NEOTYPHODIUM
IN
COOL-SEASON GRASSES

NEOTYPHODIUM IN COOL-SEASON GRASSES

Editors

Craig A. Roberts

Charles P. West

Donald E. Spiers

Blackwell Publishing

Blackwell Publishing Professional
2121 State Avenue, Ames, Iowa 50014, USA

Orders:	1-800-862-6657
Office:	1-515-292-0140
Fax:	1-515-292-3348
Web site:	www.blackwellprofessional.com

Blackwell Publishing Ltd
9600 Garsington Road, Oxford OX4 2DQ, UK
Tel.: +44 (0)1865 776868

Blackwell Publishing Asia
550 Swanston Street, Carlton, Victoria 3053, Australia
Tel.: +61 (0)3 8359 1011

Cover photo courtesy of Ho-Jong Ju and Nicholas S. Hill

First edition, 2005

Library of Congress Cataloging-in-Publication Data

Neotyphodium in cool-season grasses / editors, Craig A. Roberts, Charles P. West, Donald E. Spiers.—1st ed.
    p. cm.
Includes bibliographical references and index.
ISBN 0-8138-0189-3 (alk. paper)
1. Neotyphodium. I. Roberts, Craig A. (Craig Arthur)
II. West, Charles Patrick, 1952- III. Spiers, Donald E.

QK623.C55N465 2005
633.2'0894677—dc22
2004023764

The last digit is the print number: 9 8 7 6 5 4 3 2 1

This volume is dedicated to

Dr. Charles W. Bacon
Distinguished Senior Research Scientist
Toxicology and Mycotoxin Research Unit
USDA Agricultural Research Service

in recognition of
his seminal discovery of the link between
the tall fescue endophyte and animal toxicosis in 1977,
his vision in articulating research directions, and
his continuing output of key research on the physiology
of the *Neotyphodium*/grass symbiosis.

CONTENTS

SECTION III: ECOLOGY AND AGRONOMY

CONTRIBUTORS

John Andrae

Department of Crop and Soil Sciences
University of Georgia
Athens, GA, USA

David J. Barker

Department of Horticulture and Crop Science
The Ohio State University
Columbus, OH, USA

D. P. Belesky

USDA-ARS, Appalachian Farming Systems
Research Center
Beaver, WV, USA

Stacy A. Bonos

Department of Plant Biology & Pathology
Cook College, Rutgers University
New Brunswick, NJ, USA

Sylvie Bony

INRA-UMR Comparée des Xénobiotiques
Ecole Nationale Vétérinaire de Lyon
Marcy l'Etoile, France

Joe Bouton

Forage Improvement Division
Noble Foundation
Ardmore, OK, USA

Leah A. Brilman

Seed Research of Oregon
Corvallis, OR, USA

Gregory Bryan

AgResearch Grasslands
Palmerston North, New Zealand

Enrique J. Chaneton

IFEVA–CONICET
Facultad de Agronomía
Universidad de Buenos Aires
Buenos Aires, Argentina

Michael Christensen

AgResearch Grasslands
Palmerston North, New Zealand

Jose De Battista

INTA Estación Experimental Agropecuaria
Concepción del Uruguay
Entre Ríos, Argentina

H. Syd Easton   AgResearch Grasslands
                         Palmerston North, New Zealand

David Edwards   Primary Industries Research Victoria
                         La Trobe University
                         Bundoora, VIC, Australia

Tim J. Evans   Veterinary Medical Diagnostic Laboratory
                         University of Missouri
                         Columbia, MO, USA

Silvina A. Felitti   Primary Industries Research Victoria
                         La Trobe University
                         Bundoora, VIC, Australia

Lester R. Fletcher   AgResearch Lincoln
                         Lincoln, New Zealand

John W. Forster   Primary Industries Research Victoria
                         La Trobe University
                         Bundoora, VIC, Australia

Henry A. Fribourg   Department of Plant Sciences
                         The University of Tennessee
                         Knoxville, TN, USA

Claudio M. Ghersa   IFEVA–CONICET
                         Facultad de Agronomía
                         Universidad de Buenos Aires
                         Buenos Aires, Argentina

Nicholas S. Hill   Department of Crop and Soil Sciences
                         The University of Georgia
                         Athens, GA, USA

David E. Hume   AgResearch, Grasslands Research Centre
                         Palmerston North, New Zealand

G. C. Lewis   Institute of Grassland and Environmental Research
                         North Wyke Research Station
                         Okehampton, Devon, UK

Erica Logan   Primary Industries Research Victoria
                         La Trobe University
                         Bundoora, VIC, Australia

D. P. Malinowski

Texas A&M University System
Vernon, TX, USA

Jack W. Oliver

Department of Comparative Medicine
College of Veterinary Medicine
The University of Tennessee
Knoxville, TN, USA

Marina Omacini

IFEVA–CONICET
Facultad de Agronomía
Universidad de Buenos Aires
Buenos Aires, Argentina

Eng Kok Ong

Primary Industries Research Victoria
La Trobe University
Bundoora, VIC, Australia

Daniel G. Panaccione

Division of Plant and Soil Sciences
West Virginia University
Morgantown, WV, USA

Alison J. Popay

AgResearch, Ruakura Research Centre
Hamilton, New Zealand

K. A. Rainsford

Department of Primary Industries
Hamilton, VIC, Australia

Marc Ramsperger

Primary Industries Research Victoria
La Trobe University
Bundoora, VIC, Australia

K. F. M. Reed

Department of Primary Industries
Hamilton, VIC, Australia

Craig Roberts

Department of Agronomy
University of Missouri
Columbia, MO, USA

George E. Rottinghaus

Veterinary Medical Diagnostic Laboratory
University of Missouri
Columbia, MO, USA

Christopher L. Schardl

Department of Plant Pathology
University of Kentucky
Lexington, KY, USA

Barry Scott

Institute of Molecular BioSciences
Massey University
Palmerston North, New Zealand

C. J. Scrivener

School of Veterinary Science
The University of Melbourne
Werribee, VIC, Australia

Dwight H. Seman

J. Phil Campbell, Sr. Natural Resource
Conservation Center
USDA, Agricultural Research Service
Watkinsville, GA, USA

Takuya Shiba

National Institute of Livestock and Grassland
Science, Japan

Kate Shields

Primary Industries Research Victoria
La Trobe University
Bundoora, VIC, Australia

Daniel Singh

Primary Industries Research Victoria
La Trobe University
Bundoora, VIC, Australia

Kevin F. Smith

Department of Primary Industries
Hamilton, VIC, Australia

German C. Spangenberg

Primary Industries Research Victoria
La Trobe University
Bundoora, VIC, Australia

Donald E. Spiers

Animal Science Research Center
University of Missouri
Columbia, MO, USA

John A. Stuedemann

J. Phil Campbell, Sr. Natural Resource
Conservation Center
USDA, Agricultural Research Service
Watkinsville, GA, USA

Koya Sugawara

National Institute of Livestock and Grassland
Science, Japan

Aiko Tanaka

Institute of Molecular BioSciences
Massey University
Palmerston North, New Zealand

B. A. Tapper

AgResearch, Grasslands Research Centre
Palmerston North, New Zealand

Pei Tian

Primary Industries Research Victoria
La Trobe University
Bundoora, VIC, Australia

L. V. Walker

Department of Primary Industries
Hamilton, VIC, Australia

John C. Waller

Department of Animal Science
The University of Tennessee
Knoxville, TN, USA

W. M. Wheatley

The University of Sydney
Orange, NSW, Australia

Masayuki Yamashita

Faculty of Agriculture
Shizuoka University, Japan

Carolyn Young

Institute of Molecular BioSciences
Massey University
Palmerston North, New Zealand

Iñigo Zabalgogeazcoa

Instituto de Recursos Naturales y Agrobiología
Consejo Superior de Investigaciones Científicas
Salamanca, Spain

Eline van Zijll de Jong

Primary Industries Research Victoria
La Trobe University
Bundoora, VIC, Australia

PREFACE

Fungi belonging to the *Neotyphodium* and *Epichloë* genera are among the many types of microbes termed *endophytes* by virtue of their in planta symbiosis. *Neotyphodium* in particular is associated with several species of cool-season grasses, particularly those belonging to the *Festuca* and *Lolium* genera. The most common *Neotyphodium*-infected species of *Festuca* and *Lolium* are tall fescue and perennial ryegrass, two grasses that are widely grown in temperate climates throughout the world as forage for cattle, sheep, and horse enterprises, as well as for turf and conservation uses. The complex functions of these endophytes have received intensive study during the past three decades because of their roles in causing production losses in livestock, enhancing biotic and abiotic stress resistances in hosts, and controlling biodiversity and trophic interactions in wild populations.

This book, *Neotyphodium in Cool-Season Grasses*, presents the most recent research findings related to these endophytes. It also documents global trends in endophyte-related research and application and addresses current issues in commerce and education. *Neotyphodium in Cool-Season Grasses* was produced in conjunction with the 5th International Symposium on *Neotyphodium*/Grass Interactions, held in Fayetteville, Arkansas, USA, 23 to 26 May 2004. It was produced in lieu of standard proceedings, as symposium organizers attempted to provide a historical framework for the scientific advances reported in molecular genetics, mycology, toxicology, ecology, agronomy, and animal pathophysiology. Its intended audience includes both the researcher and the educator.

Neotyphodium in Cool-Season Grasses is divided into five sections, the first of which details *Neotyphodium* research and application in all geographical regions where these endophytes are of economic and academic importance. The next three sections present new findings while providing comprehensive research coverage from all relevant disciplines, including molecular biology, ecology and agronomy, and animal toxicoses. The final section addresses contemporary issues in commerce and public education.

The editors of *Neotyphodium in Cool-Season Grasses* are grateful to all authors for their scholarly contributions to this publication and for their unprecedented willingness to produce a book in timely fashion. We are also grateful to the associate editors who applied their expertise and volunteered their talents in reviewing the chapters. We are especially grateful to Ms. Carrie Czerwonka, Managing Editor, for her dedication in helping us honor the submission deadline of September 2004. We would also like to thank the editorial personnel at Blackwell Publishing, including Ms. Jamie Johnson, Ms. Antonia Seymor, and Ms. Dede Pedersen, for their day-to-day advice and assistance. We are also grateful for Ms. Heather Benedict for her help in the preparation of the final version of this document. We are also grateful to Ho-Jong Ju and Nicholas S. Hill for the scanning electron micrograph of *Neotyphodium coenophialum* portrayed on the cover of this book. We are grateful for major support and intellectual contribution we received from the Samuel R. Noble Foundation and the U.S. Department of Agriculture, Agricultural Research Service through the Dale Bumpers Small Farm Research Center; this work was partially supported by the U.S. Department of Agriculture, under Agreement No. 58-6227-3-016. Finally, we appreciate the generous contributions from numerous corporate donors.

It is our hope that this publication will inform the scientific community and empower science educators with the most up-to-date knowledge and definitive reviews of the state of the science. We expect this publication to facilitate a better understanding among students and practitioners regarding the agricultural, ecological, and economic roles of *Neotyphodium* endophytes in cool-season grasses.

Craig A. Roberts
Department of Agronomy
University of Missouri, USA

Charles P. West
Department of Crop, Soil, and Environmental Sciences
University of Arkansas, USA

Donald E. Spiers
Department of Animal Sciences
University of Missouri, USA

Section I

Current Trends in *Neotyphodium* Research and Application

NEOTYPHODIUM RESEARCH AND APPLICATION IN THE USA

Henry A. Fribourg and John C. Waller [1]

W e have been asked to summarize accomplishments and findings in the USA since the 4th International Symposium on *Neotyphodium*/Grass Interactions in 2000. Thus, we have limited our inclusion to publications since 1999, unless very significant and previously unreported. We have not included material or discussion that should be covered adequately in other contributions to this symposium. We have surveyed scientific publications, primarily in the USA, abstract journals, and proceedings of meetings and workshops. In particular, we have relied heavily on the annual proceedings of SERA-IEG/8, the Southern Extension and Research Activities Information Exchange Group.

Tall fescue (*Lolium arundinaceum* = *Festuca arundinacea*) toxicosis has existed in the USA ever since the 'Kentucky 31' (KY-31) cultivar was released in the 1940s and rapidly filled a vacant niche in the mid-South transition zone. Tall fescue toxicosis did not acquire its name until the mid-1970s, although earlier attempts at isolating the antiquality components did not recognize the existence of the endophytic fungus. Annual economic losses of $600 million to the U.S. cattle (*Bos* spp.) industry are probably an underestimate, covering both growth and reproduction (Allen and Segarra, 2001). In addition, the endophyte affects adversely the $60 million grass seed and hay export industries (Craig, 2003).

ENDOPHYTES AND HOST PLANTS

Recent Cultivar Developments

The most significant development in the last 4 yr has been the emergence of tall fescue cultivars in which a desired endophyte has been introduced into the host plant in the laboratory. The pioneering work in New Zealand by Latch, Easton, and co-workers was brought into the USA by Bouton and associates (Bouton et

---

[1] The University of Tennessee, Knoxville, TN, USA

al., 2002b) when they reinfected endophyte-free (E–) 'Jesup' and 'Georgia 5' tall fescues with different non-ergot alkaloid-producing endophyte strains (=nontoxic endophytes). The best combination, Jesup with AR542 (= MaxQ™), possessed greater yield and better survival than the E– entries and was equal to the fescue with a wild endophyte (E+) (*Neotyphodium coenophialum*). Animal average daily gains (ADG) were equivalent to those obtained from the E– forage and greater than those from the E+ pasture, and animals did not suffer the prolactin depression and elevated body temperature common to those grazing E+.

The AR542 endophyte was selected primarily for its lack of production of ergovaline; however, it is yet to be determined whether ergovaline is the sole or the primary cause of tall fescue toxicosis. Hill et al. (2001) investigated ergot alkaloid transport across ruminant gastric tissues, and they found that the greatest transport was for lysergic acid and lysergol rather than for other alkaloids.

A short while after Bouton et al. (2000) released Jesup/MaxQ in Georgia, the FFR Cooperative, Lafayette, Indiana, was licensed to market the cultivar 'Hi-Mag' (Sleper et al., 2002) in which had been inserted the nontoxic endophyte strain No. 4, developed by West et al. (2002) in Arkansas, currently marketed under the brand name ArkPlus™. The strain No. 4 came from wild tall fescue plants which had no detectable ergot alkaloids (Nihsen et al., 2004). It has been shown that, in response to natural selective forces, the genetic diversity within E+ and E– tall fescue cultivars leads to different genotypes with altered morphological and agronomic characteristics (Vaylay and van Santen, 2002).

Nontoxic Endophytes

It is likely that MaxQ and No. 4 are different. These two isolates are only the first releases from series of collections by New Zealand, Georgia, and Arkansas workers. Clement et al. (2001) and Hill et al. (2002a) have shown that there exists considerable diversity among *Neotyphodium* fungi from the Mediterranean basin, and that these strains can be distinguished morphologically and biochemically. The range of adaptation of E+ tall fescue extends west and north from the U.S. eastern Coastal Plain physiographic area, south from a short distance north of the Ohio River, and east of eastern Oklahoma. ArkPlus is expected to be well adapted to the Ozark region of the south-central USA and similar climatic zones to the east across the mid- to upper South (West et al., 2002).

In contrast, the GA-5/E+ cultivar was developed with a wild endophyte for adaptation to the severe summer climate and insect populations of the Coastal Plain (Gates et al., 1999). The developers of nontoxic endophyte tall fescue cultivars face the problem of determining whether the range of adaptation of these new associations coincides with the adaptation of E+ tall fescue. This challenge is complicated by the fact that mutualistic fungal endophytes often confer benefits to the hosts, but the physiological interactions involved so far have not been well characterized (Moy et al., 2002). The adaptation is influenced not only by climatic characteristics which affect drought tolerance and/or winter survival, but also by competition from other plants, management, and animal grazing effects.

Evaluation of Nontoxic Endophytes in Breeding Programs

Many assessments of the persistence of nontoxic endophyte cultivars have been done or are in progress. Numerous agronomic evaluations and grazing trials with both sheep (*Ovis aries*) (Bouton and Hopkins, 2003) and cattle (Bouton et al., 2001) have been conducted in Georgia, often using established bermudagrass (*Cynodon dactylon*) and continuous grazing to evaluate persistence under aggressive competition. Agronomic trials have included up to 14 location–years of data (Bouton et al., 2002a). Grazing studies (Parish et al., 2003a, 2003b) demonstrated the positive long-term advantage of using nontoxic endophytes to improve beef steer (*Bos* spp.) performance, even though there was more available forage produced in E+ pastures. The performance of cow–calf (*Bos* spp.) pairs was also evaluated in GA; cattle grazing MaxQ had greater weight gains and body condition scores than pairs grazing E+ (Watson et al., 2002). In Arkansas, agronomic evaluations at several locations and in three neighboring states, and grazing trials at three locations, have shown that steers grazing HiMag with nontoxic endophytes performed similarly to steers grazing E– and much better than those grazing E+ (Nihsen et al., 2004). HiMag with nontoxic endophytes had acceptable persistence for several years. Additional grazing evaluations have been conducted in several other locations (Nihsen et al., 2004).

Research in two states has used water stress as a tool for selecting persistent tall fescue/nontoxic endophyte combinations; the fungal endophyte may boost host stress tolerance through signals that amplify the production of membrane-protecting dehydrins (West et al., 2003). Irrigation levels have been used to determine seasonal distribution of yield and could be a logical criterion for selection of germplasm for use in arid but irrigated areas (Asay et al., 2001).

Tall fescue cultivars infected with different endophytes have been evaluated in small plots at several locations under grazing. In Tennessee, Fribourg et al. (2002) started in 1999, comparing 15 combinations of endophytes and host plants at three locations. Stand persistence under moderate grazing pressure was acceptable for most entries, but E– tall fescues lost more stand than tall fescues with wild or nontoxic endophytes. Similar results have been obtained at a location in Mississippi (Lang et al., 2001). At Lexington, KY, Henning et al. (2002) evaluated numerous public and commercial tall fescue cultivars, as well as many experimental host/endophyte associations, under heavy grazing pressure for several years. Nontoxic cultivars such as Jesup/MaxQ and GA-5/MaxQ persisted as well as KY-31/E+, but E– cultivars such as 'Kenhy' and Jesup were also persistent in some seedings. The Kentucky data in general indicate that both E– fescues and fescues with nontoxic endophyte may be valid alternatives to KY-31/E+ under controlled management practices. Evaluations of nontoxic endophytes in tall fescue hosts are continuing at most of these locations, and have been expanded recently to Oklahoma and Texas (Hopkins, 2003).

Seeds and Seedlings

The control and labeling for endophytic fungi in tall fescue seed are important, especially since the advent of cultivars containing nontoxic endophytes. Seed harvesting should be controlled, since more mature seed result in seedlings with greater infection by nontoxic endophytes (Hill et al., 2003). Prechilling of seed decreased germination time and uniformity (Hill et al., 2002c). The SERA-IEG/8 group was concerned about quality assurance of grass seed containing nontoxic endophytes being sold for forage (Hill et al., 2003). Diseases must be controlled when evaluating seedlings for endophyte viability; this can be done with contact fungicides (Hill and Brown, 2000). The group agreed that methods of analysis should be consistent with those of recognized state or international seed associations, that alkaloid analyses should be conducted by methods substantiated in scientific journals, and that endophyte and alkaloid analyses should be performed on statistically representative samples. Seed labels should provide information on valid grow-out analyses reflecting viable endophyte content, showing the date when the test was performed and the percentage of infected seeds which were toxin-producing off-types.

Grazing Evaluations

In the 1980s, after the realization that fungal endophytes were the cause of tall fescue toxicosis, some advocated the substitution of E+ tall fescue pastures with E- tall fescue prior to justification of this practice by reliable research data. This movement was doomed from the start and led many producers to discount future studies. In the late 1980s to early 1990s, and prior to the advent of nontoxic endophytes in the USA, the primary emphasis was to compare the productivity and persistence of E- with E+ tall fescue. An example of this approach was conducted in Tennessee (Fribourg et al., 2000a) starting in 1994. Steers grazing E- tall fescue plus clover (*Trifolium* spp.) pastures had the greatest ADG, and steers grazing E+ tall fescue had the lowest; this may be related to the reduction in forage intake and fiber digestibility observed with E+ (Humphry et al., 2002). The performance of steers grazing pastures with alternating strips of E+ and E- tall fescue was intermediate to that of steers grazing solid stands of either E- or E+. The addition of clovers improved the performance of steers on both E- and E+ pastures. Benefits obtained by renovating E+ tall fescue pastures with clover were similar to those gained by replacing half the pasture with E- tall fescue.

As seed of tall fescue cultivars containing different non-ergot alkaloid-producing endophyte strains became available for testing in the late 1990s, grazing and persistence evaluations were initiated in many states in the tall fescue belt in addition to those where the cultivars originated. An early study in Tennessee corroborated that some fungal endophyte strains were adapted and persistent in the mid-South while others were not. A few of the endophyte strains tested resulted in steer performance similar to that obtained from steers grazing E- tall fescue (Waller et al., 2001a). The performance and persistence of tall fescues with nontoxic endophytes have been evaluated in Mississippi at several locations with steers (Macoon et al., 2002) and heifers (Best et al., 2002) (*Bos* spp.). South of about 33° N latitude in the USA, dairy cows grazing E- tall fescues produced as much milk as cows grazing 'Marshall' annual ryegrass (*L. multiflorum*). Although milk production was lower for cows grazing Jesup/MaxQ than for those grazing Marshall, the economics of utilizing a perennial forage tall fescue should be attractive to dairy producers, as long as the tall fescue stand persists (Murphey et al., 2002). Winter and early spring grazing by steers in Louisiana (Alison, 2002) resulted in acceptable animal performance from Jesup/MaxQ and GA-5/MaxQ in a region marginally adapted to tall fescue, but stands of E- declined after 2 yr. Preliminary data from Oklahoma (Hopkins, 2003) indicate that nontoxic endophyte tall fescue cultivars had good persistence for 2 yr on a site with fertile soil and very good moisture holding capacity, and that weight gain was acceptable and superior to that of steers grazing E+ tall fescue. Steers that grazed E+ tall fescue entered the feedlot lighter and produced smaller carcass weights than those that grazed E– or MaxQ tall fescues, illustrating the lack of compensatory gain (Coblentz et al., 2003a; Duckett et al., 2001).

Endophyte as a Modifier of Mineral Composition of Tall Fescue

Several investigators have examined the possible relationships between the occurrence of *N. coenophialum* and minerals in tall fescue. Dennis et al. (1998) found that Cu concentrations were higher in E– than in E+ tall fescue, and increased linearly in response to N. They showed that the presence of endophyte was associated with lower Cu concentrations, possibly contributing to lowered Cu status in animals. Oliver et al. (2000b) determined that there was a relationship between Cu level and endophyte presence: Steers grazing E+ tall fescue had a serum Cu level below the generally accepted normal for cattle, but steers grazing E– tall fescue had serum Cu at the commonly accepted level. The low Cu levels could explain in part the rough haircoat usually seen on cattle grazing E+, although supplementation with Cu has not affected haircoat appearance consistently. In a 3-yr survey of Tennessee tall fescue farm pastures, presumably E+, Fisher et al. (2003) collected more than 800 plant samples. They found that Cu was marginally deficient in nearly all the samples, and that high S levels were considered marginally antagonistic to Cu utilization. Malinowski and Belesky (1999a) determined that the interaction between tall fescue genotype and endophyte status had a significant influence on mineral element uptake: Endophyte infection modified tall fescue responses to P source, suggesting that P acquisition by the plant may be determined by a specific association of fungal endophyte and tall fescue genomes. The same authors (1999b) found that Al sequestration was greater on root surfaces and in root tissues of E+ than in those of E– plants of a specific tall fescue genotype. Their results suggested that Al tolerance may contribute to the widespread adaptability and success of tall fescue/endophyte associations. Malinowski and Belesky (2000) reviewed these and other mechanisms associated with the adaptation of tall fescue to environmental stresses.

EFFECTS ON CONSUMING HERBIVORES

Whereas tall fescue toxicosis in cattle is characterized by lowered ADG, increased body temperature, rough haircoats, and reduced conception rates, mares (*Equus caballus*) exposed to E+ tall fescue exhibit increased gestation lengths, agalactia, foal mortality, and thickened placentas (Cross et al., 2000). Domperidone (Equidone®) therapy was used with considerable success as early as 1993 in South Carolina to treat tall fescue toxicosis in late gestation mares. It has evolved as a treatment of choice other than removal of the animals from E+ tall fescue for several weeks prior to parturition. Administering fluphenazine decanoate to pregnant pony mares grazing E+ tall fescue was effective in maintaining systemic relaxin and improving pregnancy outcome (Ryan et al., 2001a). It has been determined since, in Mississippi, that mares grazing Jesup/MaxQ exhibited no tall fescue toxicosis signs (Ryan et al., 2000, 2001b).

Numerous studies have shown that intake of E+ tall fescue increases hyperthermia during heat stress, which itself may induce oxidative stress. Heat challenges result in marked increase in body temperature of cattle, especially at night (Al-Haidary et al., 2001). The inability of the animal to dissipate heat significantly depletes blood glutathione (Lakritz et al., 2002). Burke et al. (2004) determined that the combination of heat stress and consumption of E+ tall fescue seed led to reduced diameter of the preovulatory dominant follicle and fewer large follicles during the estrous cycle in beef heifers. When E+ tall fescue seed, or E– seed supplemented with ergovaline, was fed to heat-stressed lambs, the ergovaline diet was not as effective as the E+ seed in producing tall fescue toxicosis signs (Gadberry et al., 2003), suggesting that alkaloids other than ergovaline may be also responsible.

Once confirmed pregnant, no differences in fetal losses of bred heifers grazing either E+ or E– tall fescue were observed (Waller et al., 2001b). However, calf birth weight was lower when dams grazed E+ tall fescue than when they grazed E– tall fescue. The difference in birth weight could be explained by decreased growth of bred heifers grazing E+ tall fescue. Follicular and luteal development and function were examined in mature lactating beef cows grazing E– or E+ tall fescue during the early postpartum period (Burke and Rorie, 2002); even though follicular dynamics were altered in cows grazing E+, follicular function was apparently not affected by ergot alkaloids. Pregnancy rate and embryonic losses tended to be different among cows between 30 and 60 d of gestation after environmental temperatures were high for 3 wk when grazing E– or E+ tall fescue (Burke et al., 2001). Pregnancy rate of yearling ewes may be reduced by E+ tall fescue, but those of older ewes were not affected (Burke et al., 2002).

The D2 dopamine receptor antagonist domperidone, so successful in treating the signs of tall fescue toxicosis in gravid mares, may have beneficial effects on beef heifers grazing E+ tall fescue, including reversal of decreased ADG and increased circulating concentrations of progesterone (Jones et al., 2003), the hormone associated with maintenance of pregnancy. The gene expression of luteal tissue from heifers fed E+ diets suggests that administration of domperidone may be beneficial in treating reproductive problems associated with tall fescue toxicosis in heifers (Jones et al., 2004).

Reproductive effects on males have not been studied extensively. Blood serum prolactin levels, which are greatly depressed in cows and steers, do not seem to be affected in bulls (*Bos* spp.) consuming E+ tall fescue pasture (Schrick et al., 2003). Male mice (*Mus musculus musculus* and *Mus musculus domesticus*) susceptible or resistant to E+ tall fescue were not affected in their reproductive responses when consuming E+ seed (Ross et al., 2004).

Several species other than cattle, sheep, horses, and mice exhibit signs of susceptibility to ergot alkaloids from tall fescue. Rabbits (*Oryctolagus cuniculus*) were affected negatively by the endophyte and were used during the search for possible vaccination against ergot alkaloids (Filipov et al., 1998). The range of meadow voles (*Microtus pennsylvanicus*) and the feed consumption by prairie voles (*M. orchrogaster*) were influenced by E+ tall fescue (Conover, 1998; Fortier et al., 2001). Consumption by hens (*Gallus gallus*) of E+ tall fescue seed resulted in reduced egg production and growth (Conover, 2003).

Several genera of invertebrates are affected, in some cases differently from mammals. Growth of earthworms (*Eisenia fetida*) fed E+ tall fescue leaves was much greater than that of those fed E– tissues (Humphries et al., 2001). Fall armyworm (*Spodoptera frugiperda*) fed E+ tall fescue responded differently at different nutrient availability levels for the plants (Bultman and Conard, 1998). Plant-parasitic nematodes may contribute to the poor growth and persistence of tall fescue on sandy soils in the southeastern USA. The presence of the wild E+ conferred resistance to feeding by some of these nematodes, notably the root-knot nematode (*Meloidogyne marylandi*) and the lesion nematode (*Pratylenchus scribneri*). It has been determined that a nontoxic endophyte did not confer feeding resistance to *P. scribneri* equivalent to that from the wild type (Elmi et al., 2000; Timper et al., 2003).

Divergent selection for ADG response to ingestion of E+ tall fescue seed by mice resulted in a favorable correlated response in survival following exposure to a chemically distinct toxin; it may be possible therefore to select livestock populations for simultaneous resistance to a variety of toxins (Wagner et al., 2000). Studies have been conducted in Missouri by Spiers, Rottinghaus, and co-workers with rats (*Rattus norvegicus*) and mice as models for cattle suffering from tall fescue toxicosis (Spiers et al., 2003a). They have exposed these animals to heat stress under controlled conditions and measured core temperatures and many physiological responses. A rat model has been used to screen experimental tall fescue cultivars with introduced strains of *N. coenophialum*; intake and ADG were consistent indicators for fungal endophytes associated with toxicity (Roberts et al., 2002b). Following Hohenboken's retirement, his work with mice in Virginia was continued in Missouri by Spiers and co-workers (Eichen et al., 2001). The research indicates that genetic variation exists in mammalian response to E+ tall fescue alkaloids and that animals might be selected for tolerance to these toxins. Unfortunately, it is not known at this time whether ruminant animals, with their extensive rumen microflora, can be modeled by rodents with a different gastrointestinal system. For this kind of research to be conducted with representative numbers of large animals would require many resources and much time.

Physiological Responses

The presence and involvement of cytochrome P450 3A (CYP3A) in the metabolism of ergot alkaloids was studied in beef liver microsomes. The data suggest that the CYP3A enzyme system in cattle may be responsible for the metabolism of ergot alkaloids produced by E+ tall fescue (Coblentz et al., 2003b). Both ergonovine and dihydroergotamine inhibited in vitro CYP3A activity, demonstrating the complexity of detoxification of liver ergot alkaloids. The CYP3A activity is significant because of its abundance in mammalian liver; it appears involved in the metabolism of a wide range of compounds (Moubarak et al., 2003). Browning et al. (2000) administered ergotamine tartrate to simulate the effect of animals ingesting alkaloids associated with tall fescue toxicosis; they found that plasma concentrations of hormones that mediate nutrient metabolism were altered, suggesting that endocrine function modification was one of the mechanisms associated with reduced animal performance.

The overall tissue ischemia and vasoconstriction known to exist in cattle that graze E+ tall fescue may be related to decreased levels of nitrate/nitrite in vascular tissues. Steers from E– tall fescue pastures had almost twice as much nitrate/nitrite per gram of tissue than those from E+. The cause of the decrease in nitric oxide formation in E+ steers may be a relative lack of tissue arginine, the substrate for nitric oxide formation (Oliver et al., 2001). Blood profiles of cattle grazing E+ indicated that the animals had decreased levels of serum arginine (Oliver et al., 2000a). Al-Tamimi and Spiers (2002) investigated the role of nitroglycerin as a nitric oxide donor and a potential vasodilator of peripheral vasculature in cattle. Although treatment with nitroglycerin failed to alleviate hyperthermia in control animals, it abrogated the more excessive hyperthermia associated with tall fescue toxicosis and returned animals to control levels. These investigators also determined that treatment with nitroglycerin overrode the persistent vasoconstrictive effects of E+ in rats, as it returned thermal circulation index values (an indicator of peripheral blood flow) to the control level. Schoning et al. (2001) demonstrated that the powerful constrictor effect of ergovaline mediated by activation of vascular receptors might explain the vascular symptoms of tall fescue toxicosis. It has been shown also that injections of ergotamine tartrate resulted in physiological responses similar to those associated with tall fescue toxicosis (Browning, 2000).

MANAGEMENT CONSIDERATIONS

Forage Systems

There have been few studies of forage systems that could decrease the impacts of tall fescue toxicosis in cattle. A grazing study in Arkansas was initiated in 2000 with the objective of reducing the long-term impacts of tall fescue toxicosis by improving longevity of overseeded, less persistent, but nontoxic forages, and by reducing exposure of calves during times of increasing toxin concentrations; results are inconclusive at this time (Coffey et al., 2003).

Another dimension relates to the management of seed fields after harvest; the need for postharvest residue burning was cultivar-specific, and mechanical residue removal was as effective for maintaining seed yields as the traditional open-field burning (Young et al., 1999).

Effects of Legumes

Several investigators at different locations have demonstrated the value of leguminous companion plants within both E+ and E– tall fescue sods. Most studies have evaluated ladino clovers (*T. repens*), but some also studied red clover (*T. pratense*), annual lespedeza (*Kummerowia stipulacea*), and alfalfa (*Medicago sativa*). More recently, mixtures of tall fescue and kura clover (*T. ambiguum*) have been promising for beef and dairy cattle in the upper Midwest (K.A. Albrecht, 2004, personal communication). The presence of these legumes in a pasture improved animal performance by 25 to 50% (Fontenot et al., 2001; Hoveland et al., 1999; Lomas et al., 1999). However, E+ level in the tall fescue may increase competition with legumes, hence seeding establishment requires careful control for seedlings and persistence of legumes. Pasture management is critical in maintaining desired legume stand density (Fribourg et al., 2000b) for positive animal response. Limited data suggest that the presence of clover reduces excretion of urinary ergot alkaloids to about half of that measured in cattle grazing E+ tall fescue (Andrae et al., 2003a). 'Durana' clover was developed recently for persistence under southern U.S. conditions; early assessment indicates it produces well in association with tall fescues containing different endophyte strains (Bouton et al., 2003).

Effects on Hay and Stockpiling

The accumulation of forage in situ is a useful management practice to substitute for the feeding of hay in late fall and winter. Endophyte-infected tall fescue hay and ammoniated hay were lower in total ergot alkaloid content than pasture or silage (Roberts et al., 2002a). Mass of stockpiled E+ tall fescue is greater per unit area than that from E– or nontoxic endophyte after autumn stockpiling. Stockpiled E+ tall fescue had substantial amounts of ergovaline in December, and these decreased by six-fold during winter. Thus, waiting until late winter to use this stockpiled E+ forage may be of benefit (Kallenbach et al., 2003). However, nutritive value of stockpiled forage was highest in mid-December and declined slowly through the winter.

Other Approaches for Reducing Tall Fescue Toxicosis

The application of a seaweed (*Ascophyllum nodosum*) extract to tall fescue pastures has alleviated some of the negative effects of tall fescue toxicosis on cattle (Allen et al., 2001, one of a series of four papers in J. Anim. Sci. vol. 79). Later research (Spiers et al., 2003b) indicates that feeding of *A. nodosum* to cattle prior to grazing E+ tall fescue reduces core temperature during heat stress and may help ADG. Since iodide, when administered to rats under heat stress, was not responsible for the protective effect associated with seaweed, other factors must be responsible (Eichen et al., 2003). Ely et al. (2004) reported improved animal performance when cattle grazing E+ tall fescue were supplemented with a modified glucomannan (FEB-200™). Aiken et al. (2001) compared protein supplementation, estrogenic implants, and removal from E+ pastures; only removal of stocker calves for 3 to 4 d alleviated signs of tall fescue toxicosis.

Conversion of E+ to Nontoxic Pastures

The development and availability of nontoxic endophytes in tall fescue cultivars have caused several investigators to consider the conversion of pastures containing E+ tall fescue. No-till glyphosate-resistant corn (*Zea mays*) and soybean (*Glycine max*), and paraquat and glyphosate fallows (Andrae et al., 2003b) are effective. Cost is a major consideration—about \$450 ha^{-1}; the use of a no-till annual cash crop may offset about half the cost (Triplett et al., 2001). Detailed protocols for successful conversions have been published (Fribourg et al., 2001; Lacefield et al., 2003).

Environmental Impacts

Franzluebbers and co-workers in Georgia have studied the spatial distribution of soil nutrients under E+ tall fescue pastures (Schomberg et al., 2000) and found that animals influence nutrient cycling within grazed systems. Endophyte levels did not affect K distribution, and affected Mg distribution only under low fertility. Grazing and pasture management were needed only occasionally to reduce potential environmental risks: Long-term cattle grazing in relatively small paddocks resulted in significant lateral and vertical changes in soil organic C and N pools (Franzleubbers et al., 2000). Clay and Holah (1999) have suggested that a host-specific endophyte with negligible biomass, such as *N. coenophialum* in tall fescue, alters plant community structure and may be reducing plant diversity. Ambient temperature, along with endophyte and host genotype, has been identified as a major factor affecting endophyte frequency (Ju et al., 2003).

LABORATORY METHODS

Several laboratory methods have been incorporated into research programs on tall fescue toxicosis. Both research programs in the USA and commercial international sales of grass seed and hay rely on accurate detection and quantification of endophyte infections and associated alkaloids. In tall fescue tissues, ELISA assays have been adapted to determine ergot alkaloids (Schnitzius et al., 2001). Monoclonal antibodies have been developed to facilitate the detection of nontoxic endophyte (Hill et al., 2002b). A simple and inexpensive thin-layer chromatographic method for the detection of ergovaline in leaf sheaths was published (Salvat and Godoy, 2001), while caution was counseled that storage and experimental conditions for purified ergopeptine alkaloid measurements should be carefully controlled (Smith and Shappell, 2002). Detection of endophyte in a pasture can be accomplished also by analysis of ergot alkaloids in urine within 2 d after exposure of the grazing animals (Stuedemann et al., 1998). Some of the 14 known isoforms of chitinase, a defense protein associated with disease resistance in tall fescue, have been shown to possess antimicrobial activity; others, related to plant growth and development, may be affected by *N. coenophialum* (Marek et al., 2000). These Missouri workers have demonstrated that chitinase could serve as a consistent marker among tall fescue cultivars across seedling stages and could help in selection for increased disease resistance.

APPLICATION OF RESEARCH FINDINGS AND FUTURE OUTLOOK

Forage and livestock producers can become aware of recent developments and their applications from extension personnel in their state, many of whom have published pamphlets on tall fescue and can provide additional information on the internet, from seed associations and companies, and other private sources. The Oregon Tall Fescue Commission (Ball et al., 2003) and the University of Arkansas (Parish et al., 2003c) have published very informative resource materials which should be useful. Increasing interest in grass-fed beef and stocker performance may result in increasing use of tall fescues with nontoxic endophytes, although the cost of seed and of converting E+ pastures is still a considerable deterrent because of the small margin of profit in, and the historically low-input nature of, beef cattle enterprises.

In 1978, while trying to convince an administrator of the urgent need for a new tall fescue research program, the senior author was asked how long it would take to resolve the tall fescue toxicosis problem. He answered: "40 years," at which there was a gasp of surprise. Early efforts identified the endophyte as a responsible agent for tall fescue toxicosis (Hoveland, 2003). It has been now about 25 years since that time, and it does appear to us that solutions should be achievable and more widely implementable within the next 15 years. The advent of E– tall fescue, and the rush to its use without supporting research data, slowed practical solutions to the problem; it is well known that "trust is hard to earn but easy to lose". More recently, the introduction of persistent, nontoxic endophytes into several adapted host genotypes promises the availability of producer-friendly tall fescue cultivars (Hopkins, 2003). This process should continue for some time, and will progress faster as better understanding of the fundamental processes and endophyte/host plant/consuming herbivore interactions become known.

REFERENCES

Aiken, G.E., E.L. Piper, and C.R. Miesner. 2001. Influence of protein supplementation and implant status on alleviating fescue toxicosis. J. Anim. Sci. 79:827–832.

Al-Haidary, A., D.E. Spiers, G.E. Rottinghaus, G.B. Garner, and M.R. Ellersieck. 2001. Thermoregulatory ability of beef heifers following intake of endophyte-infected tall fescue during controlled heat challenge. J. Anim. Sci. 79:1780–1788.

Al-Tamimi, H., and D. Spiers. 2002. Relationship between nitric oxide activity and fescue toxicosis-induced hyperthermia. *In* Proc. SERA-IEG/8 2002 Workshop.[1]

Alison, A. 2002. Evaluation of tall fescue with differing endophyte status. *In* Proc. SERA-IEG/8 2002 Workshop.[1]

Allen, V.G., K.R. Pond, K.E. Saker, J.P. Fontenot, C.P. Bagley, R.L. Ivy, R.R. Evans, C.P. Brown, M.F. Miller, J.L. Montgomery, T.M. Dettle, and D.B. Wester. 2001. Tasco-Forage. III. Influence of a seaweed extract on performance, monocyte immune cell response, and carcass characteristics in feedlot-finished steers. J. Anim. Sci. 79:1032–1040.

Allen, V.G., and E. Segarra. 2001. Anti-quality components in forage: Overview, significance, and economic impact. J. Range Manage. 54:409–412.

Andrae, J., N. Hill, and J. Bouton. 2003a. Effect of tall fescue endophyte status and white clover addition on ergot alkaloid urinary excretion and performance of stocker cattle grazing tall fescue. *In* Proc. SERA-IEG/8 2003 Workshop.[1]

Andrae, J., N. Hill, and T. Murphy. 2003b. Herbicide applications and subsequent ergot alkaloid content of renovated toxic tall fescue. *In* Proc. SERA-IEG/8 2003 Workshop.[1]

Asay, K.H., K.B. Jensen, and B.L. Waldron. 2001. Responses of tall fescue cultivars to an irrigation gradient. Crop Sci. 41:350–357.

Ball, D.M., S.P. Schmidt, G.D. Lacefield, C.S. Hoveland, and W.C. Young III. 2003. Tall fescue/endophyte concepts. Spec. Publ. 1-03. Oregon Tall Fescue Comm., Salem, OR.

Best, T.G., J.L. Howell, J.E. Huston, and R.R. Evans. 2002. Evaluation of fungus infected, fungus free and novel endophyte fescues as roughage sources for developing replacement heifers. *In* Proc. SERA-IEG/8 2002 Workshop.[1]

Bouton, J., J. Andrae, N. Hill, and C. Hoveland. 2003. Steer performance on 'Jesup' tall fescue containing different endophyte strains and inter-planted with 'Durana' white clover. *In* Proc. SERA-IEG/8 2003 Workshop.[1]

Bouton, J., J. Bondurant, N. Hill, L. Hawkins, C. Hoveland, M. McCann, F. Thompson, R. Watson, S. Easton, L. Fletcher, G. Latch, and B. Tapper. 2000. Alleviating tall fescue toxicosis problem with non-toxic endophytes. *In* Proc. SERA-IEG/8 2000 Workshop.[1]

Bouton, J., R. Gates, N. Hill, and C. Hoveland. 2002a. Agronomic trials with MaxQ tall fescue. *In* Proc. SERA-IEG/8 2002 Workshop.[1]

Bouton, J.H., R.N. Gates, and C.S. Hoveland. 2001. Selection for persistence in endophyte-free Kentucky 31 tall fescue. Crop Sci. 41:1026–1028.

Bouton, J.H., and A.A. Hopkins. 2003. Commercial applications of endophytic fungi. p. 495–516. In J.F. White et al. (ed.) Clavicipitalean fungi: Evolutionary biology, chemistry, biocontrol, and cultural impacts. Marcel Dekker, New York.

Bouton, J.H., G.C.M. Latch, N.S. Hill, C.S. Hoveland, M.A. McCann, R.H. Watson, J.A. Parish, L.L. Hawkins, and F.N. Thompson. 2002b. Reinfection of tall fescue cultivars with non-ergot alkaloid-producing endophytes. Agron. J. 94:567–574.

Browning, R., Jr. 2000. Physiological responses of Brahman and Hereford steers to an acute ergotamine challenge. J. Anim. Sci. 78:124–130.

Browning, R., Jr., S.J. Gissendanner, and T. Wakefield, Jr. 2000. Ergotamine alters plasma concentrations of glucagons, insulin, cortisol, and triiodothyronine in cows. J. Anim. Sci. 78:690–698.

Bultman, T.L., and N.J. Conard. 1998. Effects of endophytic fungus, nutrient level, and plant damage on performance of fall armyworm (Lepidoptera: Noctuidae). Environ. Entomol. 27:631–635.

Burke, J.M., W.G. Jackson, and G.A. Robson. 2002. Seasonal changes in body weight and condition, and pregnancy and lambing rates of sheep on endophyte-infected tall fescue in the south-eastern United States. Small Rumin. Res. 44:141–151.

Burke, J.M., and R.W. Rorie. 2002. Changes in ovarian function in mature beef cows grazing endophyte infected tall fescue. Theriogenology 57:1733–1742.

Burke, J.M., R.W. Rorie, E.L. Piper, and W.G. Jackson. 2001. Reproductive responses to grazing endophyte-infected tall fescue by postpartum beef cows. Theriogenology 56:357–369.

Burke, J.M., D.E. Spiers, F.N. Kojima, G.A. Perry, B.E. Salfen, S.L. Wood, D.J. Patterson, M.F. Smith, M.C. Lucy, W.G. Jackson, and E.L. Piper. 2004. Interaction of endophyte-infected fescue and heat stress on ovarian function in the beef heifer. Biol. Reprod. 65:260–268.

Clay, K., and J. Holah. 1999. Fungal endophyte symbiosis and plant diversity in successional fields. Science (Washington, DC) 285:1742–1744.

Clement, S.L., L.R. Elberson, N.N. Youssef, C.M. Davitt, and R.P. Doss. 2001. Incidence and diversity of Neotyphodium fungal endophytes in tall fescue from Morocco, Tunisia, and Sardinia. Crop Sci. 41:570–576.

Coblentz, W., K. Coffey, Z. Johnson, D. Kreider, A. Moubarak, M. Nihsen, R. Rorie, C. Rosenkrans, Jr., and C. West. 2003a. Effects of stockpiled fescue cultivars on cattle development. In Proc. SERA-IEG/8 2003 Workshop.[1]

Coblentz, W., K. Coffey, Z. Johnson, D. Kreider, A. Moubarak, M. Nihsen, R. Rorie, C. Rosenkrans, Jr., and C. West. 2003b. Hepatic metabolism of ergot alkaloids in beef cattle by cytochrome P450. In Proc. SERA-IEG/8 2003 Workshop.[1]

Coffey, K., W. Coblentz, T. Smith, J. Turner, D. Scarbrough, B. Humphry, and C. Rosenkrans, Jr. 2003. Forage systems to reduce long-term impacts of tall fescue toxicosis on cattle. In Proc. SERA-IEG/8 2003 Workshop.[1]

Conover, M.R. 1998. Impact of consuming tall fescue leaves with the endophytic fungus, Acremonium coenophialum, on meadow voles. J. Mammal. 79:457–463.

Conover, M.R. 2003. Impact of consuming tall fescue seeds infected with the endophytic fungus, Neotyphodium coenophialum, on reproduction of chickens. Theriogenology 59:5–6.

Craig, A.M. 2003. Oregon annual progress report. In Proc. SERA-IEG/8 2003 Workshop.[1]

Cross, D.L., K. Anas, W.C. Bridges, and T. Gimenez. 2000. Clinical efficacy of domperidone for treatment of fescue toxicosis in periparturient mares. In Proc. SERA-IEG/8 2000 Workshop.[1]

Dennis, S.B., V.G. Allen, K.E. Saker, J.P. Fontenot, J.Y.M. Ayad, and C.P. Brown. 1998. Influence of Neotyphodium coenophialum on copper concentration in tall fescue. J. Anim. Sci. 76:2687–2693.

Duckett, S., J. Andrae, M. McCann, and D. Gill. 2001. Performance and carcass quality of cattle finished after grazing non-toxic endophyte-infected (MaxQ™), toxic endophyte-infected, or endophyte-free tall fescue. *In* Proc. SERA-IEG/8 2001 Workshop.[1]

Eichen, P.A., S.A. Clark, M.J. Leonard, and D.E. Spiers. 2001. Identification of animal sensitivity to fescue toxicosis using mouse and cattle studies. *In* Proc. SERA-IEG/8 2001 Workshop.[1]

Eichen, P.A., M.J. Leonard, M.A. Kozma, B.M. Kronk, L.E. McVicker, D.E. Spiers, and D.P. Colling. 2003. Is iodide responsible for the heat-relief effects of *Ascophyllum nodosum*? J. Anim. Sci. 81(Suppl. 1):155.

Elmi, A.A., C.P. West, R.T. Robbins, and T.L. Kirkpatrick. 2000. Endophyte effects on reproduction of a root-knot nematode (*Meloidogyne marylandi*) and osmotic adjustment in tall fescue. Grass Forage Sci. 55:166–172.

Ely, D.G., D.K Aaron, and V. Akay. 2004. Use of a modified glucomannan to increase beef production from endophyte-infected tall fescue pasture. p. 93. *In* 20th Proc. Annu. Symp. Nutr. Biotechnol. Feed Food Ind. 24–26 May.

Filipov, N.M., F.N. Thompson, N.S. Hill, D.L. Dawe, J.A. Stuedemann, J.C. Price, and C.K. Smith. 1998. Vaccination against ergot alkaloids and the effect of endophyte-infected fescue seed-based diets on rabbits. J. Anim. Sci. 76:2456–2463.

Fisher, A.E., W.W. Gill, C.D. Lane, Jr., D.K. Joines, and J.B. Neel. 2003. Update on Tennessee tall fescue mineral study. *In* Proc. SERA-IEG/8 2003 Workshop.[1]

Fontenot, J.P., A. Brock, R.K. Shanklin, P. Peterson, and J. Fike. 2001. Grazing-tolerant alfalfa and tall fescue pastures mixtures for stocker cattle. *In* Proc. SERA-IEG/8 2001 Workshop.[1]

Fortier, G.M., M.A. Osmon, M. Roach, and K. Clay. 2001. Are female voles food limited? Effects of endophyte-infected tall fescue on home range size in female prairie voles (*Microtus orchrogaster*). Am. Midl. Nat. 146:63–71.

Franzleubbers, A.J., J.A. Stuedemann, and H.H. Schomberg. 2000. Spatial distribution of soil carbon and nitrogen pools under grazed tall fescue. Soil Sci. Soc. Am. J. 64:635–639.

Fribourg, H.A., J.C. Waller, and G.E. Bates. 2001. Methods for eliminating endophyte-infested tall fescue prior to establishing endophyte-free or nontoxic endophyte tall fescue pastures. *In* Proc. SERA-IEG/8 2001 Workshop.[1]

Fribourg, H.A., J.C. Waller, G.C.M. Latch, L.R. Fletcher, H.S. Easton, and A.E. Stratton. 2002. Evaluation under grazing of tall fescue cultivars infected with different endophytes. *In* Proc. SERA-IEG/8 2002 Workshop.[1]

Fribourg, H.A., J.C. Waller, J.H. Reynolds, K.D. Gwinn, M.A. Mueller, R.B. Graves, J.W. Oliver, and R.D. Linnabary. 2000a. Forage productivity and stand longevity of E– and E+ tall fescue pastures and steer performance at two stocking densities at the Knoxville Experiment Station. *In* Proc. SERA-IEG/8 2000 Workshop.[1]

Fribourg, H.A., J.C. Waller, M.C. Smith, and R.B. Graves. 2000b. Evaluation of clover stand persistence under grazing in tall fescue pastures at Knoxville and at Ames plantations. *In* Proc. SERA-IEG/8 2000 Workshop.[1]

Gadberry, M.S., T.M. Denard, D.E. Spiers, and E.L. Piper. 2003. Effects of feeding ergovaline on lamb performance in a heat stress environment. J. Anim. Sci. 81:1538–1545.

Gates, R.N., G.M. Hill, and J.H. Bouton. 1999. Wintering beef cattle on mixtures of 'Georgia-5' tall fescue and warm-season perennial grasses on Coastal Plain soils. J. Prod. Agric. 12:581–587.

Henning, J.C., R.F. Spitaleri, T.D. Phillips, G.D. Lacefield, C.T. Dougherty, and J.E. Roberts. 2002. Grazing tolerance of tall fescue varieties in Kentucky. *In* Proc. SERA-IEG/8 2002 Workshop.[1]

Hill, N.S., J.H. Bouton, E.E. Hiatt III, and B. Kittle. 2003. Seed maturity, germination, and endophyte relationships in tall fescue. *In* Proc. SERA-IEG/8 2003 Workshop.[1]

Hill, N.S., J.H. Bouton, F.N. Thompson, L. Hawkins, C.S. Hoveland, and M.A. McCann. 2002a. Performance of tall fescue germplasms bred for high- and low-ergot alkaloids. Crop Sci. 42:518–523.

Hill, N.S., and E. Brown. 2000. Endophyte viability in seedling tall fescue treated with fungicides. Crop Sci. 40:1490–1491.

Hill, N.S., E.E. Hiatt III, J.H. Bouton, and B. Tapper. 2002b. Strain-specific monoclonal antibodies to a nontoxic tall fescue endophyte. Crop Sci. 42:1627–1630.

Hill, N.S., H.J. Ju, and J.H. Bouton. 2002c. Germination kinetics of tall fescue seed harvested at varying stages of maturity. *In* Proc. SERA-IEG/8 2002 Workshop.[1]

Hill, N.S., F.N. Thompson, J.A. Stuedemann, G.W. Rottinghaus, H.J. Ju, D.L. Dawe, and E.E. Hiatt III. 2001. Ergot alkaloid transport across ruminant gastric tissue. J. Anim. Sci. 79:542–549.

Hill, N., C. West, R. Kallenbach, J. Henning, and C. Agee. 2003. Seed labels for non-toxic endophyte-infected grasses. *In* Proc. SERA-IEG/8 2003 Workshop.[1]

Hopkins, A. 2003. Development of non-toxic tall fescue cultivars and performance of steers grazing various tall fescue–endophyte combinations. *In* Proc. SERA-IEG/8 2003 Workshop.[1]

Hoveland, C.S. 2003. Introduction. *In* H.A. Fribourg and D. Hannaway (ed.) Tall fescue information system. Available at http://forages.oregonstate.edu/is/tfis (verified 2 Sept. 2004). Oregon State Univ., Corvallis.

Hoveland, C.S., J.H. Bouton, and R.G. Durham. 1999. Fungal endophyte effects on production of legumes in association with tall fescue. Agron. J. 91:897–902.

Humphries, S.S., K.D. Gwinn, and A.J. Stewart. 2001. Effects of endophyte status of tall fescue tissues on the earthworm (*Eisenia fetida*). Environ. Toxicol. Chem. 20:1346–1350.

Humphry, J.B., K.P. Coffey, J.L. Moyer, F.K. Brazle, and L.W. Lomas. 2002. Intake, digestion, and digestive characteristics of *Neotyphodium coenophialum*-infected and uninfected fescue by heifers offered hay diets supplemented with *Aspergillus oryzae* fermentation extract of laidlomycin propionate. J. Anim. Sci. 80:225–234.

Jones, K.L., S.S. King, K.E. Griswold, D. Cazac, and D.L. Cross. 2003. Domperidone can ameliorate deleterious reproductive effects and reduced weight gain associated with fescue toxicosis in heifers. J. Anim. Sci. 81:2568–2574.

Jones, K.L., S.S. King, and M.J. Iqbal. 2004. Endophyte-infected tall fescue diet alters gene expression in heifer luteal tissue as revealed by interspecies microarray analysis. Mol. Reprod. Develop. 67:154–161.

Ju, H.J., T. Abbott, K. Ingram, and N.S. Hill. 2003. Seasonal variation and environmental effects on endophyte transmission in tall fescue. *In* Proc. SERA-IEG/8 2003 Workshop.[1]

Kallenbach, R.L., G.J. Bishop-Hurley, M.D. Massie, G.E. Rottinghaus, and C.P. West. 2003. Herbage mass, nutritive value, and ergovaline concentration of stockpiled tall fescue. Crop Sci. 43:1001–1005.

Lacefield, G.D., J.C. Henning, and T.D. Phillips. 2003. Tall fescue. Univ. Kentucky Coop. Ext. Serv. AGR-59. Univ. of Kentucky, Lexington.

Lakritz, J., M.J. Leonard, P.A. Eichen, G.E. Rottinghaus, G.C. Johnson, and D.E. Spiers. 2002. Whole-blood concentrations of glutathione in cattle exposed to heat stress or a combination of heat stress and endophyte-infected tall fescue toxins in controlled environmental conditions. Am. J. Vet. Res. 63:799–803.

Lang, D., R. Elmore, M. Salem, A. Tokitkla, and R. Given. 2001. Persistence and preference of novel endophyte tall fescue. *In* Proc. SERA-IEG/8 2002 Workshop.[1]

Lomas, L.W., J.L. Moyer, and G.L. Kilgore. 1999. Effect of interseeding legumes into endophyte-infected tall fescue pastures on forage production and steer performance. J. Prod. Agric. 12:479–483.

Macoon, B., R.C. Vann, and F.T. Withers, Jr. 2002. Animal and forage responses on novel endophyte fescue compared with ryegrass pastures. *In* Proc. SERA-IEG/8 2002 Workshop.[1]

Malinoswki, D.P., and D.P. Belesky. 1999a. *Neotyphodium coenophialum*-endophyte infection affects the ability of tall fescue to use sparingly available phosphorus. J. Plant Nutr. 22:835–853.

Malinoswki, D.P., and D.P. Belesky. 1999b. Tall fescue aluminum tolerance is affected by *Neotyphodium coenophialum* endophyte. J. Plant Nutr. 22:1335–1349.

Malinoswki, D.P., and D.P. Belesky. 2000. Adaptations of endophyte-infected cool-season grasses to environmental stresses: mechanisms of drought and mineral stress tolerance. Crop Sci. 40:923–940.

Marek, S.M., C.A. Roberts, A.L. Karr, and D.A. Sleper. 2000. Chitinase activity in tall fescue seedlings as affected by cultivar, seedling development, and ethephon. Crop Sci. 40:713–716.

Moubarak, A., C. Rosenkrans, Jr., and Z. Johnson. 2003. Modulation of cytochrome P450 metabolism of ergonovine and dihydroergotamine. Vet. Human Toxicol. 45:6–9.

Moy, M., H.M. Li, R. Sullivan, J.F. White, Jr., and F.C. Belanger. 2002. Endophytic fungal beta-1,6-glucanase expression in the infected host grass. Plant Physiol. 130:1298–1308.

Murphey, J., T. Smith, B. Johnson, and D. Lang. 2002. A comparison of Jesup MaxQ® non-toxic endophyte-infected tall fescue and Jackson® annual ryegrass as forages for lactating dairy cows. *In* Proc. SERA-IEG/8 2002 Workshop.[1]

Nihsen, M.E., E.L. Piper, C.P. West, R.J. Crawford, T.M. Denard, Z.B. Johnson, C.A. Roberts, D.A. Spiers, and C.F. Rosenkrans, Jr. 2004. Growth rate and physiology of steers grazing tall fescue inoculated with novel endophytes. J. Anim. Sci. 82:878–883.

Oliver, J.W., H. Al-Tamimi, J.C. Waller, H.A. Fribourg, K.D. Gwinn, L.K. Abney, and R.D. Linnabary. 2001. Effect of chronic exposure of beef steers to the endophytic fungus tall fescue: Comparative effects on nitric oxide synthase activity and nitrate/nitrite levels in lateral saphenous veins. *In* Proc. SERA-IEG/8 2001 Workshop.[1]

Oliver, J.W., S.K. Cox, J.C. Waller, H.A. Fribourg, K.D. Gwinn, B.W. Rohrbach, and R.D. Linnabary. 2000a. Effect of chronic exposure of beef steers to the endophytic fungus of tall fescue: Comparative effects on serum arginine levels. *In* Proc. SERA-IEG/8 2000 Workshop.[1]

Oliver, J.W., A.E. Schultze, B.W. Rohrbach, H.A. Fribourg, T. Ingle, and J.C. Waller. 2000b. Alterations in hemograms and serum biochemical analytes of steers after prolonged consumption of endophyte-infected tall fescue. J. Anim. Sci. 78:1029–1035.

Parish, J.A., M.A. McCann, R.H. Watson, C.S. Hoveland, L.L. Hawkins, N.S. Hill, and J.H. Bouton. 2003a. Use of nonergot alkaloid-producing endophytes for alleviating tall fescue toxicosis in sheep. J. Anim Sci. 81:1316–1322.

Parish, J.A., M.A. McCann, R.H. Watson, C.S. Hoveland, L.L. Hawkins, N.S. Hill, and J.H. Bouton. 2003b. Use of nonergot alkaloid-producing endophytes for alleviating tall fescue toxicosis in stocker cattle. J. Anim Sci. 81:2856–2868.

Parish, J.A., C.P. West, J.A. Jennings, and S.M. Jones. 2003c. "Friendly" endophyte-infected tall fescue for livestock production. Univ. Arkansas Coop. Ext. Serv. FSA2140-PD-7-03N. Univ. of Arkansas, Fayetteville.

Roberts, C., R. Kallenbach, and N. Hill. 2002a. Harvest and storage method affects ergot alkaloid concentration in tall fescue [Online]. Available at www.plantmanagement network.org/cm/. Crop Manage. DOI 10.1094/cm-2002-0917-01-BR.

Roberts, C.A., D.E. Spiers, A.L. Karr, H.R. Benedict, D.A. Sleper, P.A. Eichen, C.P. West, E.L. Piper, and G.E. Rottinghaus. 2002b. Use of a rat model to evaluate tall fescue seed infected with introduced strains of *Neotyphodium coenophialum*. J. Agric. Food Chem. 50:5742–5745.

Ross, M.K., W.D. Hohenboken, R.G. Saacke, and L.A. Kuehn. 2004. Effects of feeding endophyte-infected fescue seed on reproductive traits of male mice divergently selected for resistance or susceptibility to fescue toxicosis. Theriogenology 61:651–662.

Ryan, P.L., K. Bennett-Wimbush, W.E. Vaala, and C.A. Bagnell. 2001a. Systemic relaxin in pregnant pony mares grazed on endophyte-infected fescue: Effects of fluthenazine treatment. Theriogenology 56:471–483.

Ryan, P., B. Rude, B. Warren, L. Boyd, D. Lang, R. Elmore, R. Given, D. Scruggs, and R. Hopper. 2000. Determination of pregnancy outcome of mares grazed on a toxic-free endophyte-infected tall fescue (MaxQ). *In* Proc. SERA-IEG/8 2000 Workshop.[1]

Ryan, P., B. Rude, B. Warren, L. Boyd, D. Lang, D. Scruggs, and R. Hopper. 2001b. Effects of exposing late-term pregnant mares to toxic and non-toxic endophyte-infected tall fescue pastures. *In* Proc. SERA-IEG/8 2001 Workshop.[1]

Salvat, A.E., and H.M. Godoy. 2001. A simple thin-layer chromatographic method for the detection of ergovaline in leaf sheaths of tall fescue (*Festuca arundinacea*) infected with *Neotyphodium coenophialum*. J. Vet. Diag. Invest. 13:446–449.

Schnitzius, J., N.D. Hill, C.S. Thompson, and A.M. Craig. 2001. Semi-quantitative determination of ergot alkaloids in seed, straw and digesta samples using a competitive ELISA assay. J. Vet. Diag. Invest. 13(3):230–237.

Schomberg, H.H., J.A. Stuedemann, A.J. Franzluebbers, and S.R. Wilkinson. 2000. Spatial distribution of extractable phosphorus, potassium, and magnesium as influenced by fertilizer and tall fescue endophyte status. Agron. J. 92:981–986.

Schoning, C., M. Flieger, and H.H. Pertz. 2001. Complex interaction of ergovaline with 5-HT2A, 5-HT1B/1D, and alpha1 receptors in isolated arteries of rat and guinea pig. J. Anim. Sci. 79:2202–2209.

Schrick, F.N., J.C. Waller, J.L. Edwards, M.D. Davis, H.E. Blackmon, F.N. Scenna, N.R. Rohrbach, A.M. Saxton, H.S. Adair, and F.M. Hopkins. 2003. Effects of administration of ergotamine tartrate to simulate fescue toxicosis on fertility of yearling beef bulls. J. Anim. Sci. 81(Suppl. 2):13.

Sleper, D. A., H.F. Mayland, R.J. Crawford, Jr., G.E. Shewmaker, and M.D. Massie. 2002. Registration of HiMag tall fescue germplasm. Crop Sci. 42:318–319.

Smith, D.J., and N.W. Shappell. 2002. Epimerization of ergopeptine alkaloids in organic and aqueous solvents. J. Anim. Sci. 80:1616–1622.

Spiers, D., J. Williams, G. Rottinghaus, P.A. Eichen, L. Wax, R. Settivari, and L. Thompson. 2003a. Markers of fescue toxicosis. *In* Proc. SERA-IEG/8 2003 Workshop.[1]

Spiers, D., J. Williams, G. Rottinghaus, P.A. Eichen, L. Wax, R. Settivari, and L. Thompson. 2003b. Pretreatment with *Ascophyllum nodosum* for relief of fescue toxicosis. *In* Proc. SERA-IEG/8 2003 Workshop.[1]

Stuedemann, J.A., N.S. Hill, F.N. Thompson, R.A. Fayrer-Hosken, W.P. Hay, D.L. Dawe, D.H. Seman, and S.A. Martin. 1998. Urinary and biliary excretion of ergot alkaloids from steers that grazed endophyte-infected tall fescue. J. Anim. Sci. 76:2146–2154.

Timper, P., J. Bouton, and R. Gates. 2003. Effect of tall fescue cultivar and endophyte status on reproduction of lesion (*Pratylenchus* spp.) and stubby root (*Paratrichodorus* spp.) nematodes. *In* Proc. SERA-IEG/8 2003 Workshop.[1]

Triplett, G.B., A. Rankins, Jr., M. Shankle, and R.L. Ivy. 2001. Corn and soybean production in sod to renovate tall fescue stands. *In* Proc. SERA-IEG/8 2001 Workshop.[1]

Vaylay, R., and E. van Santen. 2002. Application of canonical discriminant analysis for the assessment of genetic variation in tall fescue. Crop Sci. 42:534–539.

Wagner, C.R., T.M. Howell, W.D. Hohenboken, and D.J. Blodgett. 2000. Impacts of an endophyte-infected fescue seed diet on traits of mouse lines divergently selected for response to that same diet. J. Anim. Sci. 78:1191–1198.

Waller, J.C., H.A. Fribourg, R.J. Carlisle, K.D. Gwinn, G.C.M. Latch, L.R. Fletcher, R.J.M. Hay, H.S. Easton, and B.A. Tapper. 2001a. Evaluation of tall fescues with nontoxic endophytes. *In* Proc. SERA-IEG/8 2001 Workshop.[1]

Waller, J.C., F.N. Schrick, M.C. Dixon, A.E. Fisher, A.M. Saxton, and H.A. Fribourg. 2001b. Tall fescue based forage systems for developing beef replacement heifers. J. Anim. Sci. 79(Suppl. 1):458.

Watson, R., M. McCann, J. Parish, C. Hoveland, N. Hill, and J. Bouton. 2002. Performance of cow-calf pairs on MaxQ tall fescue. *In* Proc. SERA-IEG/8 2002 Workshop.[1]

West, C.P., R.D. Carson, B. de los Reyes, and S. Rajguru. 2003. Biochemical responses to endophyte infection and water deficit in tall fescue. *In* Proc. SERA-IEG/8 2003 Workshop.[1]

West, C.P., C.F. Rosenkrans, Jr., and E.L. Piper. 2002. Development of tall fescue with a nontoxic endophyte. *In* Proc. SERA-IEG/8 2002 Workshop.[1]

Young, W.C., III, M.E. Mellbye, and T.B. Silberstein. 1999. Residue management of perennial ryegrass and tall fescue seed crops. Agron. J. 91:671–675.

---

[1] All SERA-IEG/8 workshop proceedings available at http://animalscience.ag.utk.edu/SERA-IEG8/meetings.htm (verified 2 Sept. 2004).

NEOTYPHODIUM RESEARCH AND APPLICATION IN EUROPE

Iñigo Zabalgogeazcoa[1] and Sylvie Bony[2]

S ome of the earliest descriptions of grass endophytes were made in Europe around the beginning of the last century. Later, when the connection between animal toxicoses and consumption of endophyte-infected grasses was discovered in the USA and New Zealand, fescue and ryegrass toxicoses were still an unknown problem in Europe. Paradoxically, grass species which are hosts of *Epichloë* and *Neotyphodium* endophytes are quite common in European ecosystems. The widespread use of botanically diverse permanent pastures for livestock production seems to explain why the economic impact of endophytes in Europe is still uncertain. The use of artificial, monospecies pastures is quite limited on this continent. For instance, the land area of perennial ryegrass comprises less than 2.5% of the total pasture area in France and Spain.

The number of endophyte-focused research groups in Europe is relatively small, and this may be linked to the low incidence of endophyte toxicoses. Nevertheless, the analysis of literature databases show that the number of European research reports dealing with grass endophytes has remained steady, and perhaps increased, during the last 10 yr.

Our objective is to review the research done in Europe since the year 2000, when the 4th International *Neotyphodium*/Grass Interactions Symposium was held in Soest, Germany. In this period, more than 40 papers have been published; for reasons of limited space, all those papers could not be cited here. The main areas of study that we have identified are surveys of grass–endophyte associations, levels of alkaloids, toxin distribution in animals and cases of toxicosis, effects of endophytes on plant performance, and population biology and evolution of endophyte species.

[1] Instituto de Recursos Naturales y Agrobiología, CSIC, Salamanca, Spain
[2] INRA-UMR 1233 Mycotoxines et Toxicologie Comparée des Xénobiotiques, Ecole Nationale Vétérinaire de Lyon, Marcy l'Etoile, France

GRASS–ENDOPHYTE ASSOCIATIONS IN EUROPE

Grasses such as tall fescue (*Festuca arundinacea)* and perennial ryegrass (*Lolium perenne*), in conjunction with their *Neotyphodium* symbionts, have been the most important subjects of endophyte research. In particular, research has concentrated on forage cultivars developed in the USA and New Zealand, whose endophytes may have enhanced their adaptation to new environments. In Europe, few cultivars of tall fescue and perennial ryegrass are infected by endophytes, and in the case of perennial ryegrass, the infection rates are low (Panka and Sadowski, 2002; Saikkonen et al., 2000). However, in the permanent pastures which are so important for livestock production, and in natural ecosystems, perennial ryegrass and tall fescue are commonly infected by *Neotyphodium* endophytes, and many other grass species are also *Epichloë* hosts. In fact, the list of European grass species associated with endophytes is the largest one relative to the ones known for other continents.

The list of European grass hosts of *Epichloë* endophytes grew during the last 5 yr as a result of surveys made in Finland, France, and Spain. In a French study, seeds of 237 species of European grasses were analyzed for the presence of endophytic mycelia (Leyronas and Raynal, 2001). Seeds of 22 species were found to be infected. However, 39 species not found to be infected had already been reported elsewhere as endophyte hosts. Thus, about 25% of these 237 European grasses are potentially endophyte hosts.

In a semiarid grassland ecosystem in western Spain, 22% of 49 grass species analyzed were found to be hosts of *Neotyphodium* or *Epichloë* species (Zabalgogeazcoa et al., 2003). This ratio of grass host species is close to that from the French survey. *Festuca* and *Lolium* were the genera that encompassed the most host species in the French, Spanish, and Finnish surveys. In the latter survey, 10 grass species associated with endophytes were identified (Saikkonen et al., 2000).

Grass populations are normally thought to follow the one host–one endophyte species model. Bony et al. (2001b), however, found 12 populations of perennial ryegrass in which individual plants contained two different endophytes being either two *N. lolii* species or one *N. lolii* and a *Gliocladium*-like endophyte. Furthermore, *E. festucae* and the choke pathogen *E. typhina* have been isolated from asymptomatic plants of *Lolium* in Spain (Espadas et al., 2004). This evidence points out that, regarding the number of fungal species involved, there is more complexity in the interaction of *Lolium* with *Epichloë* and *Neotyphodium* endophytes than in most known grass–endophyte interactions.

Recent studies (Hesse et al., 2004; Leyronas and Raynal, 2001) point out that in Europe there may be a greater incidence of grass–endophyte associations in drier climates, particularly in perennial ryegrass. Studies of the endophyte species present in wild populations of this grass under different climates may reveal adaptation of different fungi to different environments.

Grass–endophyte associations have been observed from Mediterranean to Scandinavian countries. Hence, the search for new grass hosts of endophytes, and perhaps new endophyte species, is likely to yield more results in Europe. For example, the Spanish survey was limited to particular grassland ecosystems. In other parts of the country, grasses not found to be infected, or not present in these ecosystems, such as *Poa* spp. and *Festuca rubra* subsp. *pruinosa*, have been identified as potential hosts of *Epichloë* species.

ENDOPHYTE ALKALOIDS AND THEIR EFFECTS ON HERBIVORES

Most of the endophyte alkaloid measurements made in Europe concerned perennial ryegrass wild populations, and only a few cases of ryegrass staggers related to the lolitrem B content of infected forages were clearly identified during the past 4 yr.

One Spanish study examined tall fescue commercial cultivars and showed that 80% of the seedlots examined (10) were infected with incidences ranging from 1 to 72%. The ergovaline content in these seedlots ranged from 0.02 to 3.71 $\mu g\ g^{-1}$ (1.9 $\mu g\ g^{-1}$ on average), but the content in the corresponding plant tissue was much lower and averaged 0.11 $\mu g\ g^{-1}$ (Vázquez de Aldana et al., 2001). Other studies, with wild grasses from semiarid pastures in Spain confirmed that for tall fescue, the amount of ergovaline can be rather high and ranged from 0.03 to 0.85 $\mu g\ g^{-1}$ in forage and 0.28 to 3.17 $\mu g\ g^{-1}$ in seeds (Vázquez de Aldana et al., 2003). Although these ergovaline levels could be considered as inhibitory factors, it must be considered that tall fescue remains a rather uncommon species in *dehesa* pastures.

Wild ecotypes of perennial ryegrass collected in northern Spain showed high average ergovaline content (13.5 $\mu g\ g^{-1}$) despite a moderately low *Neotyphodium* infection level (40%) (Oliveira et al., 2002). Lolitrem B in seeds was about 1.1 $\mu g\ g^{-1}$, and ranged as high as 7.1 $\mu g\ g^{-1}$ in one population. These toxin levels could be considered high enough to induce episodes of ryegrass staggers or fescue toxicosis, especially when combined with heat-stress conditions, which occur readily in Spain. However, the high species diversity of pastures in northern and western Spain probably dictates against frequent acute disorders in animals.

Oliveira et al. (2003) explored the ergovaline concentration in two perennial ryegrass genotypes infected with the same lolitrem B-free endophyte during a 3-yr period. Concentration of ergovaline in total herbage showed, as already found in other countries, significant variations from a winter or early-spring minimum to a late-spring or autumn maximum. However, the range of toxin concentrations throughout this study remained rather low (0.06–0.57 $\mu g\ g^{-1}$), and suggests that agriculturally useful ryegrass–endophyte associations free from lolitrem B can be developed. In the Czech Republic, Cagaš and Flieger (2002) carried on some more investigations of ergot alkaloids (ergovaline and chanoclavine) contents in native endophyte-infected perennial ryegrass during a 3-yr period, and also found relatively low levels of toxins (range: 0.036–0.635 $\mu g\ g^{-1}$ for ergovaline and 0.07 to 0.54 $\mu g\ g^{-1}$ for chanoclavine), with a maximum during the late spring (second cut) and great variations among years related to different climatic situations.

Bony et al. (2001b) studied the fungal endophytes of 83 natural populations of perennial ryegrass in Europe to describe intraspecific variability, particularly regarding the production of mycotoxins. One-third of the isolates of *N. lolii* did not produce ergovaline, whereas only a few isolates did not produce lolitrem B. Interestingly, a clear relationship was found between ergovaline- and lolitrem B-deficient strains and morphological characteristics of colonies on potato dextrose agar medium, suggesting the possibility to select ergovaline- or lolitrem-deficient strains by examination of the morphology of isolates. This work also reported a wide range of toxin concentration data and evidence for the ubiquity of potentially toxicogenic perennial ryegrass in European natural populations. From the frequency distribution of the lolitrem B and ergovaline concentrations in the 66 whole plants harboring only one isolate, it can be estimated that 65% of the plants contained lolitrem B in the range of 2 to 6 $\mu g\ g^{-1}$, and that 50% contained ergovaline in the range of 0.5 to 3.5 $\mu g\ g^{-1}$. Thus, most of these populations had ergovaline and (or) lolitrem B concentration at least as high as the threshold values for endophyte toxicoses in horse, sheep, and cattle. Ninety percent of these associations contained peramine in the range of >2 to >40 $\mu g\ g^{-1}$, levels for which feeding deterrence is reported (Rowan et al., 1990).

In Switzerland, Leuchtmann et al. (2000) searched various grass species for loline alkaloids, which enhance host survival by deterring insect feeding. This toxin group was found in very high amounts (>5000 $\mu g\ g^{-1}$) mainly in *Festuca* hosts infected with *Neotyphodium* spp., although peramine was detected in about half of the *Neotyphodium-* and *Epichloë*-infected grasses. Interestingly, this study showed that there was a tendency for plants infected by stroma-forming endophytes to be alkaloid free, while lolines and ergovaline were associated with asymptomatic, seed-transmitted species.

Despite the rather high endophyte alkaloid levels often recorded in European grasses, no clear-cut case of ryegrass staggers or of fescue toxicosis in herbivorous animals has been recently reported in Europe, except in France. However, in many countries, some suspicious cases with animals showing strange behavior were informally recorded.

In France, the very stressful climatic conditions that occurred during late spring and summer in 2003 led to an increase in the number of ryegrass staggers episodes formally characterized compared with the previous years (Benkhelil et al., 2004). No episode was recorded with grazing animals during spring and summer, but during the next winter, some horses, mares, heifers, and bulls fed perennial ryegrass hay or straw of turf cultivars were affected at various degrees by nervous disorders. The amount of lolitrem B found in the corresponding forages was in the range of 1.8 to 4.8 µg g^{-1}. In all cases, ergovaline was also detected (0.6–2 µg g^{-1}), but no fescue toxicosis was described. Clearly, the lack of available forages due to the severe drought during late spring and summer had drawn some stockbreeders to use unusual forages for the winter feeding, like straws of turf cultivars. In two cases, forage cultivars were incriminated. For those, the unusual stressful environmental conditions probably exacerbated the toxin synthesis since the varieties involved had low infection levels (<20%) and had never been suspected before to cause disease to animals.

There has been no recent report in Europe dealing with the effects of endophyte on animal performance itself. The lack of study in this field in Europe is probably explained both by the scarcity of acute endophyte toxicosis and by the minimal effects of endophyte on production parameters recorded in feeding experiments with sheep and cattle some years ago in Germany and France, which have not fostered further studies. Animals in most European production systems are not likely to be as heavily exposed as in the USA or New Zealand. This is principally because extensive grazing systems are not as widely used, and when they are, pastures are mainly natural or artificial mixtures of many grass species and clover (*Trifolium* spp.), so that toxins contained in meadow fescue and perennial ryegrass are diluted. Additionally, European environmental conditions are not as highly stressful (heat-stress index values are usually below 23.5 throughout the year; Oliveira et al., 2003). Furthermore, rules are now applied in some European countries to limit the endophyte infection levels of commercially registered cultivars so that exposure of livestock is now rather limited.

As animal product quality and food safety are a matter of great concern in Europe, especially since the mad cow crisis, experiments were conducted to check the distribution of endophyte toxins in the animal body and to evaluate the presence of residues in milk and meat. Pharmacokinetic studies of ergovaline following an intravenous administration in various species showed that the plasma elimination half-life of the molecule was always very short (23–56 min) (Bony et al., 2001a). Goat was used as a lactating animal model to study the distribution of ergovaline and lolitrem B in milk following an intravenous injection or an intraruminal bolus. For both toxins, the amount recovered in milk was very low, and thus milk does not appear to be a major excretion route for these toxins (Grancher et al., 2004). No ergovaline and only traces of lolitrem B were found in muscle.

EFFECTS OF ENDOPHYTES ON PLANT PERFORMANCE

Knowing how endophyte infection affects plant performance is the key to understanding why high infection rates are observed in natural populations of some grasses. This information is also essential for the development of practical applications of fungal endophytes in turfgrass and forage cultivar improvement. Despite numerous studies on this subject, the nature of the interaction in terms of costs and benefits for the plant is still unclear. Chapters 7 and 8 will focus on the response of endophytic grasses to abiotic and biotic stresses. Therefore, we will briefly cite European contributions to this field.

Perennial ryegrass is the most important forage and turfgrass species in Europe, and this grass was the subject of several studies in which the responses of infected plants to abiotic stress factors such as limiting water, clipping, shading, or low soil pH were recorded (Hesse et al., 2003, 2004; Lewis, 2004). In general, plant responses varied with the genotype of the plant–endophyte combination. Thus, in some genotypes, infection increased dry weight and other characters, whereas in other genotypes, endophyte had neutral or negative effects.

In Germany, Hesse et al. (2003, 2004) noted that infected *Lolium* genotypes collected from dry areas showed increased growth or seed production compared with uninfected plants under water-limiting conditions, but showed the opposite when water supply was adequate. The stimulatory effect of endophyte under drought was not observed in a *Lolium* genotype obtained from a wet site. The authors proposed that a greater incidence of endophyte infection in dry areas may be due to endophyte-enhanced reproductive fitness of plants where water is limiting.

Genetic variability of *N. lolii* occurring in perennial ryegrass may confound the conclusions of *Lolium*–endophyte performance experiments. Host-by-endophyte genotype interactions cause different plant responses to endophyte, thereby complicating the interpretation of ecological benefits of endophytes across environments.. Equivocal responses also occur when different species of endophytic grasses are compared. Very illustrative is an intraspecific competition experiment involving two symbioses: *Brachypodium sylvaticum–E. sylvatica*, and *Bromus benekenii–E. bromicola* (Ahlholm et al., 2002). A target plant was surrounded by competitor plants, and the response of the target plant was recorded. The results showed completely different responses for each grass–endophyte association: infected target plants of *Brachypodium* had fewer tillers, less dry matter, and less seed yield than uninfected plants, while the opposite was true for infected *Bromus* plants (Brem and Leuchtmann, 2002).

Regarding responses to biotic stresses, Brem and Leuchtmann (2001) studied the performance of the lepidopteran herbivore *Spodoptera frugiperda* on *Brachypodium sylvaticum* infected by *E. sylvatica*. Insect survival and reproduction were inhibited by endophyte infection. Interestingly, no alkaloids such as lolines, peramine, or ergovaline were present in this grass–endophyte association.

STUDIES OF POPULATION BIOLOGY AND EVOLUTION

Wild plant populations in Europe are an important resource for studying the population biology and evolution of grass–endophyte associations. Quantifying genetic variability in natural populations of a seed-transmitted *Epichloë* species was the aim of a study of natural populations of *E. festucae* (Arroyo et al., 2002). *Festuca rubra* is a common species in natural grasslands of western Spain, where about 70% of the plants are infected by *E. festucae*. Grass and endophyte have coexisted for a long time in these areas, which are a center of diversity of red fescues. Whether such populations of *E. festucae* were clonal, as expected when seed transmission is the predominant dispersion mechanism, or showed substantial genetic variability as a result of sexual reproduction, was unknown. Using AFLP markers, genetic differentiation and migration were detected between populations at a 40-km distance. Within these populations, clonal lineages, but also an important amount of genetic variation, were found. Furthermore, variability within clonal lineages was detected, suggesting that genetic variation perhaps generated by nonsexual mechanisms does exist in these populations.

Mathematical modeling of grass-endophyte populations has been used to analyze conditions that could explain the coexistence of infected and uninfected plants in natural populations (Gyllenberg et al., 2002; Saikkonen et al., 2002). These papers provide a general framework for evaluating several possible explanations for this phenomenon, based on seed transmission rates and levels of mutualism.

The process of speciation in *Epichloë* has been addressed in two papers. *Epichloë bromicola* is a sexual choke pathogen of *Bromus erectus*, but isolates from *B. benekenii* and *B. ramosus* are asexual and seed-transmitted. In this system, Brem and Leuchtmann (2003) showed that isolates from each *Bromus* species are genetically differentiated from isolates of the other host species. *Epichloë* strains from *B. erectus* can infect the other two hosts, but *B. ramosus* or *B. benekenii* strains can not infect *B. erectus*. Nevertheless, isolates from different hosts are interfertile. A shift of a sexual species to a new host may have served to isolate these races.

A mechanism which could contribute to the evolution of species by isolation could be the preference of *Botanophyla* flies, which act as pollinators of stromata for a given *Epichloë* species. Bultman and Leuchtmann (2003) studied the behavior of flies on stromata of several species. Although the flies did not seem to have a particular preference for certain species, in the case of *E. clarkii*, fly visitation was uncommon relative to *E. typhina*. Interestingly, *E. clarkii* and *E. typhina* are the only known *Epichloë* species which are interfertile, but no hybrid ascospores were detected in fertilized *E. clarkii* stromata. Therefore, in this case the hypothesis of *Botanophyla* flies being a mechanism preventing hybridization seems correct. If, in addition to fungus, the plant host would also affect fly preference, then isolation by means of pollinators after a host shift could also be a mechanism of species evolution.

Wille et al. (2002) studied the joint inoculation of *Epichloë* strains on *B. erectus* plants. They observed that, in most cases, only one strain ended up infecting the host plant. However, some strains were more competitive in some host genotypes than in others. This endophyte–host interaction could explain how different strains could coexist in natural populations of grasses.

The effect of habitat fragmentation upon the incidence of choke disease caused by *E. bromicola* on *B. erectus* was studied in Switzerland by Groppe et al. (2001). In plots where habitat fragmentation was simulated by mowing the surrounding plants, the number of diseased plants increased with respect to unfragmented plots. However, the number of stems in healthy plants also increased. The increase in the number of diseased plants was at least in part due to previously asymptomatic infections reverting to a diseased phenotype.

SUMMARY

Compared with other countries where perennial ryegrass and tall fescue pastures are widely used for animal production, the number of cases of endophyte toxicosis in Europe is very low. In contrast, many different alkaloid-producing grass–endophyte associations have been identified on this continent. Perhaps the best explanation for this paradox is that high botanical diversity in European pastures dilutes endophyte toxins. Moreover, perennial ryegrass is the most important forage grass in Europe, far more than tall fescue, but most ryegrass cultivars are not infected, or their infection rates are low. Recent studies show that grass–endophyte associations of European ecosystems are a very valuable asset for the understanding of the biology of these symbioses. Furthermore, European ecosystems can be reservoirs of endophyte germplasm which could be useful for applications such as the improvement of forage and turfgrass cultivars.

REFERENCES

Ahlholm, J.U., M. Helander, S. Lehtimäki, P. Wäli, and K. Saikkonen. 2002. Vertically transmitted endophytes: Effects of environmental conditions. Oikos 99:173–183.

Arroyo, R., J.M. Martínez Zapater, B. García Criado, and I. Zabalgogeazcoa. 2002. The genetic structure of natural populations of the fungal endophyte *Epichloë festucae*. Mol. Ecol. 11:355–364.

Benkhelil, A., D. Grancher, N. Giraud, P. Bezille, and S. Bony. 2004. Intoxication par des toxines de champignons endophytes chez des taureaux reproducteurs. Rev. Med. Vet. 5:243–247.

Bony, S., A. Durix, A. Leblond, and P. Jaussaud. 2001a. Toxicokinetics of ergovaline in the horse after an intravenous administration. Vet. Res. 32:509–513.

Bony, S., N. Pichon, C. Ravel, A. Durix, F. Balfourier, and J.J. Guillaumin. 2001b. The relationship between mycotoxin synthesis and isolate morphology in fungal endophytes of *Lolium perenne*. New Phytol. 152:125–137.

Brem, D., and A. Leuchtmann. 2001. *Epichloë* grass endophytes increase herbivore resistance in the woodland grass *Brachypodium sylvaticum*. Oecologia 126:522–530.

Brem, D., and A. Leuchtmann. 2002. Intraspecific competition of endophyte infected vs uninfected plants of two woodland grass species. Oikos 96:281–290.

Brem, D., and A. Leuchtmann. 2003. Molecular evidence for host-adapted races of the fungal endophyte *Epichloë bromicola* after presumed host shifts. Evolution 57:37–51.

Bultman, T.L., and A. Leuchtmann. 2003. A test of host specialization by insect vectors as a mechanism for reproductive isolation among entomophilous fungal species. Oikos 103:681–687.

Cagaš, B., and M. Flieger. 2002. Ergot alkaloids in ecotypes of perennial ryegrass containing endophytes. Grassland Sci. Eur. 7:526–527.

Espadas, D., B.Vázquez de Aldana, M. Romo, A. García Ciudad, B. García Criado, and I. Zabalgogeazcoa. 2004. Sympatric *Neotyphodium* and *Epichloë* species in natural populations of *Lolium perenne*. *In* Proc. Int. Symp. on *Neotyphodium*/Grass Interactions, 5th, Fayetteville, AR, USA. 23–26 May 2004. Univ. of Arkansas, Fayetteville.

Grancher, D., F. Guiguen, Y. Moulard, Y. Bonnaire, P. Jaussaud, A. Durix, and S. Bony. 2004. Pharmacokinetic study of lolitrem B in goat: Distribution in milk and plasma. *In* Proc. Int. Symp. *Neotyphodium*/Grass Interactions, 5th, Fayetteville, AR, USA. 23–26 May 2004. Univ. of Arkansas, Fayetteville.

Groppe, K., T. Steinger, B. Schmid, B. Baur, and T. Boller. 2001. Effects of habitat fragmentation on choke disease (*Epichloë bromicola*) in the grass *Bromus erectus*. J. Ecol. 89:247–255.

Gyllenberg, M., D. Preoteasa, and K. Saikkonen. 2002. Vertically transmitted symbionts in structured host metapopulations. Bull. Math. Biol. 64:959–978.

Hesse, U., W. Schöberlein, L. Wittenmayer, K. Förster, W. Diepenbrock, and W. Merbach. 2004. Influence of water supply and endophyte infection (*Neotyphodium* spp.) on vegetative and reproductive growth of two *Lolium perenne* L. genotypes. Europ. J. Agron. 21:(in press).

Hesse, U., W. Schöberlein, L. Wittenmayer, K. Förster, K. Warnstorff, W. Diepenbrock, and W. Merbach. 2003. Effects of *Neotyphodium* endophytes on growth, reproduction and drought-stress tolerance of three *Lolium perenne* L. genotypes. Grass Forage. Sci. 58:407–415.

Leuchtmann, A., D. Schmidt, and L.P. Bush, 2000. Different levels of protective alkaloids in grasses with stroma-forming and seed-transmitted *Epichloë/Neotyphodium* endophytes. J. Chem. Ecol. 26:1025–1036.

Lewis, G.C. 2004. Effects of biotic and abiotic stress on the growth of three genotypes of *Lolium perenne* with and without infection by the fungal endophyte *Neotyphodium lolii*. Ann. Appl. Biol. 144:53–63.

Leyronas, C., and G. Raynal. 2001. Presence of *Neotyphodium*-like endophytes in European grasses. Ann. Appl. Biol. 139:119–127.

Oliveira, J.A., G.E. Rottinghaus, and E. González. 2003. Ergovaline concentration in perennial ryegrass infected with a lolitrem B-free endophyte in north west Spain. N. Z. J. Agric. Res. 46:117–122.

Oliveira, J.A., G.E Rottinghaus, C. Prego, and E. González. 2002. Contenido en alcaloides en semillas de poblaciones naturales de raigrás inglés del norte de España infectadas con los hongos endofitos *Neotyphodium*. Investigación Agraria. Prod. Prot. Veg. 17:247–256.

Panka, D., and C. Sadowski. 2002. Occurrence of fungal endophytes in perennial ryegrasss (*Lolium perenne*) cultivars in Poland. Grassland Sci. Eur. 7:540–541.

Rowan, D.D., J.J. Dymock, and M.A. Brimble. 1990. Effect of fungal metabolite peramine and analogs on feeding and development of Argentine stem weevil (*Listronotus bonariensis*). J. Chem. Ecol. 16:1683–1695.

Saikkonen, K., J. Ahlholm, M. Helander, S. Lehtimäki, and O. Niemeläinen. 2000. Endophytic fungi in wild and cultivated grasses in Finland. Ecography 23:346–352.

Saikkonen, K., M. Gyllenberg, and D. Ion. 2002. The persistence of fungal endophytes in structured grass metapopulations. Proc. R. Soc. London, Ser. B 269:1397–1403.

Vázquez de Aldana, B.R., A. García Ciudad, I. Zabalgogeazcoa, and B. García Criado. 2001. Ergovaline levels in cultivars of *Festuca arundinacea*. Anim. Feed Sci. Technol. 93:169–176.

Vázquez de Aldana, B.R., I. Zabalgogeazcoa, A.Garcia Ciudad, and B. Garcia Criado. 2003. Ergovaline occurrence in grasses infected by fungal endophytes of semi-arid pastures in Spain. J. Sci. Food Agric. 83:347–353.

Wille, P., T. Boller, and O. Kaltz. 2002. Mixed inoculation alters infection success of strains of the endophyte *Epichloë bromicola* on its grass host *Bromus erectus*. Proc. R. Soc. London, Ser. B 269:397–402.

Zabalgogeazcoa, I., B.R. Vazquez de Aldana, A. García Ciudad, and B. García Criado. 2003. Fungal endophytes in grasses from semiarid permanent grasslands of Western Spain. Grass Forage Sci. 58:94–97.

NEOTYPHODIUM RESEARCH AND APPLICATION IN NEW ZEALAND

Syd Easton and Brian Tapper[1]

The primary focus of endophyte research in New Zealand is perennial ryegrass (*Lolium perenne*). This species is the basis of the most productive and intensively grazed pastures in New Zealand. It contains the endophyte *Neotyphodium lolii* which causes toxicosis in grazing livestock (Fletcher et al., 1999), but it is essential to persistence of the sward (Prestidge and Ball, 1993). The most obvious symptom of toxicosis is ryegrass staggers, with livestock, particularly sheep, experiencing varying degrees of incapacitation, but thermoregulation, fecal moisture regulation, and growth rates are also affected. Enhanced persistence of the sward is mediated through tolerance and resistance to invertebrate pests (Prestidge and Ball, 1993), notably Argentine stem weevil (*Listronotis bonariensis*).

The last 4 yr have been marked by the availability and the consolidating reputation of the selected nontoxic endophyte strain, AR1, with significant research effort going into defining the best practice for its effective use. However, we have been interested in the basic biology of endophyte symbiosis, in the attributes of other endophyte strains and in the value in our environment of endophyte-infected tall fescue (*Festuca arundinacea*).

AR1 IS NOW THE BENCHMARK IN NEW ZEALAND

The AR1 endophyte produces peramine, an insect feeding deterrent, but not lolitrem B or ergovaline, the major compounds toxic to livestock (Tapper and Latch, 1999). It was intensively researched in the late 1990s (Fletcher and Easton, 2000; Hume, 1999; Popay et al., 2000) and first offered for sale in significant quantities in the southern autumn of 2002. All companies marketing proprietary ryegrass cultivars were invited to have their material inoculated with AR1. To date, seven different companies have had more than 20 cultivars

---

[1] AgResearch, Grasslands Research Centre, Private Bag 11004, New Zealand

inoculated. In 2003, the second year in which AR1 was available, 40% of all proprietary perennial ryegrass seed sold in New Zealand was infected with AR1.

On the farm, AR1 has performed to expectations. Almost all early work determining health and productivity of livestock grazing AR1 used sheep as the research animal (Fletcher and Easton, 2000). An intensively monitored dairy farm experience was reported (Keogh and Blackwell, 2001), and more recently, major dairy experiments have been undertaken and are nearing the end of their third milking season (Bluett et al., 2003; Ussher, 2003). These trials have indicated approximately 20% improved milk production through summer and autumn for cows grazing AR1-infected pasture compared with cows grazing ryegrass naturally infected with ergovaline- and lolitrem-producing endophyte, this translating into a 5 to 10% improved productivity over the whole lactation.

Evidence indicates that AR1 is robust. Research into best practice for establishing and managing AR1 pastures was widely disseminated (Hume, 1999). A trial involving several cultivars indicated that AR1 might not provide adequate protection against African black beetle (*Heteronochus arator*) (Popay and Baltus, 2001), and AR1 has been sold with a warning to farmers that in northern New Zealand, the host pasture may lack persistence in the face of black beetle attack. However, since AR1 has been on-farm in significant areas, there has not been any severe outbreak of this pest, and AR1 has provided adequate protection against what has been minor pressure. At several sites, we have monitored research plots for percent infection with AR1 and percent contamination with naturally occurring endophyte. There have been instances where an initial high incidence of contamination at establishment has increased with time, but where AR1 pasture has been successfully established, it has maintained its integrity (Bluett et al., 2001).

Seed is sold with an expectation that at least 70% of kernels are infected with live endophyte. Wholesale seed companies have had to provide retailers with protocols for dealing with seed retained at the end of the season, and farmers have been discouraged from taking possession of seed before they are ready to sow. In fact, to date, demand for seed has been such that carryover volumes have not been an issue. AgResearch has been involved with quality control, assaying certified seed lots to determine if there is any indication of contamination with seed infected with toxin-producing endemic endophyte. Of the 265 samples assayed for the 2003 harvest (seed harvested in January–February 2003), one clearly was not AR1 at all. All other samples fell well within specification, with little evidence of any contamination. Interim results for 2004 also indicate excellent compliance, with only 2 of 143 samples falling outside requirements. We consider these results to be a credit to the professionalism of the seed industry.

Clearly, this innovation has met a demand from the farming community. It is encouraging that the product has been accepted, since we had had earlier experience of having to withdraw ryegrass cultivars infected with a selected endophyte strain because these combinations proved to have unacceptably high levels of ergovaline (Fletcher and Easton, 1997). The solid body of research evidence on livestock performance and pest tolerance, the well-researched best-practice guidelines for establishment and management, and the reliability of the seed industry in meeting specifications have all contributed to this confidence.

TALL FESCUE ENDOPHYTE IN NEW ZEALAND

Tall fescue is most conspicuous in New Zealand as rampant vegetation in fertile waste places such as roadsides and by farm drains (Easton et al., 1994). It is rightly regarded by farmers as dangerous to livestock, and this of course is because it is infected with endemic endophyte (*N. coenophialum*). Tall fescue infected with endemic endophyte is never sown for pastures. In some districts it is a significant adventitious presence in pasture, but farmers seek to eradicate it. The major disorder of stock grazing tall fescue in New Zealand is *fescue foot*, impaired blood circulation in the extremities, sometimes leading to loss of the hoof.

Endophyte-free tall fescue began to assume a significant place in New Zealand pastures from 1970. Use has gradually grown (Easton et al., 1994), particularly in dry eastern regions where it provides leafy herbage more reliably in summer than does ryegrass. Endophyte-free tall fescue is tolerant of grass grub (*Costelytra zealandica*), an important root-feeding scarab. However, in northern New Zealand, where tall fescue has been shown to be valuable on farms (Milne et al., 1997), farmers have found it to lack persistence. Data from the northern regions of Northland and Waikato (Cooper et al., 2002) and from Canterbury (Easton et al., 2004) have shown infection with selected nontoxic endophyte to enhance persistence and productivity of tall fescue. Seed of four cultivars infected with AR542 was available in limited quantities in New Zealand and Australia in autumn 2004. Farmer interest is keen and we expect rapid uptake.

The situation in New Zealand (and Australia) is different to that of USA. In our farming system, tall fescue infected with endemic endophyte is not an agricultural plant. It is treated as a serious weed, farmers try to keep it off their properties, and only when they are very short of feed will they allow stock to graze it. The endophyte-infected pastures that are widespread in North America are simply not present. Therefore, the requirement for releasing tall fescue infected with selected endophyte is not simply to show that livestock response to it is better than to fescue infected with endemic endophyte, but to show conclusively that livestock health and performance are not impaired relative to endophyte-free tall fescue. As with our ryegrass work, we have more data for sheep (Fletcher et al., 2000) than for cattle.

FURTHER RESEARCH WITH RYEGRASS

Research with ryegrass–endophyte combinations has focused on characterizing bioactivity that can not be ascribed to the well-researched alkaloids, documenting further evidence of pest protection, exploring variations specific to combinations, providing more precise information on livestock response to endophyte toxins, and researching the basic biology of endophyte.

The description of the close coordination of fungal and plant growth and control of fungal metabolism (Spiering, 2000), and the unusually adapted nature of the fungal growth proposed by Mike Christensen (Hesse et al., 2004) have been among the more exciting developments in endophyte research.

For ryegrass, the enhancement of persistence and productivity ascribed to endophyte infection is mediated by enhanced resistance to pressure from invertebrate pests, complemented by protection from overgrazing by livestock (Prestidge and Ball, 1993). Research in New Zealand and North America has not found for ryegrass the direct effects on productivity and stress physiology reported for tall fescue associations. However, recent work has documented differences due to endophyte infection and strain in ryegrass root development and nutrient uptake (Popay and Bonos, Chapter 7, this volume).

Evidence that the pasture mealy bug (*Balanococcus poae*) is sensitive to endophyte led to a reevaluation of the importance of this organism in pastures (Pennell and Ball, 1999). Other pasture invertebrates have also been closely studied for their response to endophyte. This work has led to the recognition of bioactivity that can not be ascribed to the well-researched lolitrem, ergopeptine, or peramine alkaloids. The endophyte strain AR37 had been shown to produce none of the known alkaloids and was included in a trial for interest only. It was found to be very active against Argentine stem weevil, and subsequently against black beetle and other invertebrates (Popay and Bonos, Chapter 7, this volume). It has some tremorgenic activity when grazed by livestock (Fletcher, Chapter 10, this issue). Identifying the likely active factors, epoxy-janthitrems, produced by this strain (Tapper and Lane, 2004) has been the major achievement of our chemistry laboratory during the last 4 yr. The outstanding agronomic potential of this strain has driven intensive efforts to closely define the effects on live-stock. Their sporadic nature has made this difficult.

Production of bioactive alkaloids by endophyte is influenced by the genetics of the host grass (Easton et al., 2002), and there is interest in breeding host cultivars that moderate the potentially detrimental activity of otherwise valuable endophyte strains. The first selected strain exploited in New Zealand was free of lolitrem B, but produced unacceptably high levels of ergovaline (Fletcher and Easton, 1997). It was withdrawn in its combination with perennial ryegrass, but has continued to be used in association with hybrid ryegrass. This and similar strains are now being revisited in association with selected perennial ryegrass hosts.

Farmers managing livestock are often not aware of the degree of toxicosis suffered unless there is visible ryegrass staggers or obvious heat stress. Work has been undertaken in New Zealand to test the efficacy of a urine test to deter-mine exposure to ergopeptine toxins (Litherland and Fletcher, 2004, personal communication).

SEED INVENTORY AND QUALITY CONTROL

Endophyte viability declines in stored seed (Rolston et al., 1986), and historical records show that as ryegrass cultivars infected with endemic endophyte age in the market, their typical level of infection declines with some seed lines being almost endophyte-free. For new associations of cultivars with selected endo-phyte, there is a further issue of endophyte transmission and maintenance in seed being less robust than in naturally occurring associations. Seed stand management, harvest systems, and inventory management have been intensively researched to develop best practice protocols for meeting specifications of proprietary products (Rolston, 2004, unpublished data).

ENDOPHYTE GENETIC RESEARCH

Genetic research at Massey University, regularly presented at symposia and elsewhere (Scott et al., 1999), has used defined genetic strains to elucidate the close control of endophyte–host interactions (Spiering, 2000). An AgResearch group formed in the last 4 yr is also working on the genomics of ryegrass–endophyte interaction (G. Bryan, 2004, personal communication), with some projects performed in collaboration with the Massey University group. Areas of particular interest include the avoidance or subversion of the host defense mechanism (involving comparison with pathogens), use of proteomics, metabolomics and structural biology modeling to describe the systems generating active metabolites (10 different gene clusters are being analyzed), and direct and indirect effects of endophyte infection on host metabolism (there is strong evidence of effects on photosynthesis and amino acid metabolism).

NATIVE GRASSES

The New Zealand endemic grass flora includes several species of *Festuca* and other genera in the Poöidae. A clavicepitaceous endophyte was described for *Echinopogon* (Miles et al., 1998), but several studies found little of interest in *Festuca* or *Poa* species (Rolston et al., 2002). However, a recent, more thorough project using material freshly collected from natural ecosystems has shown a high incidence of infection of *Poa matthewsii* (Stewart et al., 2004). Nothing is known of the role of this endophyte in the ecosystem.

CONTROL OF VERTEBRATE PESTS

A number of bird species have been shown to be sensitive to endophyte in grasses, and work in New Zealand (Pennell and Rolston, 2003) has focused on locally important species that are inconvenient in different contexts, such as plovers at airfields, Canada geese on recreational areas, and seagulls at feedlots. Grass-feeding species are directly affected, and insect-feeding species may be indirectly affected. Management of wildlife involves attention to a complex of interacting factors, including the endophyte status of grass habitat.

CURRENT PRIORITIES

Endophyte research in New Zealand began as wholly government-funded research into an on-farm problem. It is now a mixed suite of work including research into basic biology and genomics, still funded by government, best-on-farm-practice research funded by the sheep and dairy industries, seed handling research funded by the seed industry (companies and growers), and strain development work funded from investment funds and royalty returns.

If endophyte research is to continue to promote growth in the pastoral industries, we need to provide greater confidence around the use of tall fescue infected with selected nontoxic endophyte; and we need to understand the nature and the control of the bioactive factors in endophyte strains that are currently less well understood, but which clearly have great potential for enhancing pasture persistence and productivity. The triangular interaction between livestock health, chemistry, and invertebrate pest response remains the foundation for future development. Construction of associations tailored for specific purposes not met by strains found in collections will require a sound knowledge of genomics and basic biology of the association. Basic research remains important if the powerful phenomenon of endophyte infection is to be further developed in the future.

REFERENCES

Bluett, S.J., V.T. Burggraaf, D.E. Hume, B.A. Tapper, and E.R. Thom. 2001. Establishment of ryegrass pastures containing a novel endophyte. Proc. N. Z. Grassl. Assoc. 63:259–265.

Bluett, S.J., E.R. Thom, D.A. Clark, K.A. Macdonald, and E.M.K. Minnée. 2003. Milk solids production from cows grazing perennial ryegrass containing AR1 or wild endophyte. Proc. N. Z. Grassl. Assoc. 65:83–90.

Cooper, B.M., H.S. Easton, D.E. Hume, A.J. Popay, and D.B. Baird. 2002. Improved performance in Northland New Zealand of tall fescue with a selected endophyte. p. 379–381. *In* J.A. McComb (ed.) 12th Australasian Plant Breeding Conf. of Australasian Plant Breeding Assoc., Australia.

Easton, H.S., G.C.M. Latch, B.A. Tapper, and O.J.-P. Ball. 2002. Ryegrass host genetic control of concentrations of endophyte-derived alkaloids. Crop Sci. 42:51–57.

Easton, H.S., C.K. Lee, and R.D. Fitzgerald. 1994. Tall fescue in Australia and New Zealand. N. Z. J. Agric. Res. 37:405–417.

Easton, H.S., C.G.L. Pennell, and D.E. Hume. 2004. Tall fescue cultivars and endophyte strains. *In* Proc. Int. Symp. *Neotyphodium*/Grass Interactions, 5th, Fayetteville, AR, USA. 23–26 May 2004. Univ. of Arkansas, Fayetteville.

Fletcher, L.R., and H.S. Easton. 1997. The evaluation and use of endophytes for pasture improvement. p. 209–227. *In* C.W. Bacon and N.S. Hill (ed.) *Neotyphodium*/Grass Interactions. Int. Symp. *Acremonium*/Grass Interactions, 3rd, Athens, GA, USA. 28–31 May 1997. Plenum Press, New York.

Fletcher, L.R., and H.S. Easton. 2000. Using endophytes for pasture improvement in New Zealand. p. 149–162. *In* V.H. Paul and P.D. Dapprich (ed.) The Grassl. Conf. 2000. Proc. Int. Symp. *Neotyphodium*/Grass Interactions, 4th, Soest, Germany. 27–29 Sept. 2000. Univ. Paderborn, Germany.

Fletcher, L.R., C.G. Fletcher, and B.L. Sutherland. 2000. The health and performance of sheep grazing a non-toxic tall fescue endophyte association. p. 459–464. *In* V.H. Paul and P.D. Dapprich (ed.) The Grassl. Conf. 2000. Proc. Int. Symp. *Neotyphodium*/Grass Interactions, 4th, Soest, Germany. 27–29 Sept. 2000. Univ. Paderborn, Germany.

Fletcher, L.R., B.L. Sutherland, and C.G. Fletcher. 1999. The impact of endophyte on the health and productivity of sheep grazing ryegrass-based pastures. Grassl. Res. Pract. Ser. 7:11–17.

Hesse, U., Christensen, M.J., and C.L. Schardl. 2004. Tissue specificity of endophyte development in *Epichloë/Neotyphodium* symbioses with grasses. *In* Proc. XI Molecular Plant–Microbe Interactions Congr., St. Petersburg, Russia. 19–27 July 2003. In press.

Hume, D.E. 1999. Establishing and maintaining a toxin-free pasture: A review. Grassl. Res. Pract. Ser. 7:123–132.

Keogh, R.G., and M.B. Blackwell. 2001. A three-year investigation of the performance of spring-calving dairy cows grazing ryegrass-based pastures of high or low endophyte toxin status in Northland. Proc. N. Z. Grassl. Assoc. 63:209–214.

Miles, C.O., M.E. di Menna, S.W.L. Jacobs, I. Garthwaite, G.A. Lane, R.A. Prestidge, S.L. Marshall, H.H. Wilkinson, C.L. Schardl, O.J.P. Ball, and G.C.M. Latch. 1998. Endophytic fungi in indigenous Australasian grasses associated with toxicity to livestock. Appl. Environ. Microbiol. 64:601–606.

Milne, G.D., R. Shaw, R. Powell, B. Pirie, and J. Pirie. 1997. Tall fescue use on dairy farms. Proc. N. Z. Grassl. Assoc. 59:163–167.

Pennell, C.G., and O.J.-P. Ball. 1999. The effects of *Neotyphodium* endophytes in tall fescue on pasture mealy bug (*Balanococcus poae*). Proc. N. Z. Plant Prot. Soc. 52:259–263.

Pennell, C.G.L., and M.P. Rolston. 2003. The effect of grass-endophyte associations on feeding of Canada geese (*Branta canadensis*). Proc. N. Z. Grassl. Assoc. 65:239–243.

Popay, A.J., and J.G. Baltus. 2001. Black beetle damage to perennial ryegrass infected with AR1 endophyte. Proc. N. Z. Grassl. Assoc. 63:267–271.

Popay, A.J., J.G. Baltus, and C.G.L. Pennell. 2000. Insect resistance in perennial ryegrass infected with toxin-free *Neotyphodium* endophytes. p. 187–193. *In* V.H. Paul and P.D. Dapprich (ed.) The Grassl. Conf. 2000. Proc. Int. Symp. *Neotyphodium*/Grass Interactions, 4th, Soest, Germany. 27–29 Sept. 2000. Univ. Paderborn, Germany.

Prestidge, R.A., and O.J.P. Ball. 1993. The role of endophytes in alleviating plant biotic stress in New Zealand. p. 141–151. *In* D.E. Hume et al. (ed.) Proc. Int. Symp. *Acremonium/Grass Interactions*: Plenary Papers, 2nd, Palmerston North, NZ. AgResearch Grasslands, Palmerston North.

Rolston, M.P., M.D. Hare, K.K. Moore, and M.J. Christensen. 1986. Viability of *Lolium* endophyte fungus in seed stored at different moisture contents and temperatures. N. Z. J. Exp. Agric. 14:297–300.

Rolston, M.P., A.V. Stewart, G.C.M. Latch, and D.E. Hume. 2002. Endophytes in New Zealand seeds: Implications for conservation of grass species. N. Z. J. Bot. 40:365–372.

Scott, D.B., C.L. Young, and L. McMillan. 1999. Molecular biology of *Epichloë* endophyte toxin biosynthesis. Grassl. Res. Pract. Ser. 7:77–83.

Spiering, M.J. 2000. Distribution of *Neothyphodium lolii*-endophyte metabolic activity in perennial ryegrass (*Lolium perenne* L.) and its implications for alkaloid distribution and photosynthesis. Ph.D. thesis. Massey Univ., Palmerston North, NZ.

Stewart, A., D. Hume, G. Latch, B. Tapper, W. Simpson, and P. Matthewsii. 2004. The first endemic New Zealand grass with an endophyte. *In* Proc. Int. Symp. *Neotyphodium/Grass Interactions*, 5th, Fayetteville, AR, USA. 23–26 May 2004. Univ. of Arkansas, Fayetteville.

Tapper, B.A., and G.A. Lane. 2004. Janthitrems found in a *Neotyphodium* endophyte of perennial ryegrass. *In* Proc. Int. Symp. *Neotyphodium*/Grass Interactions, 5th, Fayetteville, AR, USA. 23–26 May 2004. Univ. of Arkansas, Fayetteville.

Tapper, B.A., and G.C.M. Latch. 1999. Selection against toxin production in endophyte-infected perennial ryegrass. Ryegrass endophyte: An essential New Zealand symbiosis. Grassl. Res. Pract. Ser. 7:107–112.

Ussher, G. 2003. Northlands pasture toxin project. p. 62–64. *In* New Zealand Large Herds Assoc. 34th Ann. Conf. New Zealand Large Herds Assoc., Paihia.

NEOTYPHODIUM RESEARCH AND APPLICATION IN AUSTRALIA

K. F. M. Reed,[1,2] C. J. Scrivener,[3] K. A. Rainsford,[1] and L. V. Walker[1]

Perennial ryegrass (*Lolium perenne*) occupies more than 6 million ha in Australia, where much of it experiences four months of hot, dry conditions during summer. In the cool, temperate, winter–spring rainfall zone, perennial ryegrass is commonly dominant in pasture. In the warm, summer rainfall zone, perennial ryegrass is less dominant. There it is often sown at only 1 to 3 kg ha^{-1} mixed with other grass seed, and on the coast it competes with warm-season grasses.

Endophyte (*Neotyphodium lolii*)-infected perennial ryegrass, relative to endophyte-free plants, may exhibit greater seedling vigor, tillering, yield, and persistence (Quigley, 2000). In old, naturalized populations, *N. lolii* frequency averaged 90% (Reed et al., 2000). Concentrations of ergovaline and lolitrem B in grass peak in summer–autumn (Reed et al., 2001; Woodburn et al., 1993). Both are widespread in southwestern Victoria, and may exceed critical levels on >30% of autumn pastures (Reed et al., 2004).

Ryegrass staggers in sheep is observed nearly every year in some regions. Horses fed perennial ryegrass seed cleanings exhibited intense neurological symptoms and tenesmus (Munday et al., 1985) and lolitrem B and ergovaline have been associated with poor ewe fertility and lamb rearing (Foot et al., 1994). There have been several anecdotal reports of nervous behavior in cattle grazing perennial ryegrass in summer–autumn in southwestern Victoria. Reduced milk yield and meat quality problems have been associated with these observations. Some case studies have been supported by toxin analysis. In New South Wales, reduced production and health of dairy cows fed on perennial ryegrass silage were associated with a sample containing 1.78 mg kg^{-1} ergovaline (Lean, 2001). There have been isolated reports from producers about marked improvements in the milk production, pregnancy rate, and behavior of cattle when endophytic,

[1] Department of Primary Industries, PB 105, Hamilton, VIC 3300, Australia
[2] Cooperative Research Centre for Plant-Based Management of Dryland Salinity
[3] School of Veterinary Science, The University of Melbourne, Werribee, VIC 3030, Australia

perennial ryegrass pasture has been replaced by endophyte-free (usually hybrid/Italian, *L. multiflorum*) ryegrass. Despite research recording a 12% loss of autumn milk due to the subclinical effect of wild endophyte (Valentine et al., 1993), 80% of the perennial ryegrass seed sold contains wild endophyte. Hay and straw is sold or exported without requirement for toxin testing.

We have summarized the development of grass endophyte technology occurring post-1999 and described a recent epidemic of perennial ryegrass toxicosis that illustrates the risks inherent in relying on perennial ryegrass containing wild endophyte.

USE OF GRASS ENDOPHYTE TECHNOLOGY

Seed merchants estimate that approximately 15% of the seed of tall fescue (*Festuca arundinacea*) pasture cultivars sold this year will contain select endophyte. 'Quantum', 'Jessup', 'Vulcan', 'Advance', and two winter-active, summer-dormant Mediterranean cultivars ('Flecha' and 'Resolute') containing MaxP endophyte were released in 2003. Tests made in high rainfall regions demonstrated that in areas with African black beetle (*Heteronychus arator*), summer–autumn production was vastly improved (i.e., >200%) by the select endophyte (Wheatley et al., 2003); this may reflect the benefit of lolines. Both concepts, Mediterranean-type cultivars and safe endophyte, may aid tall fescue persistence in marginal districts.

For 2004, merchants estimate that approximately 10% of the certified perennial ryegrass seed sold for pasture will be endophyte free, and 10% will contain select endophyte. AR1 is available in four cultivars and AR6 in two. The inoculated cultivars are of New Zealand origin and are suited primarily to dairy farms. Persistence is a more serious problem for the lower rainfall districts' sheep–beef farms, where nil endophyte perennial ryegrass often does not persist as well as wild endophyte material (e.g., Quigley, 2000). Two years after sowing, New Zealand cultivars with wild endophyte maintained a markedly greater frequency of tillers when compared with the same cultivars containing a select endophyte, AR1/NEA2 ($P < 0.01$). The results from this on-farm research conducted at Bega were associated with accumulated stress from drought, African black beetle, and pasture scarabs (*Sericesthis* spp.) (H.W. Kemp, NSW Agriculture, 2004, personal communication). Perennial ryegrass with resistance to native, root-feeding pests such as *Adorphorus couloni*, *Scitala sericans*, and *Sericesthis* spp. remains an important goal.

Confusion surrounding a microscopic organism that combines positive benefits for the grass with negative effects for livestock, lack of field research, and conflicting messages from merchants has not encouraged a strong demand for select endophyte ryegrass.

While there is no requirement for marketers to monitor endophyte in seed lines, and no standards are set for seed sold as low/high endophyte status, some companies are invoking measures to minimize seed moisture and set safe sow-by dates to ensure that select endophyte cultivars deliver benefits after leaving the cold store.

PERENNIAL RYEGRASS TOXICOSIS 2002

Diagnosis

Ryegrass staggers can be seen in most years, and there have been three serious epidemics in the past 20 yr, the most recent in February to April 2002. Following good spring and early summer rain, ryegrass toxicosis occurred during late summer and autumn 2002, in the 600- to 700-mm rainfall zone across the four southern states. In that season, many southern Victorian sheep exhibited the usual head tremors and nystagmus. Stock introduced to toxic pasture exhibited symptoms within a few days, while on removal, symptoms persisted up to 3 wk. Crucial husbandry and marketing operations were badly disrupted, with many sheep becoming recumbent. Trembling sheep displayed poor coordination and a stiff-legged gait. Often, sheep went down on their hocks, and those seriously affected became recumbent with legs stiffly extended and in tetanic spasms. Losses were common. Many sheep remained prostrate for 2 to 3 d and died or required euthanasia. Variable scouring, increased respiratory rate, and severe muscle twitching over the head, trunk, and legs were observed, with increased internal body temperature (> 41°C) in some sheep. Dehydration, hyperthermia, starvation, hypoglycemia, and predators also contributed to subsequent death. The staggers condition was aggravated by disturbances such as predators, mustering, delivery of supplements, and fertilizer spreading. Several farmers observed mass drowning events as heat-stressed sheep flocked to open water. Losses were diagnosed as severe perennial ryegrass toxicosis on the basis of botanical dominance, clinical symptoms, toxin determinations, and postmortem examination.

Differential diagnoses were made, including polioencephalomalacia, entero-toxemia, phalaris toxicity, annual ryegrass toxicity, and acute liver disease. In April, four affected Merino sheep were examined postmortem at The University of Melbourne. This revealed severely reduced trabecular bone density and severe loss of trabecular bone on the long bones of all sheep. Histopathology examination of neuronal cells in the brain revealed low numbers of Purkinje axonal swellings scattered throughout the granular layer of the cerebellum, sometimes in association with moderate vacuolar changes at the interface between the molecular and granular layers, features consistent with longstanding perennial ryegrass toxicosis (Mason, 1968; Radostits et al., 2000). Blood parameter changes were variable and compounded with secondary effects such as starvation and flystrike. Creatinine phosphate kinase was elevated; blood calcium levels were normal; magnesium and potassium were sometimes slightly reduced. Creatinine, alkaline phosphatase, gamma globulins, total protein, and albumin were low.

Farm Reports

Regional Department of Primary Industries (DPI) staff asked farmers to describe livestock problems and pasture associated with staggers in questionnaires distributed by advisors, media, and newsletters. Two hundred and twenty-four replies were returned from affected farms. From these, we attributed the death of 29 109 sheep, 448 cattle, 140 deer, and four horses to perennial ryegrass toxicosis. Conservatively, the respondents who completed the questionnaire represented no more than one-third of the sheep depastured in the affected climatic zone where perennial ryegrass-dominant pasture is common. Up to 30% mortality occurred in some groups, the majority of losses being weaned lambs or young sheep.

Ill-thrift, staggers, and death of sheep and cattle from the epidemic in the summer–autumn months was common in Victoria (Table 1.4.1), and most severe in the 600- to 700-mm rainfall zone. Twelve farms lost >500 head. Problem pastures were heavily dominated by perennial ryegrass. The main legume in this zone is the annual, subterranean clover (*Trifolium subterraneum*). Invariably, it is desiccated and only present as seed burr during late summer and early autumn. Various management regimes were implicated. Examples included regrowth after mowing, pastures offering abundant feed (i.e., pasture mass > 3 Mg ha^{-1} dry wt.), and closely grazed and cell-grazed pasture; 19 varieties were named, with the long-used Victorian ecotype the most common.

Table 1.4.1. Number of farms that voluntarily reported effects of perennial ryegrass toxicosis in 2002 and their loss of livestock.

Region†	No. of reports	Sheep deaths	Cattle deaths
SE South Australia	6	179	3
Tasmania	11	3 232	336
NE Victoria	11	74	3
SE Victoria	21	2 294	17
SW Victoria	175	23 330	89
Total	224	29 109	448

† The epidemic also caused losses of young dairy cattle in southwestern West Australia (Dr. D. Forshaw, Agriculture West Australia, 2002, personal communication) and significant loss of sheep on Kangaroo Island, South Australia (Dr. G. Johnson, 2002, personal communication).

'Ellett', 'Victoca', 'Yatsyn 1', and 'Grasslands Impact' were the next most frequent.

Farmers reported that sheep and cattle were in good condition after an excellent pasture season (livestock prices were at a 30-yr high). As the epidemic spread, farmers commented that live-weight losses exceeded normal seasonal effects. Surviving sheep that had displayed clinical symptoms and caused management difficulties contributed to further economic losses. Supplementary feeding and fertilizer spreading were postponed. High loss of ewes and weaned lambs seriously disrupted breeding goals. Shearing and marketing of stock was delayed. Lambing was reduced and protracted due to the difficulty in removing rams. Delayed crutching and drenching led to severe intestinal worm and flystrike problems. Wool cut was reduced and tenderness increased. Lameness in sheep and cattle and a high incidence of pregnancy toxemia were reported. Fox and crow preyed on weak and recumbent stock.

Many prime lamb dams failed to conceive. Their performance on perennial ryegrass at DPI's Hamilton farm was not atypical for autumn joined flocks on other affected farms. At scanning (June), the 4-yr-old ewe flock (which had shown clinical symptoms) had an 18% empty rate to natural mating; the 3-yr-old flock had a 24% empty rate (staggers symptoms less severe). Their scanned twinning rates were 54 and 61% respectively. Five-year-old ewes mated 3 wk later were more normal with an empty rate of 3% and a twinning rate of 69% (L.J. Cummins, DPI, 2004, personal communication). Aside from the direct losses in autumn, veterinary scientists estimated that a similar number of sheep died in winter due to intestinal worms. Most of the latter loss was associated with the missed anthelmintic treatment normally given in late summer. No treatment appeared reliable for reversing the clinical condition. For sheep, some veterinarians tried the use of promazine, diazepan, metoclopramide, and domperidone for short periods. No substantial improvement was noticed. The latter two (dopamine antagonists) showed some promise. Treatment with calcium-boroglutonate alone or with glucose, magnesium and phosphorus, or magnesium sulfate solution gave no beneficial response other than an initial stimulus to sheep that had just gone down. Some valuable stock were housed, harnessed, and nursed with stomach tubing to feed electrolytes and energy supplements.

On-Farm Toxin Monitoring

Ten farms were monitored for endophyte and toxins in April and May. These farms lost a total of 7847 head in autumn (9.8% of their flocks). Perennial ryegrass from 2 to 5 paddocks on each monitor farm was sampled once for endophyte detection. The sample set included two paddocks that were only lightly affected. For toxin analysis, three successive samples were taken with sampling repeated approximately three and six weeks after the initial sampling. Pastures and livestock on a further 10 affected farms were sampled on one occasion only. Twenty plants were sampled to determine endophyte frequency (three tillers per plant were cut below the crown). Perennial ryegrass was cut below the crown from 60 sites for toxin analysis. Toxin samples were separated into green and dead fractions then freeze dried. Fecal samples were collected from 3 to 6 recumbent/staggering animals. Individual samples were freeze dried and ground, and for each paddock a composite sample was made.

Neotyphodium lolii presence was determined for each of the three tillers by ELISA (Musgrave and Fletcher, 1986; Guy, 1992). Endophyte concentration was rated on a 1-to-5 scale. Antiserum and alkaline-phosphatase conjugates were used. Alkaloids were determined in grass and feces using HPLC: lolitrem B (Gallagher et al., 1984) and ergovaline (Shelby and Flieger, 1997—modified as in Reed et al., 2004). Dry-fraction samples of perennial ryegrass from 14 paddocks were tested for the possible presence of the corynetoxins associated with annual ryegrass toxicity; all tested negative.

The toxin concentration in the green grass fraction was approximately two- to fourfold that observed in the (dominant) dry fraction. Ergovaline exceeded 0.45 mg kg^{-1} in every sample of green grass and with lolitrem B, all but three samples exceeded 2.25 mg kg^{-1}. Affected paddocks contained a high perennial ryegrass content and, on average, approximately 3 Mg dry matter ha^{-1} in early April (Table 1.4.2). The green component accounted for 5 to 15% of the total DM. Median endophyte frequency was 93%. Cultivars encountered were mainly Victorian (ecotype), with Ellett, Lincoln, Grasslands Impact, and Aries represented. During April through May, the endophyte (e) measurements made for 21 pastures accounted for 61.8% ($P < 0.001$) of the variation in lolitrem B in green grass (mean across three sampling dates): Lolitrem B = $-0.516 + 0.024$ (± 0.005) endophyte frequency (%) + 0.014 (± 0.007) endophyte concentration (% of max. possible rating for endophyte-infected plants). However, a similar relationship between endophyte infection and ergovaline was poor, accounting for only 25.8% ($P = 0.027$) of the variation in ergovaline among affected pastures. Toxins were readily detected in fecal samples. On one farm, three paddocks returned high fecal lolitrem B concentrations of 11.5, 4.9, and 2.9 mg kg^{-1} dry wt.

Surviving stock exhibited poor live-weight gain, wool growth, wool strength, and reproduction extending through the winter and spring, as reported in spring by nine out of 10 farmers whose pastures had been monitored. Live-weight gain was retarded for much of the year, especially in young stock. Intestinal worm problems and scouring were severe relative to previous years. Increased flystrike was attributed to increased breech soiling in down sheep. Some respondents observed a 15% drop in fleece weight. Merino lambing percentages for May and June were 14% on one farm and 40 to 50% for some others. For meat–lamb flocks, reduced rates of gain and difficulties in moving lambs to better feed led to reduced and delayed income.

Table 1.4.2. Pasture details, endophyte presence, and toxin concentration in the green perennial ryegrass fraction and feces of affected livestock from pastures monitored in southern Victoria, April to May 2002.

	No. of samples	Range	Median	Mean ± SD
Herbage dry mass, 1st sampling, Mg ha^{-1}	31	1.0–6.3	3	3.2 ± 1.3
Perennial ryegrass content, % dry wt.	31	15–95	90	80 ± 23
Age of pasture, yr	31	1–50	15	16 ± 12
Endophyte frequency, %†	21	15–100	93	85 ± 20
Endophyte concentration, %‡	21	8–90	57	57 ± 19
Alkaloid concentrations	——————mg kg^{-1} dry wt.——————			
Ergovaline in grass§	31	0–4.30	1.51	1.75 ± 0.99
Ergovaline in feces	30	0–0.55	0.10	0.18 ± 0.16
Lolitrem B in grass§	31	1.10–4.60	3.06	3.02 ± 0.87
Lolitrem B in feces	31	0.10–11.50	0.78	1.36 ± 2.15

† Proportion of 20 plants infected.
‡ Estimated concentration in endophyte infected plants (ELISA rated 1 = low, 5 = high).
§ Maximum concentration where serial samples were analyzed.

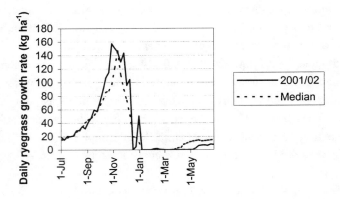

Fig. 1.4.1. *Growth rate of perennial ryegrass (dry wt.) at Hamilton, Victoria, simulated weekly by the GRASSGRO model (Moore et al., 1997) from meteorological data for 2001–2002 and from the median data for 1965 through 2002.*

Climatic and Genetic Factors

Seasons in which clinical effects are widespread usually have heavy late-spring rain and a hot, dry summer–autumn. In 2001–2002, the first of these conditions resulted in strong pasture and aftermath growth across southwestern Victoria (Fig. 1.4.1). The summer was mild and favored survival of more worm eggs than usual, and so aggravated the problem of delayed anthelmintic treatment caused by staggers. In Caramut, the worst-affected district, 200 mm of rain fell in summer. Stock in this district suffered chronic toxicosis for more than three months. The unseasonal rain caused seed germination. Seedlings can be a concentrated source of alkaloids, causing ryegrass staggers (Aasen et al., 1969). Although the autumn drought was prolonged and effective growth delayed until the end of May, most sheep deaths occurred in March to April, prior to the peak drought stress. From monitor farms, toxin concentration in the whole plant was computed from fractions and while well above critical levels (Hovermale and Craig, 2001) lolitrem B was similar to concentrations measured in Victoria in small surveys conducted in 1999 and 2000 when the usual symptoms were observed but the acute condition was uncommon. Ergovaline was only slightly higher than in the earlier surveys (Reed et al., 2004). Possibly several other known tremorgens produced by the ryegrass endophyte (Lane, 1999) may have contributed directly or synergistically (Reid et al., 1978) to the high mortality in 2002; no determination for these was made.

The uniqueness of the 2002 season may perhaps be the extended summer period for consumption of fresh ryegrass rather than an unusually high toxin concentration. Coincidently, in an unusual step, live sheep exports from Portland, southwestern Victoria, were temporarily suspended by the federal government on 1 Oct. 2002. This moratorium was brought about by the excessive losses. In winter shipments to the Middle East, 17 000 sheep died. The problem was peculiar to sheep boarding from southeastern Australia. The main cause of death was given as heat stress and vulnerability to infection, such as *Salmonella*. A residual effect of ergot alkaloid ingestion possibly contributed to an excess heat load in animals under stress. Ergot alkaloids can accumulate in fat tissue (Realini et al., 2004). The retention and slow release of potent alkaloids from this reservoir during an autumn feed gap/stress may therefore impact on livestock entering feedlots or embarking on ships. Fescue toxicosis symptoms in cattle have been observed in feedlots long after their removal from toxic feed (Duckett et al., 2001).

High toxin associations may arise from host genotypes that encourage intense infection and mycelial mass. Given the restricted genetic diversity of *N. lolii* in the naturalized populations and cultivars of perennial ryegrass in Australasia (Van Zijll de Jong et al., 2002), the wide diversity in toxin production when old populations are grown under uniform conditions (Reed et al., 2000) may reflect the importance of host genetic control.

RESEARCH

Independent applied research on agronomic performance and alkaloid profiles of cultivars within the major regions is required to enable farmers to objectively appraise the option of select endophyte perennial ryegrass cultivars. Use of select endophyte may avert toxicosis provided that unsafe seed can be removed from the environment. Comparative animal production studies involving select endophyte perennial ryegrass pasture are required to assess benefits. Optimum management guidelines for old pasture are needed in order to minimize the impact of toxins on the productivity and welfare of those flocks and herds that will depend on it for many years.

Summary

Recent field research on grass–endophyte interactions in Australia is limited to small-plot cultivar trials. Tall fescue cultivars containing MaxP *Neotyphodium coenophialum* endophyte have overcome insect predation in isolated field trials. Together with the release of summer dormant cultivars, MaxP may help to expand the area for which tall fescue is accepted as a useful forage. While endophyte-free perennial ryegrass has not persisted as well as plants with wild endophyte, cultivars containing the select AR1 *Neotyphodium lolii* endophyte have also shown low persistence relative to wild endophyte at the one site from which results are available.

No endophyte–host–consuming herbivore interaction studies or reports of animal production measurements have been published since the last international symposium. Clinical symptoms of perennial ryegrass toxicosis are seen in livestock in autumn, and occasionally severe epidemics occur. Following good spring and early-summer rain, an epidemic of perennial ryegrass toxicosis occurred during late summer and autumn 2002 in the 600- to 750-mm rainfall zone, across the four southern states. Crucial husbandry and marketing operations were badly disrupted with many sheep becoming recumbent. Livestock losses in the tens of thousands were reported. Up to 30% mortality occurred in some groups, the majority of losses being weaned lambs or young sheep. Necropsy of sheep revealed histological changes in the cerebellum consistent with perennial ryegrass toxicosis.

For perennial ryegrass pastures monitored on affected properties, the mean endophyte frequency was 85%. Ergovaline and lolitrem B in green perennial ryegrass averaged 1.75 and 3.02 mg kg^{-1} dry matter and the feces of affected animals contained mean concentrations of 0.18 and 1.36 mg kg^{-1}, respectively. Surviving sheep exhibited poor growth rate. For young sheep especially, this effect lasted well into spring on some properties. Aside from the direct deaths in autumn, a similar scale loss occurred in winter due to heavy burdens of intestinal worms. The unusually high winter mortality could in part be attributed to sheep being unavailable for anthelmintic treatment in late summer–early autumn as a result of ryegrass staggers. The new select endophyte technology offers solutions; research is required to quantify the benefits and guide farmers and advisers in its application.

ACKNOWLEDGMENTS

The authors thank the Grassland Society of Southern Australia for travel assistance and for distributing questionnaires; Mr. Peter Cross, Department of Primary Industries, Tasmania, for endophyte detection tests; Dr. Steven Colegate, CSIRO, for analysis of corynetoxins; Dr. Leo Cummins, DPI Hamilton, and Dr. Pat Kluver, Dr. Angus Campbell, and Dr. Janine Sandy, School of Veterinary Science, The University of Melbourne, for postmortem studies; and Mr. Stephen Clark and Mr. Dion Borg, DPI Hamilton, for assistance.

REFERENCES

Aasen, A.J., C.C.J. Culvenor, E.P. Finnie, A.W. Kellock, and L.W. Smith. 1969. Alkaloids as a possible cause of ryegrass staggers in grazing livestock. Aust. J. Agric. Res. 20:71–86.

Duckett, S.K., J.A. Bondurant, J.G. Andrae, J. Carter, T.D. Pringle, M.A. McCann, and D. Gill. 2001. Effect of grazing tall fescue endophyte types on the subsequent feedlot performance and carcass quality. J. Anim. Sci. (Suppl. 1) 79:221.

Foot, J.Z., O.J. Woodburn, J.R.Walsh, and P.G. Heazlewood. 1994. Responses in grazing sheep to toxins from perennial ryegrass/endophyte associations. p. 375–380. *In* S.M. Colegate and P.R. Dorling (ed.) Plant-associated toxins—Agricultural, phytochemical and ecological aspects. CAB International, Wallingford, UK.

Gallagher, R.T., A.D. Hawkes, P.S. Steyn, R. Vleggaar. 1984. Tremorgenic neurotoxins from perennial ryegrass causing ryegrass staggers disorder of livestock: Structure elucidation of lolitrem B. J. Chem. Soc. Chem. Comm. 614–616.

Guy, P.L. 1992. Enzyme-linked immuno absorbent assays (ELISA) to detect perennial ryegrass endophyte (*Acremonium lolii*) and *Metarhizium anisopliae*, a potential bio-control agent. p.321–325. *In* E.S. Delfosse (ed.) Pests of pastures weed, invertebrate and disease pests of Australian sheep pastures. CSIRO, Melbourne, Australia.

Hovermale, J.T., and A.M. Craig. 2001. Correlation of ergovaline and lolitrem B in endophyte-infected perennial ryegrass (*Lolium perenne*). J. Vet. Diag. Invest. 13:323–327.

Lane, G.A. 1999. Chemistry of endophytes: Pattern and diversity. p. 85–94. *In* D.R. Woodfield and C. Matthew (ed.) Ryegrass endophyte: An essential New Zealand Symbiosis. Research and Practice Ser. No. 7. N. Z. Grassl. Assoc., Palmerston North.

Lean, I.J. 2001. Association between feeding perennial ryegrass (*Lolium perenne* cultivar Grasslands Impact) containing high concentrations of ergovaline, and health and productivity in a herd of lactating dairy cows. Aust. Vet. J. 79:262–264.

Mason, R.W. 1968. Axis cylinder degeneration associated with ryegrass staggers in sheep and cattle. Aust. Vet. J. 44:428.

Moore, A.D., J.R. Donnelly, and M. Freer. 1997. GRAZPLAN: Decision support systems for Australian grazing enterprises. III. Pasture growth and soil moisture sub-models and the GrassGro DSS. Agric. Syst. 55:535–582.

Munday, B.L., I.M. Monkhouse, and R.T. Gallagher. 1985. Intoxication of horses by lolitrem B ryegrass seed cleanings. Aust. Vet. J. 62:207.

Musgrave, D.R., and L.R. Fletcher. 1986. Optimisation and characterisation of enzyme-linked immunosorbent assay for detecting *Acremonium* endophyte. Molec. Gen. Genet. 233:1–9.

Quigley, P. 2000. Effects of *Neotyphodium lolii* infection, and sowing rate of perennial ryegrass, on dynamics of ryegrass/subterranean clover swards. Aust. J. Agric. Res. 50:47–56.

Radostits, O.M., C.C. Gay, D.C. Blood, and K.W. Hinchcliff. 2000. Veterinary medicine, a textbook on the diseases of cattle, sheep, pigs, goats and horses. 9th ed. Saunders, Sydney, Australia.

Realini, C.E., S.K. Duckett, N.S. Hill, C.S. Hoveland, B.G. Lyon, J.R. Sackman, and M.H. Gillis. 2004. Effect of endophyte type on carcass traits, meat quality and fatty acid conmposition of beef cattle grazing tall fescue. J. Anim. Sci. 82:(in press).

Reed, K.F.M., A. Leonforte, P.J. Cunningham, J.R. Walsh, D.I. Allen, G.R. Johnstone, and G. Kearney. 2000. Incidence of ryegrass endophyte (*Neotyphodium lolii*) and diversity of associated alkaloid concentrations among naturalised populations of perennial ryegrass. Aust. J. Agric. Res. 51:569–578.

Reed, K.F.M., N.M. McFarlane, and J.R. Walsh. 2001. Seasonal alkaloid concentration in perennial ryegrass dairy pasture. Proc. Aust. Agron. Conf., 10th, Hobart, 29–31 Jan. 2001 [Online]. Available at www.regional.org.au/au/asa/2001/ [verified 22 Aug. 2004]. Australian Society of Agronomy.

Reed, K.F.M., J.R. Walsh, N.M. McFarlane, P.A. Cross. 2004. Ryegrass endophyte (*Neotyphodium lolii*) alkaloids and mineral concentrations in perennial ryegrass (*Lolium perenne*) from south West Victoria. Aust. J. Exp. Agric. 44(8) (in press).

Reid, C.S.W., R.G. Keogh, G.C.M. Latch, and R.T. Gallagher. 1978. Ryegrass staggers: A role for fungal tremorgens. Proc. N. Z. Soc. Anim. Prod. 38:53–57.

Shelby, R., and M. Flieger. 1997. Improved method of analysis for ergovaline in tall fescue by high performance liquid chromatography. J. Agric. Food Chem. 45:1797–1800.

Valentine, S.C., B.D. Bartsch, and P.D. Carroll. 1993. Production and composition of milk by dairy cattle grazing high and low endophyte cultivars of perennial ryegrass. p. 138–141. *In* D.E. Hume et al. (ed.) Proc. 2nd Int. Symp. *Acremonium*/grass Interactions, Palmerston North, NZ. 4–6 Feb. 1993. AgResearch, Palmerston North.

Van Zijll de Jong, E., N.R. Bannan, J. Batley, K.M. Guthridge, G.C. Spangenberg, K.F. Smith, and J.W. Forster. 2002. Genetic diversity in the perennial ryegrass fungal endophyte *Neotyphodium lolii*. p. 155–164. *In* A. Hopkins et al. (ed.) Molecular breeding of forage and turf. Kluwer Academic Press, Boston.

Wheatley, W.M., D.E. Hume, H.W. Kemp, M.S. Monk, K.F. Lowe, A.J. Popay, D.B. Baird, and B.A. Tapper. 2003. Effects of fungal endophyte on the persistence and productivity of tall fescue at three sites in eastern Australia. Proc. Aust. Agron. Conf., 11th, Geelong, VIC, Australia [Online]. Available at www.regional.org.au/au/asa/2003/ [verified 22 Aug. 2004]. Australian Society of Agronomy.

Woodburn, O.J., J.R. Walsh, J.Z. Foot, and P.G. Heazlewood. 1993. Seasonal ergovaline concentrations in perennial ryegrass cultivars of differing endophyte status. p. 100–102. *In* D.E. Hume et al (ed.) Proc. 2nd Int. Symp. *Acremonium*/grass Interactions, Palmerston North, NZ. 4–6 Feb. 1993. AgResearch, Palmerston North.

NEOTYPHODIUM RESEARCH AND APPLICATION IN JAPAN

Koya Sugawara,[1] Takuya Shiba,[1] and Masayuki Yamashita[2]

As a part of East Asia, the Japanese Islands are located in the northwestern side of the Pacific Ocean, of similar latitude to the islands of New Zealand (Fig. 1.5.1). Vegetation and land use, however, are very different between the two island groups owing to different seasonal temperatures and Japan's large population of 126 million.

Fig. 1.5.1. *Latitude comparison of the East Asia region with other parts of the world. Each area is displayed in Mercator's projection.*

The presence of *Neotyphodium* endophytes in Japan has been recognized during the past 20 yr, starting with reports in imported forage grass species (Koga et al., 1993; Sato et al., 1995). The first nationwide survey included native grass species and recorded several new endophyte–grass associations (Enomoto et al., 1998). Searches for endophyte–grass associations have continued (Saiga et al., 2003a; Yamashita et al., 2003; Yanagida et al., 2004) (Table 1.5.1). Detailed descriptions and phylogenetics of these endophytes are not yet well documented, but Yanagida et al. (2004) reported results of phylogenetical analysis of *tub2* sequences indicating that isolates collected from *Agropyron ciliare* var. *minus* (syn. *Elymus racemifer*) and *A. tsukusiense* var. *transiens* (syn. *Elymus tsukushiensis*) are grouped apart from the known *Epichloë–Neotyphodium* species.

[1] National Institute of Livestock and Grassland Science, Japan
[2] Faculty of Agriculture, Shizuoka University, Japan

There may be many hidden genetic resources of *Neotyphodium* endophytes in East Asia, including Japan.

Table 1.5.1. *Neotyphodium*-infected plants reported in Japan.

Name of plant	Reference
Achnatherum extraorientale	Enomoto et al., 1998
Agrostis clavata	Enomoto et al., 1998
Agrostis clavata spp. *matsumurae*	Enomoto et al., 1998
Agrostis dimorpholemma	Enomoto et al., 1998
Avena fatua	Enomoto et al., 1998
Brachypodium sylvaticum	Enomoto et al., 1998
Bromus catharticus	Enomoto et al., 1998
Bromus pauciflorus	Enomoto et al., 1998
Diarrhena fauriei	Enomoto et al., 1998
Diarrhena japonica	Enomoto et al., 1998
Elymus dahuricus	Enomoto et al., 1998
Elymus racemifer	Enomoto et al., 1998;
(syn. *Agropyron ciliare* var. *minus*)	Yanagida et al., 2004
Elymus tsukushiensis	Enomoto et al., 1998;
(syn. *Agropyron tsukushiensis* var. *transiens*)	Yanagida et al., 2004
Festuca parvigluma	Enomoto et al., 1998
Festuca pratensis	Takai et al., 2001
Festuca rubra	Enomoto et al., 1998
Glyceria ischyroneura	Enomoto et al., 1998
Lolium multiflorum	Yamashita et al., 2003;
	Sugawara et al., 2004a, 2004b
Lolium perenne	Koga et al., 1993; Sato et al., 1995;
	Enomoto et al., 1998
Melica onoei	Enomoto et al., 1998
Poa annua	Enomoto et al., 1998
Poa nipponica	Enomoto et al., 1998
Poa sphondylodes	Enomoto et al., 1998
Poa trivialis	Enomoto et al., 1998
Trisetum bifidum	Enomoto et al., 1998

National surveys of endophyte occurrences reveal new *Neotyphodium* species and associations of academic and agricultural interest. Surveys also reveal the potential for the occurrence of endophyte toxicosis problems as observed in New Zealand and the USA. The causes of such toxicoses in those two countries are tall fescue (*Festuca arundinacea*) infected with *N. coenophialum* and perennial ryegrass (*Lolium perenne*) infected with *N. lolii,* which were imported and used for meadows, sports turf, and soil conservation in Japan dating from the start of the Meiji Restoration (i.e., opening up of the country, late 1800s). Saiga et al. (2003a) reported that *Festuca* and *Lolium* populations in northern Japan (Hokkaido Island and northern part of Honshu Island) may include plants infected with the endophytes which produce ergovaline and (or) lolitrem B.

Although plants producing those alkaloids do not comprise the majority of plant populations, and toxicosis problems in grazing animals have not been reported, effective methods for observation and identification of those infected plants have been sought. Ohkubo et al. (2000) reported that the endophytes can be more easily identified in plant tissue when using acid fuchsin stain in lactic acid (0.1%, w/v) than the stains currently used, such as rose bengal and aniline blue. Majewska-Sawka and Nakashima (2004) documented the seed transmission process in perennial ryegrass–*N. lolii* associations by microscopic observation that utilized immunological staining. Sugawara et al. (2004b) reported that the endophytes can be easily seen with differential interface contrast microscopy in immature ovaries of host plants that have been cleared with lactic acid in glycerol (lactic acid–glycerol–water ratio = 1:2:1, v/v/v), and suggested that this method provides an accurate and easy way to detect *Neotyphodium* in flowering grasses (Fig. 1.5.2).

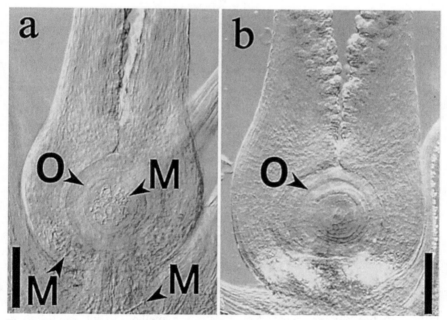

Fig. 1.5.2. *Neotyphodium endophyte observed in chemically cleared immature ovary of Lolium multiflorum. (a) An ovary of infected L. multiflorum. Mycelia of N. occultans* (M) *are present in and around the ovule* (O). *Differential interference contrast microscopy. Bar = 100 μm.* (b) *An ovary of noninfected L. multiflorum in similar developing stage as the ovary in* (a). *No mycelium is seen in and around the ovule* (O). *Bar = 100 μm.*

It has been assumed in Japan that the humid climate and the large biodiversity within plant populations may limit any fitness benefits of infected plants, so that there has been no proliferation of infected plants and thus no animal toxicosis problems. However, recent studies revealed that naturalized Italian ryegrass (*L. multiflorum*) plants infected with *N. occultans* dominate populations in many parts of the country (Sugawara et al., 2004a; Yamashita et al., 2003) (Fig. 1.5.3).

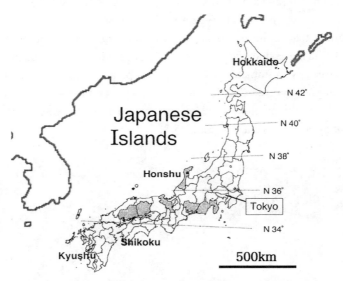

Fig. 1.5.3. *Regions (prefectures) where Lolium populations infected with Neotyphodium occultans were recorded in Japan in 2003 (shadowed).*

The Italian ryegrass/*N. occultans* symbiosis has received little investigation, as no problems or benefits resulting from the symbiosis have been recognized. The relatively small biomass of hyphae in plants has perhaps been another reason why the presence of the endophyte has been largely overlooked. However, the dominance of infected plants in many Italian ryegrass populations in Japan strongly suggests that some benefit does result from the symbiosis and further studies are warranted. Novel detection methods using host flowers (Sugawara et al., 2004b, Fig. 1.5.2) may help accelerate the process.

Toxicosis problems in Japan have been caused by imported endophyte-infested feed. Because of the small land area available for livestock farming, owing to high population density and other social factors, Japan has been a huge importer of food and animal feed, including straw. Beef cattle and milking cows comprise the majority of the grazing livestock in Japan, numbering 2 838 000 and 1 726 000, respectively, according to a 2001 survey (Ministry of Agriculture, Forestry and Fishery, 2004), and are producing 459 000 Mg of beef (35% of national demand) and 8 312 000 Mg of milk (68% of national demand) (in 2002, Ministry of Agriculture, Forestry and Fishery, 2004). The livestock industry in Japan imports approximately 2.3 million Mg of hay per year, of which 80% is from the USA (in 2001, Ministry of Agriculture, Forestry and Fishery, 2003). Miyazaki et al. (2001) reported that there were at least 26 cases of cattle toxicosis and three cases of horse toxicosis, from 1997 to 1999, caused by the endophyte in imported straw. Two major animal toxins, ergovaline and lolitrem B, were detected in straw samples from those cases, but the concentration of these alkaloids did not always exceed the threshold levels proposed by Aldrich-Markham et al. (2003). Feeding experiments using infected straw were carried out, and the results strongly implied that Japanese Black cattle, the breed mainly grown for meat in Japan, are highly sensitive to endophyte toxins, and can be affected by those toxins in lower concentrations than the proposed threshold (Miyazaki, 2003; Miyazaki et al., 2001). Now that Japanese Black cattle are grown not only in Japan but also in other parts of the world, further study on the impact of these toxins on their health may be needed to prevent expansion of the problem.

To elucidate factors relating to the endophyte–grass symbiosis, techniques have been established to inoculate plants, to eradicate the endophyte from tillers of mature plants, and to kill the endophyte in seed. Saiga et al. (2003b) tested several fungicides to eradicate the endophyte in grass tillers, and reported that soaking in 2% benomyl solution was effective for treating tall fescue and perennial ryegrass. Sato and Takada (2002) demonstrated that heat treating infected seeds of perennial ryegrass at 80°C for 2 to 8 d eradicated the endophyte without loss of seed viability. Although seedling inoculation is the main way that novel endophyte–grass associations are obtained, some novel associations have been made by crossing experiments. Crossings between infected *Lolium* species and tall fescue have generated intergeneric hybrids (×*Festulolium* spp.) that were stably infected with the *Neotyphodium* endophyte derived from perennial ryegrass (Sugawara et al., 2003).

Artificial inoculation with endophytes in Japan has been carried out mainly to increase the growth and insect tolerance of turf and forage grasses. For forage grasses, *N. uncinatum*, an endophyte from meadow fescue (*F. pratensis*), has been used as a candidate for inoculation. Meadow fescue was imported about a century ago and has been selected and used for pasture in the northern part of Japan. Takai et al. (2001) reported that neither ergovaline nor lolitrem B was detected in naturally infected meadow fescue cultivars, and thus this species is unlikely to cause problems with grazing animals. Following this finding, endophyte strains from meadow fescue cultivars were isolated and inoculated into meadow-type cultivars of *L. multiflorum*, the most popular pasture grass in Japan, to increase the growth and insect resistance of host plants (Kasai et al., 2004). The infected ryegrasses made stable symbiotic associations with the endophyte (Kasai et al., 2004) and were reported as having enhanced insect resistance (Shiba et al., 2004). Commercial production of those infected ryegrass cultivars is now in preparation.

Fig. 1.5.4. *Kernel spotting of rice ("spotted rice" or "pecky rice") and one of the major causes of the symptom, Trigonotylus caelestialium (Heteroptera: Miridae, rice leaf bug). (a) Spotted rice, (b) an adult of T. caelestialium, (c) a larva of T. caelestialium, Bar = 1 mm , (d) scanning electron micrograph of a larva of T. caelestialium, bar = 200 μm.*

The insect-repelling effects of endophytes are now getting attention in Japan as insect pests in meadows are causing problems in adjacent rice (*Oryza sativa*) paddies. In Japanese farming regions, a variety of crops are planted like patchwork on small land areas. Meadows, on which little or no measures for pest control are applied, are sometimes considered a hotbed of pests, especially insects. One of the most serious insect problems related to grass production is kernel spotting of rice grains (*spotted rice* or *pecky rice*, Fig 1.5.4a), caused by rice leaf bugs [*Trigonotylus caelestialium* (Heteroptera: Miridae); Fig. 1.5.4b–1.5.4d].

Grasslands, especially those planted with ryegrass species, are considered one of the major sources of the propagation of the rice leaf bugs (Takada et al., 2000). They propagate on meadows and move to nearby rice paddies following the harvest of grasses. Contamination by spotted grains results in a severe downgrading of the quality and hence the price received for the rice crop. The dark grains are visually offensive because, in Japan, rice is usually served after simple boiling or steaming. Shiba et al. (2004) reported that perennial ryegrass infected with *Neotyphodium* endophytes, and Italian ryegrass inoculated with *N. uncinatum* (Kasai et al., 2004), expressed increased tolerance to this major species causing the spotted rice in northern Japan. Studies are under way to identify the deterrent factor(s) and other novel chemicals present in these endophyte–infected grasses. Shiba et al. (2004) detected loline, but no peramine, ergovaline, or lolitrem B in Italian ryegrass inoculated with *N. uncinatum*. Araya et al. (2003) identified two ergosterols known to have antibacterial activity from extracts of *N. lolii* mycelium.

SUMMARY

Neotyphodium endophytes have been found in many native grass species and in some exotic species in Japan, but no toxicosis problems in grazing animals have been reported. Toxicosis problems observed in Japan have been caused only by imported straw. Studies indicate that Japanese Black cattle, the cattle breed mainly grown for meat in Japan, tend to be highly sensitive to endophyte toxins. Meadows can be sources of pests for neighboring crop fields, especially rice. Use of forage grasses infected with endophyte has been considered a potential solution to insect pest problems in and around forage grass fields in Japan. There are real and potential problems associated with endophytes in Japan, and hidden genetic resources are waiting to be revealed. Japanese research has expanded in recent years into more comprehensive surveys of endophyte occurrence and novel techniques for endophyte detection and transfer.

ACKNOWLEDGEMENTS

Authors are grateful for helpful advice given by M.J. Christensen, AgResearch Grassland Research Centre, New Zealand, Professor C.L. Schardl, University of Kentucky, and Professor C.P. West, University of Arkansas, USA. We are also grateful for help and advice from E. Kasai and T. Sasaki, Japan Grassland Farming and Forage Seed Association, and Y. Mikoshiba and H. Ohkubo, National Institute of Livestock and Grassland Science, Japan, in the preparation of the manuscript.

REFERENCES

Aldrich-Markham, S., G. Pirelli, and A.M. Craig. 2003. Endophyte toxins in grass seed fields and straw effects on livestock [Online]. Oregon State Univ. Ext. and Exp. Stn. No. EM 8598. Available at http://eesc.orst.edu/agcomwebfile/edmat/EM8598.pdf [revised Aug. 2003; cited 27 May 2004; verified 22 Aug. 2004]. Oregon State Univ., Corvallis, OR, USA.

Araya, H., M. Kobayashi, M. Ebina, H. Nakagawa, E. Nishihara, and Y. Fujimoto. 2003. Identification of two $5\alpha,8\alpha$-epidioxyergosta-3β-ols from the endophyte, *Neotyphodium lolii*. Biochem. System. Ecol. 31:1337–1339.

Enomoto, T., T. Tsukiboshi, and T. Shimanuki. 1998. The gramineous plants in which *Neotyphodium* endophytes were found. (In Japanese.) J. Weed Sci. Technol. 43(suppl.):76–77.

Kasai, E., T. Sasaki, and H. Okazaki. 2004. Artificial infection of Italian ryegrass (*Lolium multiflorum* Lam.) with *Neotyphodium uncinatum*. (In Japanese with English summary.) Grassl. Sci. 50:180–186.

Koga H., T. Kimigafukuro, T. Tsukiboshi, and T. Uematsu. 1993. Incidence of endophytic fungi in perennial ryegrass in Japan. Ann. Phytopathol. Soc. Jpn. 59:180–184.

Majewska-Sawka, A., and H. Nakashima. 2004. Endophyte transmission via seeds of *Lolium perenne* L: Immunodetection of fungal antigens. Fungal Genet. Biol. 41:534–541.

Ministry of Agriculture, Forestry and Fishery. 2003. 2002 Ryutsu Shiryou Binran (Feed Trade Handbook 2002). (In Japanese. Translated title by the present authors.) Nourin Toukei Kyoukai, Tokyo.

Ministry of Agriculture, Forestry and Fishery. 2004. Pocket Chikusan Toukei Heisei 14 nendo ban (Pocket almanac of livestock production, 2002 edition). (In Japanese. Translated title by the present authors.) Nourin Toukei Kyoukai, Tokyo.

Miyazaki, S. 2003. Current subjects on endophyte toxicosis. Mycotoxins 53:57–62.

Miyazaki, S., M. Fukumura, M. Yoshioka, and N. Yamanaka. 2001. Detection of endophyte toxins in the imported perennial ryegrass straw. J. Vet. Med. Sci. 63:1013–1015.

Ohkubo, H., Tsukiboshi, T., Sugawara, K. and Shimanuki, T. 2000. Comparison of detection of *Neotyphodium* endophytes from examining seeds and plants of tall fescue (*Festuca arundinacea*) and perennial ryegrass (*Lolium perenne*). p. 102. *In* Abstracts of Asian Mycological Congress 2000, Hong Kong, China, July 9–13.

Saiga, S., T. Inoue, H. Nakashima, A. Maejima, S. Yoshid, and M. Tsuki. 2003a. Incidence of *Neotyphodium* endophytes among naturalized perennial ryegrass and tall fescue plants in northern Japan and alkaloid concentration of the seeds from infected plants. Grassl. Sci. 49:444–450.

Saiga S, Y. Kodama, H. Takahashi, and M. Tsuiki. 2003b. Endophyte removal by fungicides from ramets of perennial ryegrass and tall fescue. Grassl. Sci. 48:504–509.

Sato, N., T. Tagawa, T. Kitamori, and N. Akiyama. 1995. Incidence of endophyte fungi on forage grasses in the grassland in northern and central region of Hokkaido and preventive methods of infected seeds. J. Hokkaido Soc. Grassl. Sci. 29:78–84.

Sato, N., and Y. Takada. 2002. Sterilization of endophyte in perennial ryegrasses (*Lolium perenne* L.) seed by heat treatment. (In Japanese with English summary.) J. Hokkaido Soc. Grassl. Sci. 36:39–42.

Shiba, T., T. Sasaki, E. Kasai, K. Sugawara, and K. Kanda. 2004. Resistance for rice leaf bug (*Trigonotylus caelestialium*) observed in endophyte infected forage grasses. (In Japanese. Translated title by the present authors.) Grassl. Sci. 50(suppl.):226–227. (abstr.)

Sugawara, K., M. Fujimori, M. Ebina, S. Sugita, H. Ohkubo, and Y. Mikoshiba. 2003. Seed transmission of *Neotyphodium* endophyte to intergeneric hybrid grasses. p. 322. *In* Abstracts of Offered Papers, 8th Int. Congr. Plant Pathol. 2–7 Feb. 2003, ICPP, Christchurch, NZ.

Sugawara, K., T. Inoue, M. Yamashita, S. Saiga, H. Nakashima, H. Ohkubo, and Y. Mikoshiba. 2004a. Occurrence of *Neotyphodium* endophyte in naturalized *Festuca* and *Lolium*, and their ergovaline contents. (In Japanese.) Jpn. J. Phytopathol. 70:226 (abstr.).

Sugawara, K., H. Ohkubo, M. Yamashita, and Y. Mikoshiba. 2004b. Flowers for *Neotyphodium* endophyte detection: A new method for detection from grasses. Mycoscience 45:222–226.

Takada, M., H. Tanaka, and T. Chiba. 2000. Occurrence of the spotted rice in Iwate prefecture in 1999. (In Japanese.) Ann. Rep. Plant Prot. N. Jpn. 51:165–169.

Takai, T., S. Nakayama, Y. Terada, S. Hojito, H. Daido, H. Araki, K. Mizuno, S. Sugita, and K. Ito. 2001. Breeding of 'Harusakae' meadow fescue and its characteristics (in Japanese with English summary). Res. Bull. Hokkaido Natl. Agric. Exp. Stn. 173:47–62.

Yamashita, M., H. Tobina, K. Sugawara, and H. Sawada. 2003. Distribution of maternally transmitted fungal endophyte in naturalized ryegrass (*Lolium* spp.) in western Japan. p. 40. *In* Abstracts of Int. Symp. of the Society for the Study of Species Biology (SSSB), Sapporo, Japan. 16–17 Oct. 2003.

Yanagida, N., C. Teramoto, T. Irie, and A. Tajimi. 2004. Difference between choke pathogens isolated from *Agropyron ciliare* var. *minus* and *A. tsukusiense* var. *transiens*. (In Japanese. Translated title by the present authors.) Grassl. Sci. 50(suppl.):236–237.

NEOTYPHODIUM RESEARCH AND APPLICATION IN SOUTH AMERICA

Jose De Battista [1]

The agronomic and economic importance of *Neotyphodium* endophytes in South America is related to adaptation areas of their hosts, tall fescue (*Festuca arundinacea* = *Lolium arundinaceum*) and perennial ryegrass (*Lolium perenne*), whose zones comprise the humid and subhumid temperate regions of southern Brazil, Uruguay, Argentina, and Chile. Evidence that Italian ryegrass (*Lolium multiflorum*), an extensively used winter annual forage, is commonly infected with a *Neotyphodium*-like endophyte (probably *N. occultans*) (De Battista, 2002) has extended the interest of the impact of these associations in animal agriculture. This chapter will cover the incidence of endophyte infection on economically important grasses, the frequency and importance of animal toxicoses, and general recommendations to deal with the problems. The potential use of nontoxic endophyte technology will be addressed. Finally, current endophyte research efforts will be described.

INCIDENCE OF ENDOPHYTES AND ANIMAL TOXICOSES

Regional incidence of endophyte infection in tall fescue and perennial ryegrass was reviewed by De Battista et al. (1997), showing wide dissemination of *N. coenophialum* and *N. lolii* in seedlots and pastures. De Battista et al. (2001) extended a previous survey on *N. coenophialum* incidence in commercial tall fescue seedlots for the 1995–2000 period. As was shown in the previous period (1987–1994) (De Battista et al., 1995), endophyte incidence continued to decline across time; however, the frequency of potentially toxic samples (>5% infection level) varied between 11 and 19%, indicating that there is still a risk of seeding toxic pastures.

Occurrences of animal disorders grazing tall fescue at the farm level are not very common. Usually they occur in old pastures that were seeded during the

[1] INTA Estación Experimental Agropecuaria Concepción del Uruguay, Casilla de Correo No. 6, (3260) C. del Uruguay, Entre Ríos, Argentina

1980s and have not been replanted (E. Odriozola, 2004, personal communication). This low frequency of fescue toxicosis is probably related to farmers' awareness of the problem and official regulations that have set a standard level (<5%) for certified tall fescue seed. In addition, tall fescue is usually seeded mixed with legumes and replanted every 4 to 6 yr, resulting in a dilution effect of toxins that could be present in tall fescue forage (De Battista et al., 1997).

The impact of endophyte infection on tall fescue plant growth and persistence has been well documented in the international literature (Bouton et al., 1993; West et al., 1993) and in local conditions (De Battista et al., 1997). Usually endophyte-infected tall fescue provides increased forage production and persistence, although this response varies with environment and genotypic composition of the association. This indicates that there would be an advantage of using endophyte-infected tall fescue, providing that animal toxicoses are prevented. Since tall fescue mixed pastures are mainly used for steer fattening, management practices favor forage quality over pasture production and persistence. Therefore, recommendations are aimed at establishing endophyte-free pastures.

Outbreaks of perennial ryegrass toxicosis are much less frequent than fescue toxicosis, and the few detected cases of ryegrass staggers were associated with close summer grazing of infected seed production pastures (E. Odriozola, 2004, personal communication). Endophyte infection seems not to affect perennial ryegrass productivity or persistence in its areas of adaptation (De Battista et al., 1997; Tourne et al., 2002). Similar results were obtained in cool, moist environments in New Zealand (Eerens et al., 1997). No local information is available on the possibility that endophyte infection can extend the perennial ryegrass adaptation area to warmer regions. As with tall fescue, recommendations promote the establishment of endophyte-free pastures. However, since no regulations have been established concerning endophyte levels in perennial ryegrass seed, both highly and lowly infected cultivars are marketed by seed companies.

Italian ryegrass is a widespread, naturalized grass species in rangelands and crops in the humid and subhumid areas of Argentina, Chile, Uruguay, and southern Brazil. It is also cultivated as a winter forage crop, producing excellent forage quantity and quality during the critical autumn–winter period. Its use area extends into warmer regions than those of tall fescue. Italian ryegrass can harbor an endophytic fungus, recently named as *N. occultans* (Moon et al., 2000). Local surveys have shown that most naturalized populations in Entre Ríos Province, Argentina, are highly infested by a *Neotyphodium*-like endophyte

(Medvescigh et al., 2004c). When the survey was extended to populations collected in Buenos Aires Province, similar results were obtained (De Battista, 2002). These results show the extremely high incidence of *Neotyphodium* infection in spontaneous populations of Italian ryegrass, suggesting a strong selective advantage of infected plants under natural reseeding conditions.

Endophyte infection in commercial Italian ryegrass cultivars has also been studied. Results showed that most tetraploid cultivars are endophyte-free, and diploid cultivars can be infected at high, medium, or zero levels (J. De Battista, 2003, unpublished data). This variation in infection level among commercial cultivars could have arisen through manipulation of seedlots during breeding and (or) multiplication processes.

Despite the extensive use of Italian ryegrass, no reports of animal disorders related to endophyte infection have been documented locally. A similar situation was described by Nelson and Ward (1990) in the USA. Intra- and interspecific variation in toxin production have been found in *Neotyphodium*–grass associations. Tepaske et al. (1993) reported presence of lolines in infected *L. multiflorum*, but no ergot alkaloids as found in tall fescue and perennial ryegrass. Lack of ergot alkaloids in infected Italian ryegrass could explain the apparent lack of toxicity with cattle. No specific recommendations with respect to endophyte infection currently exist for seeding Italian ryegrass in the region.

NONTOXIC ENDOPHYTE–GRASS ASSOCIATIONS

The importance of using nontoxic endophytes in tall fescue and perennial ryegrass to capitalize on benefits conferred to the plant by the fungus has long been recognized. Selection of fungal strains from natural grass–endophyte associations with nontoxic characteristics has so far been the most effective method of development (Fletcher and Easton, 2001). Two endophyte strains, AR542 and AR1 for tall fescue and perennial ryegrass, respectively, showed nontoxic characteristics for livestock and maintained insect deterrence and (or) enhanced plant persistence (Bouton, 2000; Fletcher and Easton, 2001).

As for the potential use of these so-called safe endophytes, tall fescue would probably be the most promising species for Argentina, Brazil, and Uruguay, and perennial ryegrass for Chile. Local seed companies have shown interest in introducing this technology, provided local information on the economic benefits justify the higher seed cost. This will require experiments assessing plant and animal performance across a range of environmental conditions. Another need will be reliable and practical assays to distinguish nontoxic from wild endophytes to overcome restrictions on seed-borne endophytes and assure the user about the identity of the nontoxic fungus.

CURRENT ENDOPHYTE RESEARCH

Most endophyte research activities in South America are concentrated in Argentina. The main objectives can be summarized as (i) survey and characterization of *Neotyphodium* endophytes in native and naturalized grasses, (ii) ecological implications of endophyte infection in Italian ryegrass, and (iii) agronomic impact of *Neotyphodium* infection on Italian ryegrass and other forage species.

During the past 10 yr, D. Cabral and co-workers have studied the occurrence of endophytes in native cool-season grasses, mainly from semiarid rangelands of Patagonia and northwest Argentina. *Neotyphodium*-like endophytes were isolated from several toxic grasses (*Festuca argentina*, *Poa huecu*, and *Festuca hyeronymi*) and nontoxic grasses (*Poa rigidifolia*, *Festuca magellanica*, *Melica stuckertii*, *Bromus setifolius*, and *B. auleticus*, among others) (Bertoni et al., 1993; Cabral and Bertoni, 1991; Cabral and Lugo, 1994; Ianonne, 2003; White et al., 1996, 2001). Several endophyte isolates were characterized in culture and, using isozymes and DNA markers, were compared with Northern Hemisphere endophytes. As a result, a new species was named *N. tembladerae* (Cabral et al., 1999). Despite the extensiveness of surveys, no sexual stages (*Epichloë* spp.) have been found in Argentina or reported elsewhere in the Southern Hemisphere (D. Cabral, 2004, personal communication). Ongoing research, in collaboration with C. Schardl (University of Kentucky), addresses the evolutionary origins of Argentine grass endophytes.

Using seed from germplasm banks, A. Peretti and co-workers carried out a survey of endophytes of grasses from the humid Pampas. Sixty-eight accessions of 13 species were evaluated, representing genera like *Briza*, *Dactylis*, *Thinopyrum*, *Lolium*, *Avena*, *Bromus*, *Festuca*, and *Poa*. In contrast with Cabral's findings, very few populations of *Bromus* and *Lolium* were found to harbor *Neotyphodium*-like endophytes (Colabelli et al., 2001a, 2001b). Local research on ecological implications of endophyte infection in Italian ryegrass is addressed in Chapter 6 of his volume by M. Omacini.

Since Italian ryegrass is a very important forage for the region, and its spontaneous populations are highly infested, there is strong research interest in the agronomic impact of *Neotyphodium* infection. Because of the annual nature of the species, research has focused mainly on the effect of endophyte on early stages of development and reproductive ability. Endophyte presence enhanced seed germination in non-water-stressed and well-watered environments. Final germination percentage was greater and germination rate was faster in infected than endophyte-free seed (Medvescigh et al., 2004a). Endophyte infection produced seedlings with fewer but heavier tillers and larger leaves (Medvescigh et al., 2004b). Other research includes the study of reproductive ability of infected and noninfected Italian ryegrass and studies of vertical transmission of endophyte.

Bromus auleticus is a very promising cool-season native grass that has recently been domesticated. Natural populations have shown high frequency of a *Neotyphodium*-like endophyte (Iannone, 2003). Studies on the impact of endophyte infection showed that infected seed was lighter but germinated more and faster. Infected seedlings produced less biomass at 10 d of age, but afterwards produced more tillers and longer leaves than endophyte-free seedlings (Iannone, 2003).

SUMMARY

Endophyte infection is fairly common in tall fescue and perennial and Italian ryegrass seedlots and pastures. However, toxicoses are infrequent in the field, probably because of farmers' awareness and the typical ways of using these species. General recommendations usually encourage the use of endophyte-free pastures. Nontoxic-endophyte technology has potential applications in South America, but its adoption will depend on local research data demonstrating their agronomic and economic benefits. New *Neotyphodium*–grass associations have recently been identified.

REFERENCES

Bertoni, D.M., D. Cabral, and N. Romero. 1993. Endófitos fúngicos en especies sudamericanas de *Festuca* (Poaceae). Boletín de la Asociación Argentina de Botánica 2:25–34.

Bouton, J.H. 2000. The use of endophytic fungi for pasture improvement in the USA. p. 163–168. *In* P. Volker and D. Dapprich (ed.) The Grassland Conference 2000. 4th Int. *Neotyphodium*/grass Interactions Symp., Soest, Germany. 27–29 Sept. 2000. Univ. Paderborn, Germany.

Bouton, J.H., R.N. Gates, D.P. Belesky, and M. Owsley. 1993. Yield and persistence of tall fescue in the southeastern Coastal Plain after removal of its endophyte. Agron. J. 85:52–55.

Cabral, D., and D.M. Bertoni. 1991. Una nueva especie de *Acremonium*: Endófito de *Festuca argentina*. Resúmenes de las Jornadas Argentinas de Botánica, XXIII. 14–18 Oct. 1991. San Carlos de Bariloche, Sociedad Argentina de Botánica, Esperanza, Argentina.

Cabral, D., M.J. Cafaro, O.B. Saidman, A.M. Lugo, V.P. Reddy, and F.J. White, Jr. 1999. Evidence supporting the occurrence of a new species of endophyte in some South American grasses. Mycologia 91:315–325.

Cabral, D., and M.A. Lugo. 1994. Los endófitos fúngicos en gramíneas de Argentina. p. 106. *In* R.H. Fortunato and N.M. Bacigalupo (ed.) Congr. Latinoamericano de Botánica, VI. 2–8 Oct. 1994. Mar del Plata. Libro de resúmenes. Missouri Botanical Garden Press, St. Louis.

Colabelli, M.N., M.S. Torres, A. Peretti, and M. Clausen. 2001a. Presencia de hongos endófitos (*Neotyphodium* spp.) en gramíneas nativas y naturalizadas de los géneros *Briza, Dactylis, Elytrigia* y *Lolium*. Boletín de la Sociedad Argentina de Botánica 36:112 (abstr.).

Colabelli, M.N., M.S. Torres, A. Peretti, and M. Clausen. 2001b. Relevamiento de endófitos en semillas de especies nativas y naturalizadas de los géneros *Avena, Bromas, Festuca* y *Poa* en la Argentina. Informativo Abrates 11:314 (abstr. 530).

De Battista, J.P. 2002. El endófito de raigrás anual. Actas de la Reunión Anual de Forrajeras [CD-ROM]. El pasto: El recurso más barato. INTA EEA Pergamino, Buenos Aires.

De Battista, J., N. Altier, D.R. Galdames, and M. Dall'Agnol. 1997. Significance of endophyte toxicosis and current practices in dealing with the problem in South America. p. 383–388. *In* C. Bacon and N. Hill (ed.) *Neotyphodium*/Grass Interactions. Proc. Int. Symp. *Acremonium*/Grass Interactions, 3rd, Athens, GA, USA. 28–31 May 1997. Plenum Press, New York.

De Battista, J., A. Peretti, A. Carletti, A. Ramirez, M. Costa, and I. Schultz. 1995. Evolución de la incidencia de la infección de *Acremonium coenophialum* en la oferta de semilla de festuca alta en Argentina. Rev. Arg. Prod. Anim. 15:300–302.

De Battista, J., A. Peretti, M. Sala, I. Schultz, M. Costa, E. Picardi, A. Salvat, J. Medevecigh, and D. Bacigaluppi. 2001. Evolution of the incidence of *Neotyphodium coenophialum* infection on tall fescue seed market in Argentina. Period 1995–2000. p. 16. *In* Proc. Seed Symp. ISTA Congr., 26th, Angers, France. 18–20 June 2001.

Eerens, J.P.J., H.S. Easton, R.J. Lucas, J.G.H. White, and K.B. Miller. 1997. Influence of the ryegrass endophyte on pasture production and composition in a cool-moist environment. p. 157–159. *In* C. Bacon and N. Hill (ed.) *Neotyphodium*/Grass Interactions. Proc. Int. Symp. *Acremonium*/Grass Interactions, 3rd, Athens, GA, USA. 28–31 May 1997. Plenum Press, New York.

Fletcher, L.R., and H.S., Easton. 2001. Advances in endophyte research. Progress and priorities in temperate areas. *In* Proc. of the XIX International Grassland Congr., Sao Pedro, Sao Paulo, Brazil [CD-ROM]. 11–21 Feb. 2001.

Iannone, L.J. 2003. Caracterización del endófito de *Bromus auleticus* y sus efectos sobre la biología del hospedante. Tesis de Licenciatura. Facultad de Ciencias Exactas y Naturales, Univ. de Buenos Aires.

Medvescigh, J., J. De Battista, O.R. Del Longo, and M. Costa. 2004a. Effect of *Neotyphodium occultans* infection and substrate water potential on annual ryegrass seed germination. Proc. 5th Int. Symp. *Neotyphodium*/Grass Interactions. Fayetteville, AR, USA. 23–26 May 2004. Univ. Arkansas, Fayetteville.

Medvescigh, J., J. De Battista, O.R. Del Longo, and M. Costa. 2004b. Effect of endophyte infection and water stress on annual ryegrass seedling growth. Proc. 5th Int. Symp. *Neotyphodium*/Grass Interactions. Fayetteville, AR, USA. 23–26 May 2004. Univ. Arkansas, Fayetteville.

Medvescigh, J., A. Maidana, J. De Battista, R. Sabatini, and M. Costa. 2004c. Incidence and infection level of endophyte infection on *Lolium multiflorum* Lam. naturalized populations in Entre Ríos Province, Argentina. Proc. 5th Int. Symp. *Neotyphodium*/Grass Interactions, Fayetteville, AR, USA. 23–26 May 2004. Univ. Arkansas, Fayetteville.

Moon, C.D., D.B. Scott, C.L. Schardl, and M.J. Christensen. 2000. The evolutionary origins of *Epichloë* endophytes from annual ryegrasses. Mycologia 92:1103–1118.

Nelson, L.R., and S.L. Ward. 1990. Presence of fungal endophyte in annual ryegrass. p. 41–43. *In* S.S. Quisenberry and R.E. Joost (ed.) Proc. Int. Symp. *Acremonium*/Grass Interactions. 5–7 Nov. 1990. Louisiana Agric. Exp. Stn., Baton Rouge.

TePaske, M.R., R.G. Powell, and S.L. Clement. 1993. Analyses of selected endophyte-infected grasses for the presence of loline-type and ergot-type alkaloids. Agric. Food Chem. 41:2299–2303.

Tourne, S., S.G. Assuero, and J. Castaño. 2002. Acumulación de materia seca de raigrás perenne con y sin endófito bajo distintas condiciones de manejo. Prod. Anim. 22 Sup. 1:169–170.

West, C.P., E. Izekor, E. Turner, and A. Elmi. 1993. Endophyte effects on growth and persistence of tall fescue along a water-supply gradient. Agron. J. 85:264–270.

White, F.J., Jr., I.T. Martin, D. Cabral. 1996. Endophyte-host associations in grasses. XXIV. Conidia formation by *Acremonium* endophytes on the phylloplanes of *Agrostis hiemalis* and *Poa rigidifolia*. Mycologia 88:174–178.

White, F.J., Jr., F.R. Sullivan, A.G. Balady, J.T. Gianfagna, Q. Yue, A.W. Meyer, D. Cabral. 2001. A fungal endosymbiont of the grass *Bromus setifolius*: Distribution in some populations, identification and examination of beneficial properties. Symbiosis 31:241–257.

Molecular Biology of *Neotyphodium*

BIOSYNTHESIS OF ERGOT AND LOLINE ALKALOIDS

Christopher L. Schardl[1] and Daniel G. Panaccione[2]

G rasses that possess symbiotic *Epichloë* and *Neotyphodium* species may benefit in various ways, particularly by enhanced resistance to potential abiotic and biotic stresses (Clay and Schardl, 2002; Malinowski and Belesky, 2000). Among the biotic factors from which some endophytes provide demonstrable protection are grazing vertebrate herbivores, insects, and nematodes. Although the basis for antinematode activity remains to be elucidated, it is clear that alkaloids play a major role in averting, sickening, or even killing insect and mammalian herbivores. Four classes of alkaloids have been associated with endophyte presence and effects: Ergot alkaloids cause toxicoses in grazing mammals, ranging from stupor and appetite suppression to reproductive problems, dry gangrene, and death. The indolediterpenes are tremorgenic and cause staggers in mammals. Peramine is an insect feeding deterrent, and lolines are insecticidal. This chapter focuses on biosynthesis and molecular genetics of loline alkaloids and ergot alkaloids, and Chapter 3 discusses the indolediterpenes.

ERGOT ALKALOIDS

History of Ergot Alkaloids

Ergotism has played an important role throughout history (Matossian, 1989; Panaccione and Schardl, 2003). Ergots have been used for millennia to aid in labor or promote abortion. Hallucinations and physical symptoms (St. Anthony's fire) were often interpreted as a state of bewitchment, and very likely led to accusations triggering the Salem Witch Trials in Massachusetts and similar events in Europe. An epidemic of ergot poisoning in the Russian army thwarted the designs of Peter the Great on the Ottoman Empire, and widespread ergotism

[1] Department of Plant Pathology, University of Kentucky, Lexington, KY
[2] Division of Plant and Soil Sciences, West Virginia University, Morgantown, WV

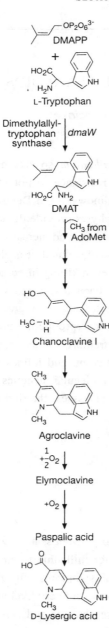

Fig. 2.1. *Biosynthesis of clavines and lysergic acid.*

is a proposed cause of the first crusades. The most modern manifestations of ergot alkaloids are the recreational and experimental uses of lysergic acid diethylamide (LSD), and therapeutic uses of other ergot alkaloid derivatives. Because of the medicinal uses and industrial production of ergot alkaloids, there has been considerable interest in the biosynthetic pathway and regulation of their expression.

The *Claviceps* spp., particularly *C. purpurea* on rye (*Secale cereale*), are the main culprits in historical ergot alkaloid poisoning. These fungi infect grass florets and produce dense, hard resting structures (sclerotia), commonly called ergots, which are difficult to remove with the chaff. Modern milling techniques help remove most ergots, but the potential remains for livestock poisoning if they eat the seedheads of grasses in fence rows or other pasture areas that might escape mowing. Such poisonings could be associated with fescues (*Festuca* spp.) or ryegrasses (*Lolium* spp.) (Shelby et al., 1997), which could thereby vector ergot alkaloids both from *C. purpurea* and their related endophytic fungi (Lyons et al., 1986). Other fungi outside the Clavicipitaceae can produce ergot alkaloids (Vining, 1973). Furthermore, species of the plant family Convolvulaceae are known to accumulate ergot alkaloids, although new evidence suggests that the source in these plants is actually fungal. Accumulation of ergot alkaloids in *Ipomea asarifolia* correlates with the presence of an epiphytic fungus in association with secretory glands on the leaves, and fungicide treatment of these plants can cure them of these fungi and render them free of ergot alkaloids (Kucht et al., 2004).

Despite the widespread occurrence of ergot-alkaloid-producing fungi, the endophytes in pasture grasses remain the main concern. Shortly after the release in 1942 of the tall fescue (*L. arundinaceum* = *F. arundinacea*) cultivar Kentucky 31, an association was noted between tall fescue pastures and toxicoses in livestock. Such toxicoses mimicked ergot alkaloid poisoning, yet evidence was lacking for *Claviceps* spp. as the causal agents. Further study focused on relatives of *Claviceps* spp., particularly *Balansia* spp., for which fruiting structures were evident on grasses in some pastures. However, no *Balansia* sp. was known to infect tall fescue. Eventually, Bacon et al. (1977) identified an endophytic symbiont of tall fescue, and referred the endophyte to *E. typhina*, another member of the family Clavicipitaceae. The presence of this endophyte, now called *N. coenophialum*, was also correlated with accumulation in the grass of ergot alkaloids, particularly the ergopeptine ergovaline. Furthermore, *N. coenophialum* could be induced to express ergovaline in culture (Porter et al., 1979, 1981).

Despite the fact that ergot alkaloid biosynthetic enzymes had been studied for decades (Gröger and Floss, 1998), only two enzymes of the pathway have been purified and partially sequenced. Identification of the genes for these enzymes ensued, and culminated in mutagenesis experiments that verified involvement of those genes and the corresponding enzymatic steps in the pathway. These studies are summarized in the following sections.

Ergot Alkaloid Biosynthesis Pathway

Early investigations indicated that tryptophan and mevalonate are precursors to the ergot alkaloids. Feeding *C. purpurea* fermentation cultures with labeled mevalonate resulted in labeling patterns that suggested isoprene as an intermediate. The enzyme, 4-γ,γ-dimethylallyltryptophan synthase (Fig. 2.1), was purified (Gebler and Poulter, 1992) and the gene was subsequently cloned from *C. fusiformis* (Tsai et al., 1995).

The second step in ergot alkaloid synthesis is α-N-methylation (Otsuka et al., 1980). Subsequent oxidation/oxygenation steps on the prenyl chain are believed first to form a diene, which is then epoxidized, promoting spontaneous cyclization of the C ring with a simultaneous release of the α-carboxyl group (Kozikowski et al., 1993) (Fig. 2.1, 2.2). This yields chanoclavine-I, the first clavine in the pathway, which accumulates to some extent in fermentation cultures as well as in some endophyte-infected grasses. Cyclization of the D ring, catalyzed by chanoclavine-I cyclase (possibly a FAD-containing oxidoreductase), likely involves an aldehyde intermediate (Floss et al., 1974). Isomerization of the substrate in the active site would be required to bring the aldehyde in proximity to the amine and facilitate their condensation to form an imine bond (Schiff base), reduction of which would generate the D ring (Floss et al., 1974). The product, agroclavine, is hydroxylated at the methyl carbon 17 to form elymoclavine. Further oxidations yield paspalic acid, and a shift of the D ring unsaturation yields lysergic acid.

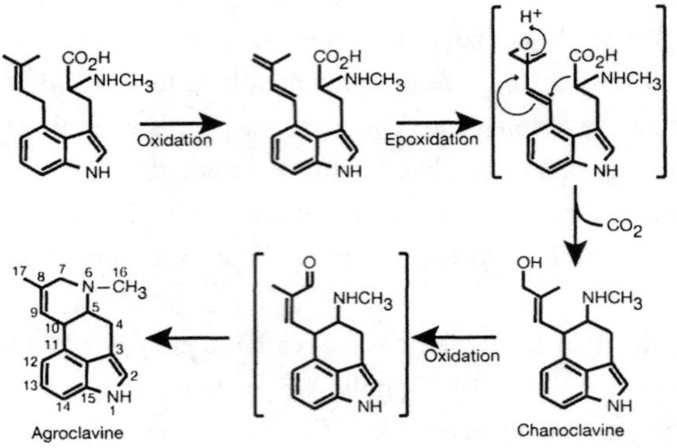

Fig. 2.2. *Cyclization of the C and D rings in agroclavine biosynthesis.*

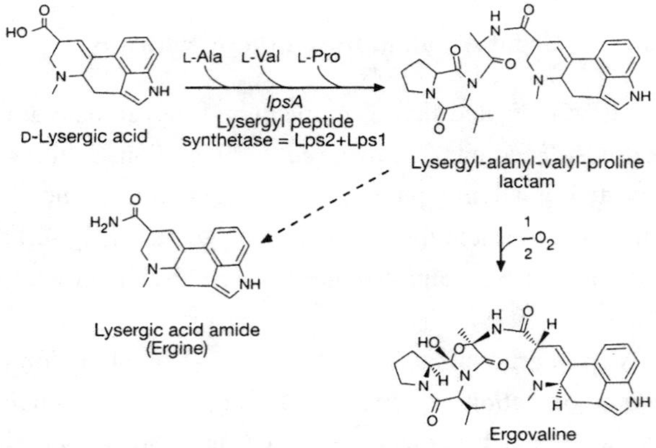

Fig. 2.3. *Synthesis of the ergopeptine, ergovaline, from lysergic acid. A possible origin of ergine is also suggested (dashed arrow).*

Many clavicipitaceous fungi produce several amide derivatives of lysergic acid, of which the most abundant and acutely toxic are the ergopeptines. These are formed by the condensation of lysergic acid with three amino acids (Fig. 2.3), and variations among the amino acids determine the different ergopeptines (Table 2.1). Synthesis involves one or more lysergyl peptide synthetases (Lps)

(Riederer et al., 1996; Tudzynski et al., 2001). The first steps of the multifunctional Lps are to activate lysergic acid to form a thiol ester with the Lps2
subunit, and similarly to establish a thiol-ester bond with amino acid I in the first
module of the Lps1 subunit. Lysergic acid is condensed with amino acid I, and a
similar process in module 2 of Lps1 extends the chain with amino acid II.
Amino acid III, generally proline, forms a thiol ester with Lps1 module 3.
Amino acids II and III form a cyclic lactam, probably during release of the
peptide from the Lps complex. The lactam is further cyclized, probably by a
monooxygenase, to form the ergopeptine (Fig. 2.3).

Table 2.1. Composition of ergopeptine alkaloids. Each of the ergopeptines listed has
L-proline as amino acid III.

L-amino acid I	L-amino acid II			
	Val	Phe	Leu	Ile
Ala	ergovaline†	ergotamine	ergosine	β-ergosine
Val	ergocornine	ergocristine	ergokryptine	β-ergokryptine
2-aminobutyric acid	ergonine	ergostine	ergoptine	β-ergoptine

† Ergovaline is the only ergopeptine known from *Epichloë* and *Neotyphodium* spp. The
related ergopeptine, ergobalansine, has Ala, Val, and Ala at amino acid positions I, II,
and III, respectively.

Ergot Alkaloid Biosynthesis Genes

The 1990s saw a flourish of molecular advances in ergot alkaloid studies. The
complete purification and characterization of *C. fusiformis* dimethylallyltryptophan (DMAT) synthase was published by Gebler and Poulter (1992), who
treated the protein with CNBr to generate fragments for partial peptide sequencing. On the basis of this information, the first ergot alkaloid synthesis gene,
dmaW, was cloned (Tsai et al., 1995). The function of the cloned gene was
confirmed by expression of the cDNA in yeast (*Saccharomyces cerevisiae*). The
transformed yeast yielded activity that transferred the prenyl group from [^{3}H]-
dimethylallyl diphosphate (DMAPP) to L-tryptophan. Acid hydrolysis of
DMAPP generates a volatile allyl alcohol, which is driven out of solution by
boiling. Tryptophan-dependent incorporation of the tritium label into a nonvola-
tile form is a convenient activity assay for DMAT synthase. Yeast with cDNA
(but not the antisense-cDNA control) expressed from the inducible *GAL4*
promoter generated this activity. The enzyme expressed in yeast was authenti-
cated by H. Wang and C.D. Poulter, who characterized the product by HPLC,
UV spectrometry, and mass spectrometry (Tsai et al., 1995).

Initial attempts to use the cloned *C. fusiformis dmaW* as a probe for the endophyte homologs were unsuccessful. Therefore, Wang et al. (2004) took an alternative approach, whereby they identified and sequenced the *C. purpurea* gene in order to identify segments of the enzyme that might be conserved. On the basis of comparison of the *C. fusiformis* and *C. purpurea* homologs, four putatively conserved sequences were identified, and PCR primers were designed based on those sequences. With these primers, PCR generated the expected fragments from various endophytes, including *N. coenophialum*, *E. typhina* × *N. lolii* isolate Lp1, and *Balansia obtecta*. When the *C. purpurea* and *B. obtecta* homologs (modified to remove introns) were expressed in yeast they directed DMAT synthase expression, but the endophyte genes did not (Wang et al., 2004). Nevertheless, J. Wang and C. Machado used marker exchange mutagenesis to disrupt the *dmaW* gene of Lp1. This mutant was introduced into several perennial ryegrass (*L. perenne*) plants, and ergot alkaloids of the resulting symbiota were analyzed (Wang et al., 2004). No ergovaline, lysergic acid amide, or chanoclavine-I was detected in plants with the *dmaW* disruption mutant. The mutation was complemented by a construct consisting of the *C. fusiformis* cDNA under control of the transcription promoter from the β-tubulin gene of *E. typhina*. These genetic experiments demonstrated that *dmaW* was a key gene for clavine and ergot alkaloid biosynthesis, thus supporting the long-held hypothesis that DMAT synthase is the determinant step in the pathway.

A mystery remains as to why the *Neotyphodium* spp. *dmaW* genes were not expressed or active in the yeast system. The possibility was considered that the pathway might be slightly different in endophytes. A reasonable conjecture was that tryptophan methylation preceded prenylation. So, *N*-methyl-L-tryptophan was used as a substrate for DMAT synthase assays, but no significant activity with this substrate was detected from yeast expression systems with *dmaW* of *Claviceps* spp. or *Neotyphodium* spp. (J. Wang and C.L. Schardl, unpublished data). The problem with the yeast system for the endophyte genes remains to be elucidated.

The other enzyme that has been well characterized biochemically and genetically is the lysergyl peptide synthetase. U. Keller and co-workers (Riederer et al., 1996) purified the *C. purpurea* lysergyl peptide synthetase complex, which they characterized as a two-subunit complex. They obtained partial peptide sequence information from the larger subunit, Lps1. Meanwhile, Tudzynski et al. (1999) identified the *C. purpurea* homolog of *C. fusiformis dmaW*, and also identified a cluster of genes closely linked with the homolog. In the cluster was a large gene encoding a predicted trimodular peptide synthetase, with peptide sequence closely matching that of Lps1. While this work was under way, J. Wang, C.L. Schardl, and D.G. Panaccione identified from *C. purpurea* ATCC 20102 three cosmid clones containing both *dmaW* and the gene for a peptide synthetase, a portion of which had been cloned previously (Panaccione, 1996). Then, C. Young in D.B. Scott's laboratory used the peptide synthetase gene fragment as a probe to isolate clones from the perennial ryegrass endophyte *N. lolii*. The endophyte gene, *lpsA*, was sequenced by P. Damrongkool and D.G. Panaccione, the latter of whom also generated a marker-exchange mutant in another perennial ryegrass endophyte, *E. typhina* × *N. lolii* isolate Lp1. The *lpsA* mutant was introduced into perennial ryegrass, where it proved incapable of ergovaline production but capable of producing lysergic acid and chanoclavine-I, as well as other clavine alkaloids (Panaccione et al., 2001, 2003). These results supported the proposed function of the trimodular peptide synthetase in ergopeptine synthesis.

Interestingly, the simple lysergyl amide, ergine, was also undetectable in plants with the *lpsA* mutant, as was lysergyl alanine (Panaccione et al., 2003). These results suggest that some simple amide derivatives of lysergic acid are derived from ergopeptines, the lysergyl peptide lactam intermediate (Fig. 2.3), or by other intermediates generated by Lps activity.

The trimodular peptide synthetase identified in *C. purpurea* differs by approximately 50% from that of *N. lolii* (P. Damrongkool et al., unpublished data). Furthermore, analysis of the cosmid clones from *C. purpurea* identified a second trimodular peptide synthetase gene downstream of the first (Machado, 2004). The putative proteins encoded by these linked genes share 89% identity. Southern-blot analysis suggests that *C. purpurea* might have additional homologs of this gene (Panaccione et al., 2001). The identification of multiple Lps1 homologs in *C. purpurea* is intriguing in light of the fact that the species produces several ergopeptines (Table 2.1), almost certainly because of variations in the amino acids I and II added by the Lps1 subunit. In contrast, *Neotyphodium* sp. Lp1 has only the one Lps1 homolog, and produces ergovaline and a dehydro-derivative (described as didehydroergovaline by Shelby et al., 1997) as its only ergopeptines. It is tempting to speculate that different Lps1 homologs might differ in specificity, accounting for the diversity of ergopeptines typifying *Claviceps* species and isolates. Alternatively, the adenylation domains of *C. purpurea* Lps1 may have lower specificity for their amino acid substrates than does Lps1 produced by endophytes.

The role of the monomodular subunit, Lps2, was established by P. Tudzynski's and U. Keller's groups, who conducted genetic knockout in *C. purpurea* of a predicted monomodular peptide synthetase gene from the ergot alkaloid cluster (Correia et al., 2003). Activity assays were conducted on extracts from wild-type *C. purpurea*, isolate P1, and its Lps2-gene-deletion mutant. The assays were for protein-thiol-ester formation with radiolabeled substrates D-[^{3}H]dihydrolysergic acid, L-[^{14}C]valine, and L-[^{14}C]phenylalanine. As expected, D-[^{3}H]dihydro-lysergic acid labeled the Lps2 subunit in the wild type, but no such labeled subunit was identified in the mutant. The L-[^{14}C]-amino acids labeled Lps1 in the wild type and the mutant. Heterologous expression in *Escherichia coli*, assayed by a D-lysergic acid-dependent ATP-pyrophosphate exchange assay, provided further evidence that the gene encoded Lps2.

Progress toward Modifying *Neotyphodium coenophialum*

Southern-blot hybridization analysis indicated that *N. coenophialum* has two *dmaW* homologs designated *dmaW-1* and *dmaW-2* (Wang, 2000). Each homolog was cloned and sequenced, and the two predicted proteins share 95% identity. The *dmaW-2* product is identical to the Lp1 DMAT synthase, so *dmaW-2* was targeted first for disruption. The marker exchange mutation of *dmaW-2* (Machado, 2004) was much like that used to disrupt *dmaW* in *Neotyphodium* sp. Lp1 (Wang et al., 2004). The mutant was introduced into tall fescue plants, which accumulated ergovaline to levels comparable with plants with wild-type *N. coenophialum* (D.G. Panaccione and C. Machado, unpublished data). Thus, a subsequent mutation of *dmaW-1* will be attempted to eliminate ergovaline production by this endophyte. If successful, such a modified endophyte might be an appropriate substitute for wild-type endophytes in tall fescue cultivars bred for livestock forage.

LOLINE ALKALOIDS

History and Background of Lolines

The loline alkaloids possess a saturated pyrrolizidine ring structure with an ether linkage between C-2 and C-7, and a 1-amine (Fig. 2.4). The 1-amine can bear formyl, acetyl, and/or methyl groups. Lolines have been identified in the following grass-endophyte symbiota: *L. giganteum* (= *F. gigantea*) with *E. festucae* (Leuchtmann et al., 2000; Siegel et al., 1990), *L. arundinaceum* with *N. coeno-phialum* (Leuchtmann et al., 2000; Siegel et al., 1990) and with the endophytes tentatively classified as FaTG-3, *L. pratense* (= *F. pratensis*, meadow fescue) with *N. uncinatum* (Leuchtmann et al., 2000; Siegel et al., 1990) and with *N. siegelii* (Craven et al., 2001), *L. multiflorum*, *L. rigidum*, and *L. temulentum* with *N. occultans* (TePaske et al., 1993), *Echinopogon ovatus* with *N. aotearoae* (Moon et al., 2002), *Agrostis hyemalis* with *E. amarillans* (J. Faulkner, J.D. Blankenship, C.D. Moon, and C.L. Schardl, unpublished data), endophyte-infected *Poa autumnalis* (Siegel et al., 1990), *F. argentina* (Casabuono and Pomilio, 1997), and *Achnatherum robustum* (TePaske et al., 1993). *N*-formylloline (NFL) and sometimes *N*-acetylloline (NAL) tend to accumulate to the highest levels in most of the symbiota, although tall fescue with FaTG-3 endophytes and *A. hyemalis* with *E. amarillans* have only *N*-acetylnorloline (NANL). The only nongrass plants found to have lolines are *Adenocarpus* spp. (Powell and Petroski, 1992) and *Argyreia mollis* (Tofern et al., 1999).

In 1898, Hofmeister isolated from *L. temulentum* an alkaloid he named temuline. In 1955, loline was reisolated from seeds of a ryegrass species (*L. cuneatum*) by Yunosov and Akramov (1955), who named it loline. Later, Yates and Tookey (1965) isolated loline (which they called festucine) from tall fescue. The structures of loline alkaloids were determined by several groups (Aasen and Culvenor, 1969; Bates and Morehead, 1972; Yunosov and Akramov, 1960), and Petroski et al. (1989) published the proton and ^{13}C nuclear magnetic resonance (NMR) spectra. Until recently, no similar alkaloids were identified outside of the grasses, and other plants that have them remain few. The lolines were among the first alkaloids, other than the ergopeptines, to be identified in *N. coenophialum*-infected tall fescue (Robbins et al., 1972). Although they were used as a proxy for infection levels in studies of animal toxicosis, there is little indication of any antimammalian activity of loline alkaloids other than possible immunosuppression (Dew et al., 1990). In contrast, lolines exhibit a broad range of antiinsect activities when applied as water or acetone solutions to insects (Riedell et al., 1991).

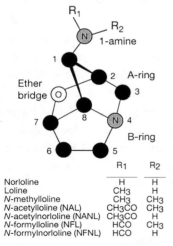

	R₁	R₂
Norloline	H	H
Loline	CH₃	H
N-methylloline	CH₃	CH₃
N-acetylloline (NAL)	CH₃CO	CH₃
N-acetylnorloline (NANL)	CH₃CO	H
N-formylloline (NFL)	HCO	CH₃
N-formylnorloline (NFNL)	HCO	H

Fig. 2.4. *Structure of loline alkaloids. Carbons of the saturated pyrrolizidine ring system are numbered.*

Loline alkaloids are toxic to many insects, including *Rhopalosiphum padi* (bird-cherry-oat aphid) (Siegel et al., 1990; Wilkinson et al., 2000), *Schizapus graminum* (greenbug aphid)

(Siegel et al., 1990; Wilkinson et al., 2000), and larvae of *Haematobia irritans* (horn fly) (Dougherty et al., 1998), *Popillia japonica* (Japanese beetle) (Patterson et al., 1991), *Spodoptera frugiperda* (fall armyworm) (Riedell et al., 1991), and *Ostrinia nubilalis* (European corn borer) (Riedell et al., 1991). In a genetic test of loline activity against aphids (Wilkinson et al., 2000), *E. festucae* siblings segregating for loline alkaloid production were introduced into meadow fescue, and the infected plants were challenged with greenbug and bird-cherry-oat aphids. The results confirmed the protective effect of lolines at levels ranging from 30 to 300 μg g^{-1} dry wt., far lower than is typical of tall fescue with *N. coenophialum*.

Until recently, the lolines were the only alkaloid class associated with endophytes but never to have been observed in endophyte fermentation cultures. This was surprising, considering that they are by far the most abundant of the alkaloid classes in endophyte-infected tall fescue and meadow fescue (Bush et al., 1997). Therefore, the question remained whether these were fungal alkaloids, plant metabolites induced by endophytes, or some product of a joint plant–fungus pathway. The possibility remained that signals or nutrient status of the plant apoplast induces the endophyte to produce lolines. This was a reasonable hypothesis, given that plant pathogens restricted to apoplastic spaces—such as *Cladosporium fulvum* and many species of bacteria—require low-nutrient conditions to express some genes in the plant milieu. The Schardl group attempted to manipulate culture conditions to promote loline alkaloid expression by the meadow fescue endophyte, *N. uncinatum*. This endophyte was chosen because *L. pratense–N. uncinatum* symbiota have the highest reported levels of lolines (ca. 3000–20 000 μg g^{-1} dry wt.). As expected, complex nutrient media supported good growth but no loline alkaloid expression, whereas defined minimal-salts media supported less growth but detectable production of NFL and NANL (Blankenship et al., 2001). Growth kinetics indicated that loline alkaloid production rates were greatest as the fungus went into stationary phase. Interestingly, loline alkaloid production was actually promoted by abundant sources of nitrogen—particularly amino acids or urea—and abundant sucrose as a carbon source. Curiously, no loline alkaloid production by *N. coenophialum* has been observed in similar culture conditions (J.D. Blankenship, M.J. Spiering, and J. Faulkner, unpublished data), indicating that different endophytes may respond to different signals from their host environments.

Expression of lolines by fermentation cultures of *N. uncinatum* has facilitated studies of the biosynthetic pathway. At the same time, the *E. festucae* genetic system and *N. uncinatum* culture system have facilitated identification of genes likely to be involved in loline alkaloid biosynthesis. These studies are described below.

Loline Alkaloid Pathway Studies

Several amino acids have been fed to fermentation cultures of *N. uncinatum* to investigate the fates of those labels in NFL (Blankenship, 2004). Labels from amino acids in the L-glutamate family, including L-ornithine and L-proline, greatly enriched NFL. In radioisotope feeding studies, the best of these was L-[^{14}C]proline, from which approximately threefold more label was incorporated compared with L-[^{14}C]glutamate and L-[^{14}C]ornithine. Similar experiments were conducted with L-[1,2-^{13}C$_2$]ornithine, L-[5-^{13}C]ornithine, and universally labeled L-[U-^{13}C$_5$,^{15}N]proline, all of which enriched the B ring of NFL. The ring nitrogen of L-[U-^{13}C$_5$,^{15}N]proline was also observed to enrich the pyrrolizidine-ring nitrogen of NFL. The ^{15}N and four ^{13}C atoms from the L-proline ring were incorporated together, as evidenced by mass spectrometry and ^{13}C-NMR.

The involvement of the L-glutamate family of amino acids raised the question whether a polyamine intermediate might be involved in the biosynthetic pathway. If so, the B ring would most likely be derived from putrescine (Bush et al., 1993). Therefore, cultures were fed with [^{14}C$_4$]putrescine, [1,4-^{13}C$_2$]putrescine, and [2,3-^{13}C]putrescine. In none of these experiments was there any indication of label incorporated in NFL or other loline alkaloids.

Incorporation of [^{14}C$_4$] putrescine into spermidine confirmed that the label entered the cells in those experiments. Further evidence against involvement of polyamines was that the label from L-[5-^{13}C]ornithine chased specifically into C-5, and the label from L-[1,2-^{13}C$_2$]ornithine chased specifically into C-8 of NFL (in the latter case, it is expected that the carboxyl 1-[^{13}C] was lost as CO$_2$ in the pathway, so that the 2-[^{13}C] label contributed the 8-[^{13}C] of NFL). In the polyamine pathway, the chemically asymmetrical ornithine is decarboxylated to form the chemically symmetrical putrescine. There is no metabolic means to distinguish between C-1 and C-4 of putrescine. Therefore, a hypothetical polyamine-based pathway would predict that L-[5-^{13}C]ornithine would distribute its label equally between C-5 and C-8 of NFL, and a similar distribution should be observed from feeding L-[1,2-^{13}C$_2$]ornithine. The fact that the labels from L-ornithine were specific was additional evidence against a polyamine pathway to lolines, and also established that the α-carbon of proline (derived from the α-carbon of ornithine) became the C-8 atom of NFL.

Experiments in which either L-[U-^{13}C$_5$] methionine or L-[6-^{13}C]methionine was fed to cultures indicated that both the *N*-methyl and *N*-formyl carbons were derived from the *S*-methyl carbon of L-methionine, almost certainly from *S*-adenosylmethionine via a classical methyltransferase. The L-[U-^{13}C$_5$] methionine feeding gave no indication of a polyamine pathway, in that the A ring was not labeled.

Similar feeding experiments established that the A ring was derived from L-aspartic acid, with an L-homoserine intermediate. L-[4-^{13}C]aspartate enriched the C-4 carbon of NFL. Feeding experiments with L-[^{15}N] aspartate indicated that the label incorporated at N-1 was derived from the α-amine of L-aspartate. When cultures were fed L-[4,4-^{2}H$_2$]homoserine, and the NFL was analyzed by mass spectrometry, significant enrichment in the +2 peaks for the parent compound and several of the fragments indicated that both deuterium atoms were incorporated together. This ruled out a scheme whereby L-homoserine or L-aspartate might have been converted to the 4-aldehyde, which could then form an imine bond with the ring nitrogen of L-proline. Such an aldehyde intermediate would bear only a single hydrogen at C-4, so the two deuterium atoms would not be expected to remain together in the labeled NFL. Although L-homoserine might not be a proximate precursor, it is reasonable to hypothesize that an activated form such as *O*-acetyl-L-homoserine is a direct precursor. In fungi, *O*-acetyl-L-homoserine is a primary metabolite and serves as an L-methionine precursor. *O*-acetyl-L-homoserine is a substrate for either of the related enzymes, cystathionine γ-synthase and homocysteine synthase, which catalyze reactions in alternative routes to L-methionine (Brzywczy et al., 1993). Molecular genetic studies described below suggest that an enzyme related to these may indeed be involved in the loline alkaloid pathway. A plausible biosynthetic route to lolines is shown in Fig. 2.5.

Loline Alkaloid Biosynthesis Genes

Two approaches were taken in an effort to identify genes that may be involved in the loline alkaloid pathway. One was to identify molecular markers—amplified fragment length polymorphisms (AFLPs)—that might cosegregate with loline alkaloid expression among *E. festucae* progeny in test crosses (Wilkinson et al., 2000). The other was to identify in *N. uncinatum* sequences of genes expressed under conditions promoting loline alkaloid expression, but not under conditions that suppressed expression (Spiering et al., 2002). These two approaches converged on a cluster of at least nine genes that appear to be unique to loline-alkaloid-producing endophytes. Two of the cluster genes, *lolA* and *lolC*, were expressed at higher levels when lolines were expressed in a minimally defined medium, but at extremely low levels when expression of loline alkaloids was suppressed by growth in a complex medium. Furthermore, the AFLP marker linked with the expression phenotype was localized to the *E. festucae* homolog of *lolC*.

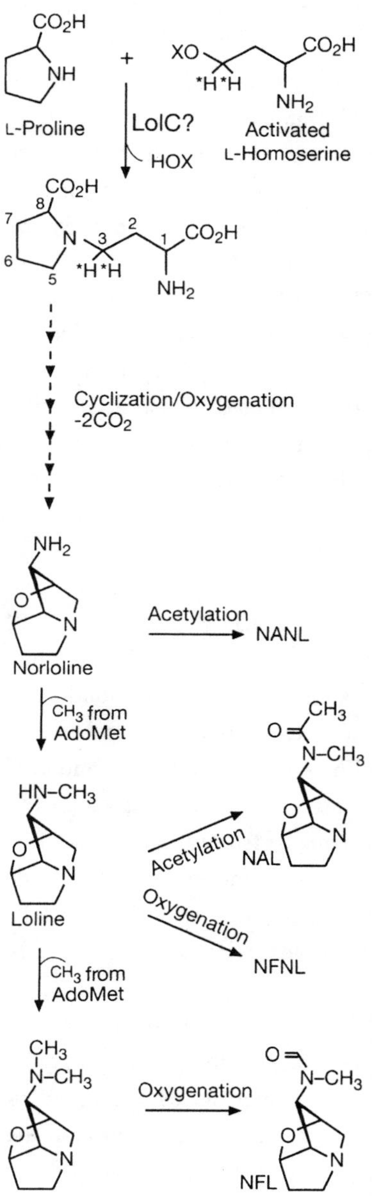

At least nine cluster genes were found to be duplicated in *N. uncinatum* (M.J. Spiering, C.D. Moon, and C.L. Schardl, unpublished data). In *LOL* locus 1, the genes occur in the order *lolF*, *lolC*, *lolD*, *lolO*, *lolA*, *lolU*, *lolP*, *lolT*, and *lolE*. *LOL* locus 2 contains homologs of all of the genes from *lolE* to *lolC* in the same order. Also duplicated is *lolF*, though the possible linkage of the duplicate *lolF* to locus 2 has not been determined. Analysis of mRNA both from cultures and from grass–*N. uncinatum* symbiota indicated that all nine genes are expressed under conditions of loline alkaloid production. Homology searches and conserved-domain-database searches suggest the following relationships (highly significant): *lolF*, FAD-containing monooxygenase; *lolC*, γ-class of pyridoxal phosphate (PLP)-containing enzymes; *lolD* and *lolT*, α-class PLP enzymes; *lolO*, nonheme-Fe oxidoreductase; *lolA*, allosteric domain of L-aspartyl kinase; *lolP*, cytochrome P450;

Fig. 2.5. *Biosynthesis of lolines. A possible determinant step is suggested in which LolC catalyzes condensation of L-proline and an activated L-homoserine. X indicates a substituent that provides a leaving group for the condensation reaction, and *H*H are the two hydrogen atoms from homoserine that were observed to be maintained in the pathway. Carbons in the proposed first intermediate are numbered according to their expected positions in the lolines.*

and *lolE*, epoxidase. In addition, a DNA-binding signature was identified in the predicted product of *lolU*.

Thus, seven of the nine genes are predicted to encode biosynthetic enzymes—three with PLP cofactors (LolC, LolD, and LolT) and four likely to catalyze redox reactions (LolO, LolP, LolF, and LolE)—and one gene in the cluster (*lolU*) may encode a regulatory protein. The seven biosynthesis genes appear adequate to carry out the synthesis of norloline, considering the number of oxidation steps and the cyclizations that must occur (Fig. 2.5).

The possible role of *lolA* remains elusive. The predicted LolA protein has a highly significant match to the *N*-terminal region of fungal L-aspartyl kinases, but this region does not include the active site (Spiering et al., 2002). Exhaustive analysis of *lolA* cDNA confirms that it lacks the predicted kinase active site. This relationship is intriguing, given the role of the L-homoserine, which in primary metabolism is derived from L-aspartate by a pathway that involves the kinase. A possibility is that LolA regulates the primary pathway to enhance production of L-homoserine, but this is a matter of conjecture.

Direct evidence for a role of LolC in loline alkaloid biosynthesis was garnered by an RNA-interference (RNAi) experiment (M.J. Spiering and C.L. Schardl, unpublished data). A clone was generated that was predicted, when expressed in *N. uncinatum*, to generate a double-stranded RNA corresponding in sequence to the first exon of *lolC*. This construct was introduced into *N. uncinatum*, and two transformants were tested for loline alkaloid production and *lolC* mRNA levels. Positive controls were transformants with the cloning vector but lacking the RNAi construct. The RNAi transformants had approximately 25% of *lolC-1* and *lolC-2* mRNAs compared with the controls, and produced approximately 50% lower levels of total loline alkaloids compared with the controls. The differences in mRNA levels and loline alkaloid levels were statistically significant ($P < 0.05$).

The reason for targeting *lolC* in the RNAi experiments was based on the conjecture that the determinant—therefore, regulated—step in the pathway may be catalyzed by LolC (Fig. 2.5). The LolC homologs, such as cystathionine γ-synthase and homocysteine synthase, catalyze a nucleophilic substitution at the γ-carbon of homoserine (usually activated by acetylation, succinylation, or phosphorylation). Therefore, we speculate that an analogous process might result in condensation of L-homoserine with L-proline as shown in Fig. 2.5. Although such a reaction appears unprecedented in the literature, it seems the most obvious proposal to explain both the gene relationships and the results of precursor feeding studies, including the retention of both deuterium atoms from L-[4,4-^{2}H$_2$]homoserine.

SUMMARY

Ergot alkaloids and loline alkaloids are two of the three alkaloid classes known to be produced by the tall fescue endophyte *N. coenophialum*, and both are potent toxins with neurotropic activities. However, the former present a major problem to livestock producers because of their antimammalian activities, whereas the latter are much more specific against insects and are therefore desirable for forage grasses. In the past decade, genes and gene clusters have been implicated in the biosynthetic pathways for these alkaloids.

Parallel studies in *Claviceps* spp. and *Neotyphodium* spp. have identified and characterized the gene for dimethylallyltryptophan synthase (DMAT synthase), the determinant step in the clavine and ergot alkaloid pathways. Similarly, genes have been identified for the two subunits (Lps1 and Lps2) of lysergyl peptide synthetase, which catalyzes the penultimate step in ergopeptine synthesis. The roles of these genes have been verified by genetic tests, whereby the DMAT synthase and Lps1 genes were disrupted in a *Neotyphodium* sp., and the Lps2 gene was disrupted in *C. purpurea*. As expected, clavines, lysergic acid, and ergopeptines are not produced by DMAT-synthase-gene disruptants, and only the lysergic acid amides and ergopeptines are lacking in the Lps1- and Lps2-gene disruptants.

Recent studies of loline alkaloids have established that they are fungal metabolites mainly derived from L-proline and L-homoserine, and that their production is most likely directed by a cluster of genes common to all loline alkaloid producers. Precursor-feeding experiments and an RNA-interference (RNAi) experiment indicate a key role for LolC, an apparent homolog of the L-methionine biosynthesis enzymes, homocysteine synthase, and cystathionine γ-synthase.

It is estimated that at least 11 enzymatic steps are involved in ergopeptine synthesis, and at least eight steps are involved in loline biosynthesis. Reasonable conjectures can be made as to which genes encode enzymes for which steps, but these will remain speculative until thorough characterizations of those genes and enzymes are also conducted.

ACKNOWLEDGMENTS

Research by the authors has been supported by the United States Department of Agriculture National Research Inititative, and by the U.S. National Science Foundation.

REFERENCES

Aasen, A.J., and C.C.J. Culvenor. 1969. Abnormally low vicinal coupling constants for O-CH-CH in a highly strained five-membered-ring ether; the identity of loline and festucine. Austral. J. Chem. 22:2021–2024.

Bacon, C.W., J.K. Porter, J.D. Robbins, and E.S. Luttrell. 1977. *Epichloë typhina* from toxic tall fescue grasses. Appl. Environ. Microbiol. 34:576–581.

Bates, R.B., and S.R. Morehead. 1972. Absolute configurations of pyrrolizidine alkaloids of the loline group. Tetrahedron Lett. 17:1629–1630.

Blankenship, J.D. 2004. Loline alkaloid biosynthesis in *Neotyphodium uncinatum*, a fungal endophyte of *Lolium pratense*. Ph.D. diss. Univ. of Kentucky, Lexington.

Blankenship, J.D., M.J. Spiering, H.H. Wilkinson, F.F. Fannin, L.P. Bush, and C.L. Schardl. 2001. Production of loline alkaloids by the grass endophyte, *Neotyphodium uncinatum*, in defined media. Phytochemistry 58:395–401.

Brzywczy, J., S. Yamagata, and A. Paszewski. 1993. Comparative studies on O-acetylhomoserine sulfhydrylase: Physiological role and characterization of the *Aspergillus nidulans* enzyme. Acta Biochim. Pol. 40:421–428.

Bush, L.P., F.F. Fannin, M.R. Siegel, D.L. Dahlman, and H.R. Burton. 1993. Chemistry, occurrence and biological effects of saturated pyrrolizidine alkaloids associated with endophyte–grass interactions. Agric. Ecosyst. Environ. 44:81–102.

Bush, L.P., H.H. Wilkinson, and C.L. Schardl. 1997. Bioprotective alkaloids of grass-fungal endophyte symbioses. Plant Physiol. 114:1–7.

Casabuono, A.C., and A.B. Pomilio. 1997. Alkaloids from endophyte-infected *Festuca argentina*. J. Ethnopharmacol. 57:1–9.

Clay, K., and C. Schardl. 2002. Evolutionary origins and ecological consequences of endophyte symbiosis with grasses. Am. Nat. 160:S99–S127.

Correia, T., N. Grammel, I. Ortel, U. Keller, and P. Tudzynski. 2003. Molecular cloning and analysis of the ergopeptine assembly system in the ergot fungus *Claviceps purpurea*. Chem. Biol. 10:1281–1292.

Craven, K.D., J.D. Blankenship, A. Leuchtmann, K. Hignight, and C.L. Schardl. 2001. Hybrid fungal endophytes symbiotic with the grass *Lolium pratense*. Sydowia 53:44–73.

Dew, R.K., G.A. Boissonneault, N. Gay, J.A. Boling, R.J. Cross, and D.A. Cohen. 1990. The effect of the endophyte (*Acremonium coenophialum*) and associated toxin(s) of tall fescue on serum titer response to immunization and spleen cell flow cytometry analysis and response to mitogens. Vet. Immunol. Immunopathol. 26:285–295.

Dougherty, C.T., F.W. Knapp, L.P. Bush, J.E. Maul, and J. Van Willigen. 1998. Mortality of horn fly (Diptera: Muscidae) larvae in bovine dung supplemented with loline alkaloids from tall fescue. J. Med. Entomol. 35:798–803.

Floss, H.G., M. Tcheng-Lin, C.-j. Chang, B. Naidoo, G.E. Blair, C.I. Abou-Chaar, and J.M. Cassady. 1974. Biosynthesis of ergot alkaloids. Studies on the mechanism of the conversion of chanoclavine-I into tetracyclic ergolines. J. Am. Chem. Soc. 96:1898–1909.

Gebler, J.C., and C.D. Poulter. 1992. Purification and characterization of dimethylallyltryptophan synthase from *Claviceps purpurea*. Arch. Biochem. Biophys. 296:308–313.

Gröger, D., and H.G. Floss. 1998. Biochemistry of ergot alkaloids: Achievements and challenges. Alkaloids 50:171–218.

Kozikowski, A.P., C. Chen, J.-P. Wu, M. Shibuya, C.-G. Kim, and H.G. Floss. 1993. Probing ergot alkaloid biosynthesis: Intermediates in the formation of ring C. J. Am. Chem. Soc. 115:2482–2488.

Kucht, S., J. Gross, Y. Hussein, T. Grothe, U. Keller, S. Basar, W.A. König, U. Steiner, and E. Leistner. 2004. Elimination of ergoline alkaloids following treatment of *Ipomoea asarifolia* (Convolvulaceae) with fungicides. Planta 219:619–625.

Leuchtmann, A., D. Schmidt, and L.P. Bush. 2000. Different levels of protective alkaloids in grasses with stroma-forming and seed-transmitted *Epichloë/Neotyphodium* endophytes. J. Chem. Ecol. 26:1025–1036.

Lyons, P.C., R.D. Plattner, and C.W. Bacon. 1986. Occurrence of peptide and clavine ergot alkaloids in tall fescue grass. Science (Washington, DC) 232:487–489.

Machado, C. 2004. Studies of ergot alkaloid biosynthesis genes in clavicipitaceous fungi. Ph.D. diss. Univ. of Kentucky, Lexington.

Malinowski, D.P., and D.P. Belesky. 2000. Adaptations of endophyte-infected cool-season grasses to environmental stresses: Mechanisms of drought and mineral stress tolerance. Crop Sci. 40:923–940.

Matossian, M.K. 1989. Poisons of the past: Molds, epidemics, and history. Yale Univ. Press, New Haven, CT.

Moon, C.D., C.O. Miles, U. Järlfors, and C.L. Schardl. 2002. The evolutionary origins of three new *Neotyphodium* endophyte species from grasses indigenous to the Southern Hemisphere. Mycologia 94:694–711.

Otsuka, H., F.R. Quigley, D. Gröger, J.A. Anderson, and H.G. Floss. 1980. In vivo and in vitro evidence for *N*-methylation as the second pathway-specific step in ergoline biosynthesis. Planta Med. 40:109–119.

Panaccione, D.G. 1996. Multiple families of peptide synthetase genes from ergopeptine-producing fungi. Mycol. Res. 100:429–436.

Panaccione, D.G., R.D. Johnson, J.H. Wang, C.A. Young, P. Damrongkool, B. Scott, and C.L. Schardl. 2001. Elimination of ergovaline from a grass–*Neotyphodium* endophyte symbiosis by genetic modification of the endophyte. Proc. Nat. Acad. Sci. USA 98:12 820–12 825.

Panaccione, D.G., and C.L. Schardl. 2003. Molecular genetics of ergot alkaloid biosynthesis. p. 399–424. *In* J.F. White, Jr. et al. (ed.) Clavicipitalean fungi: Evolutionary biology, chemistry, biocontrol and cultural impacts. Vol. 19. Marcel-Dekker, New York.

Panaccione, D.G., B.A. Tapper, G.A. Lane, E. Davies, and K. Fraser. 2003. Biochemical outcome of blocking the ergot alkaloid pathway of a grass endophyte. J. Agric. Food Chem. 51:6429–6437.

Patterson, C.G., D.A. Potter, and F.F. Fannin. 1991. Feeding deterrency of alkaloids from endophyte-infected grasses to Japanese beetle grubs. Entomol. Exp. Appl. 61:285–289.

Petroski, R.J., S.G. Yates, D. Weisleder, and R.G. Powell. 1989. Isolation, semi-synthesis, and NMR spectral studies of loline alkaloids. J. Nat. Prod. 52:810–817.

Porter, J.K., C.W. Bacon, and J.D. Robbins. 1979. Ergosine, ergosinine, and chanoclavine I from *Epichloë typhina*. J. Agric. Food Chem. 27:595–598.

Porter, J.K., C.W. Bacon, J.D. Robbins, and D. Betowski. 1981. Ergot alkaloid identification in Clavicipitaceae systemic fungi of pasture grasses. J. Agric. Food Chem. 29:653–657.

Powell, R.G., and R.J. Petroski. 1992. The loline group of pyrrolizidine alkaloids. p. 320–338. *In* Alkaloids: Chemical and biological perspectives. Vol. 8. Springer Verlag, Berlin.

Riedell, W.E., R.E. Kieckhefer, R.J. Petroski, and R.G. Powell. 1991. Naturally occurring and synthetic loline alkaloid derivatives: Insect feeding behavior modification and toxicity. J. Entomol. Sci. 26:122–129.

Riederer, B., M. Han, and U. Keller. 1996. D-lysergyl peptide synthetase from the ergot fungus *Claviceps purpurea*. J. Biol. Chem. 271:27524–27530.

Robbins, J.D., J.G. Sweeny, S.R. Wilkinson, and D. Burdick. 1972. Volatile alkaloids of Kentucky 31 tall fescue seed (*Festuca arundinacea* Schreb.). J. Agric. Food Chem. 20:1040–1043.

Shelby, R.A., J. Olsovska, V. Havlicek, and M. Flieger. 1997. Analysis of ergot alkaloids in endophyte-infected tall fescue by liquid chromatography electrospray ionization mass spectrometry. J. Agric. Food Chem. 45:4674–4679.

Siegel, M.R., G.C.M. Latch, L.P. Bush, F.F. Fannin, D.D. Rowan, B.A. Tapper, C.W. Bacon, and M.C. Johnson. 1990. Fungal endophyte-infected grasses: Alkaloid accumulation and aphid response. J. Chem. Ecol. 16:3301–3315.

Spiering, M.J., H.H. Wilkinson, J.D. Blankenship, and C.L. Schardl. 2002. Expressed sequence tags and genes associated with loline alkaloid expression by the fungal endophyte *Neotyphodium uncinatum*. Fung. Genet. Biol. 36:242–254.

TePaske, M.R., R.G. Powell, and S.L. Clement. 1993. Analyses of selected endophyte-infected grasses for the presence of loline-type and ergot-type alkaloids. J. Agric. Food Chem. 41:2299–2303.

Tofern, B., M. Kaloga, L. Witte, T. Hartmann, and E. Eich. 1999. Phytochemistry and chemotaxonomy of the Convolvulaceae: Part 8—Occurrence of loline alkaloids in *Argyreia mollis* (Convolvulaceae). Phytochemistry 51:1177–1180.

Tsai, H.-F., H. Wang, J.C. Gebler, C.D. Poulter, and C.L. Schardl. 1995. The *Claviceps purpurea* gene encoding dimethylallyltryptophan synthase, the committed step for ergot alkaloid biosynthesis. Biochem. Biophys. Res. Commun. 216:119–125.

Tudzynski, P., T. Correia, and U. Keller. 2001. Biotechnology and genetics of ergot alkaloids. Appl. Microbiol. Biotechnol. 57:593–605.

Tudzynski, P., K. Hölter, T. Correia, C. Arntz, N. Grammel, and U. Keller. 1999. Evidence for an ergot alkaloid gene cluster in *Claviceps purpurea*. Mol. Gen. Genet. 261:133–141.

Vining, L.C. 1973. Physiological aspects of alkaloid production by *Claviceps* species. p. 405–419. *In* Z. Vanek et al. (ed.) Genetics of industrial microorganisms. II. Actinomycetes and fungi. Vol. II. Elsevier Publ., Amsterdam.

Wang, J. 2000. *dmaW* encoding tryptophan dimethylallyltransferase in ergot alkaloid biosynthesis from clavicipitaceous fungi. Ph.D. diss. Univ. of Kentucky, Lexington.

Wang, J., C. Machado, D.G. Panaccione, H.-F. Tsai, and C.L. Schardl. 2004. The determinant step in ergot alkaloid biosynthesis by an endophyte of perennial ryegrass. Fung. Genet. Biol. 41:189–198.

Wilkinson, H.H., M.R. Siegel, J.D. Blankenship, A.C. Mallory, L.P. Bush, and C.L. Schardl. 2000. Contribution of fungal loline alkaloids to protection from aphids in a grass–endophyte mutualism. Mol. Plant–Microbe Interact. 13:1027–1033.

Yates, S.G., and H.L. Tookey. 1965. Festucine, an alkaloid from tall fescue (*Festuca arundinacea* Schreb.): Chemistry of the functional groups. Aust. J. Chem. 18:53–60.

Yunosov, S.Y., and S.T. Akramov. 1955. Alkaloids of seeds of *Lolium cuneatum*. Zh. Obshch. Khim. 25:1813–1820.

Yunosov, S.Y., and S.T. Akramov. 1960. Investigation of alkaloids of *Lolium cuneatum* II. Zh. Obshch. Khim. 30:677–682.

MOLECULAR AND GENETIC ANALYSIS OF LOLITREM AND PERAMINE BIOSYNTHETIC PATHWAYS IN *EPICHLOË FESTUCAE*

Barry Scott,[1] Carolyn Young,[1] Aiko Tanaka,[1] Michael Christensen,[2] Brian Tapper,[2] and Gregory Bryan[2]

Fungal endophytes of the *Epichloë–Neotyphodium* group of clavicipitaceous fungi (Clavicipitaceae, Ascomycota) are symbionts of temperate grasses (subfamily Pooideae), to which they confer a number of fitness benefits (Clay, 1990; Schardl and Clay, 1997). These biotrophic fungi systemically colonize the intercellular spaces of leaf primordia, leaf sheaths, and leaf blades of vegetative tillers and the inflorescence tissues of reproductive tillers.

At least nine different sexual species are recognized, including *E. typhina*, a broad host range species (Craven et al., 2001a; Schardl and Wilkinson, 2000), and *E. festucae*, a natural symbiont of *Festuca* spp. (Leuchtmann et al., 1994) that is also capable of forming compatible associations with perennial ryegrass, *Lolium perenne* (Christensen et al., 1997). *Neotyphodium lolii*, the predominant endophyte of *L. perenne*, is a haploid asexual derivative of *E. festucae* (Christensen et al., 1993; Schardl et al., 1994). Nearly all other asexual *Neotyphodium* spp. are interspecific hybrids (Craven et al., 2001b; Moon et al., 2000; Schardl et al., 1994; Tsai et al., 1994).

Endophyte benefits to the host include resistance to mammalian and insect herbivory (Clay, 1990; Clay and Schardl, 2002), drought tolerance (Arachevaleta et al., 1989; West, 1994), resistance to nematodes (Kimmons et al., 1990; West et al., 1988) and some fungal pathogens (Gwinn and Gavin, 1992), various growth enhancements (Malinowski and Belesky, 2000), and greater field persistence (Hill et al., 1990; West et al., 1988). Host benefits to the endophyte include provision of nutrients and a means of dissemination through the seed.

While the biochemical and physiological mechanisms underlying many of the endophyte benefits to the host are unknown, fungal synthesis of a range of

[1] Centre for Functional Genomics, Institute of Molecular BioSciences, Massey University, Palmerston North, New Zealand
[2] AgResearch Grasslands, Palmerston North, New Zealand

bioprotective metabolites are thought to be responsible for the antiinsect and antimammalian effects. Peramine and lolines are potent antifeeding compounds, and the ergot alkaloids and indole-diterpenes have a range of antimammalian activities. This review provides an overview of recent developments on understanding the biosynthesis of peramine and lolitrems.

ENDOPHYTE SYNTHESIS OF BIOPROTECTIVE METABOLITES

Peramine, classified as a pyrrolopyrazine alkaloid, appears to be a unique metabolite to the *Epichloë–Neotyphodium* genera of fungi (Clay and Schardl, 2002). Production of this metabolite in cultures of *N. lolii* (Rowan, 1993) and *E. typhina* (Schardl et al., 1999), albeit in low amounts compared with the levels found in endophyte-infected grass tissue, confirms that it is a fungal metabolite. To date, there is no experimental information on the biosynthesis of this compound. Analysis of its structure (Fig. 3.1) would suggest that it is the product of a reaction catalyzed by a two-module nonribosomal peptide synthetase (NRPS) that utilizes proline and arginine as substrates (Schardl et al., 1999). Peramine is a potent antifeeding deterrent against adult Argentine stem weevil, *Listronotus bonariensis*, a major pest of perennial ryegrass (Prestidge et al., 1985; Rowan and Gaynor, 1986; Rowan et al., 1990).

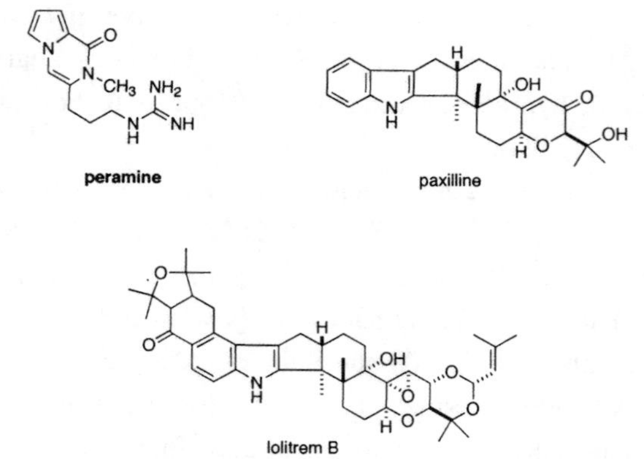

peramine

paxilline

lolitrem B

Fig. 3.1. *Structure of peramine, paxilline, and lolitrem B.*

Indole-diterpenes are a structurally diverse group of metabolites principally found in filamentous fungi of the genera *Penicillium*, *Aspergillus*, *Claviceps*, and *Epichloë* (Steyn and Vleggaar, 1985). This group of metabolites is characterized by the presence of a cyclic diterpene skeleton derived from four isoprene units, and an indole moiety derived from tryptophan or a tryptophan precursor such as anthranilic acid (Byrne et al., 2002). While biosynthetic schemes have been proposed for the synthesis of this class of compounds (Mantle and Weedon, 1994; Munday-Finch et al., 1996), very little is known about the nature of the intermediates or the enzymology of their biosynthesis. Many of these compounds are potent mammalian tremorgens, while others are known to confer antiinsect activity (Parker and Scott, 2004).

Lolitrems are an important subgroup of indole-diterpene metabolites that are relatively abundant in leaf sheath tissue and seeds of perennial ryegrass and some tall fescue (*Festuca arundinacea*) grasses containing *E. festucae*, *N. lolii*, and some interspecific hybrids of these and other *Epichloë–Neotyphodium* species (Bush et al., 1997; Christensen et al., 1993; Siegel et al., 1990). Lolitrems, in particular lolitrem B (Fig. 3.1), are thought to be responsible for the livestock disorder known as ryegrass staggers (Fletcher and Harvey, 1981; Gallagher et al., 1984). There have been two reports of indole-diterpene biosynthesis in axenic cultures of *Neotyphodium* spp., confirming that these compounds are fungal metabolites (Penn et al., 1993; Reinholz and Paul, 2001). Penn et al. (1993) detected indole-diterpenes in cultures of *N. lolii*, *N. coenophialum* (FaTG-1), *Neotyphodium* spp. (FaTG-2), *Neotyphodium* spp. (FaTG-3), *E. festucae*, and *N. uncinatum*. Reinholz and Paul (2001) reported the presence of lolitrem B in agar plate cultures of *N. lolii*. The preferential synthesis of these metabolites in planta suggests that the genes for lolitrem biosynthesis are symbiotically regulated.

EPICHLOË FESTUCAE AS A MODEL EXPERIMENTAL SYSTEM FOR GENETIC ANALYSIS OF ENDOPHYTES

As discussed by Schardl (2001), *E. festucae* is an attractive experimental system to use for molecular and genetic analysis of grass endophytes. *Epichloë festucae* is relatively fast growing compared with most *Neotyphodium* spp., taking 2 wk, compared with 4 to 6 wk, to form a 1.5-cm colony on potato dextrose agar plates. *Epichloë festucae* is relatively easy to protoplast and gives high transformation rates. Targeted replacements in the genome can be generated at frequencies of between 1 to 5% (C. Young, A. Tanaka, and B. Scott, unpublished results). The fungus is naturally haploid with a genome size of approximately 29 Mb (Kuldau et al., 1999) and has a heterothallic mating system (Leuchtmann et al., 1994). The natural hosts for *E. festucae* include *Festuca*, *Lolium*, and *Koeleria* spp. (Schardl, 2001). Artificial associations of *E.*

festucae with several of these species can be readily established, thereby allowing manipulation of the endophyte in culture and subsequent reintroduction into the host to study the symbiotic phenotype. All four main classes of alkaloids are synthesized by *E. festucae* (Schardl, 2001). Because of the above attributes, we have adopted *E. festucae* and *L. perenne* as our model experimental system to study the symbiotic interaction.

GENETICS AND MOLECULAR CLONING OF A PERAMINE BIOSYNTHESIS GENE CLUSTER

Genetic analysis of crosses between peramine-positive and peramine-negative strains of *E. typhina* indicated that peramine expression is under the control of a single genetic locus (Schardl et al., 1999). Segregation of peramine production, as determined by protection of the plant from greenbug aphid (*Schizaphis graminum*) herbivory, in F1 and BC1 generations was not significantly different from a ratio of 1:1. This Mendelian test supports the postulated role of peramine as a plant bioprotectant molecule.

As discussed above, peramine is predicted to be encoded by a two-module NRPS. Each module of a NRPS is a semiautonomous unit that recognizes, activates, and modifies a single amino acid residue of the final polypeptide (Marahiel et al., 1997; von Döhren et al., 1997). The minimal repeating unit is comprised of condensation (C), adenylation (A), and thiolation (T) domains. The versatility of NRPSs is enhanced by the presence of additional domains such as those for *N*-methylation (M) and epimerization (E). The C-terminus of the last module in bacterial NRPSs usually contains a thioesterase (TE) domain for release, and sometimes cyclization, of the polypeptide. While fungal NRPSs are known to have a TE domain, more commonly this activity is part of the C domain, or alternatively there is an additional reductase (R) domain (Keating et al., 2001; Wiest et al., 2002).

The conservation of polypeptide sequence in the A domain has allowed PCR strategies to be used to clone NRPSs from clavicipitalean fungi (Panaccione, 1996; Panaccione et al., 2001). Using a similar strategy, candidate sequences for a peramine NRPS were amplified by RT-PCR using total RNA from endophyte-infected perennial ryegrass (A. Tanaka and B. Scott, unpublished results). Three unique NRPS products were identified. Analysis of RT-PCR, with specific primers designed to each of the three sequences, revealed that the transcripts of two of them were significantly up-regulated in planta. One of these clones cross hybridized to known peramine-producing strains. This probe was used to isolate a genomic clone from a pMOcosX cosmid library to *E. festucae*, strain Fl1. Sequence analysis of a representative cosmid identified a two module NRPS, which we have designated as *perA*, with functions predicted for the biosynthesis of peramine.

A replacement construct of *perA* was prepared and recombined into the genome of Fl1. A PCR screen of the transformants identified one mutant containing the desired gene replacement event. This was confirmed by Southern analysis. Associations established between perennial ryegrass and the *perA* mutant lacked detectable levels of peramine (A. Tanaka, M. Christensen, B. Tapper, and B. Scott, unpublished results). By contrast, the levels of lolitrem B and ergovaline were similar to those found in wild-type associations. These results confirm that *perA* is a peptide synthetase for peramine biosynthesis.

MOLECULAR CLONING AND GENETIC ANALYSIS OF A GENE CLUSTER FOR PAXILLINE BIOSYNTHESIS

The cloning from *Penicillium paxilli* of a cluster of genes for the biosynthesis of paxilline (Fig. 3.1) has provided for the first time an insight into the biochemistry of indole-diterpene biosynthesis (Young et al., 2001). A combination of gene deletion and chemical complementation studies have confirmed that at least five genes are required for paxilline biosynthesis (McMillan et al., 2003; Young et al., 2001). Key genes identified in this cluster include a geranylgeranyl diphosphate (GGPP) synthase (*paxG*), a FAD-dependent monooxygenase (*paxM*), a prenyl transferase (*paxC*), and two cytochrome P450 monooxygenases, *paxP* and *paxQ*. PaxG is proposed to catalyse the determinant step in paxilline biosynthesis. PaxM and PaxC are proposed to catalyse the addition of indole-3-glycerol phosphate to GGPP and subsequent cyclization to form the first stable indole-diterpene, possibly paspaline (Parker and Scott, 2004). PaxP and PaxQ are proposed to catalyze the conversion of paspaline to paxilline via 13-desoxypaxilline (McMillan et al., 2003). At least one additional gene is required for oxygenation at the C-10 position, to convert β-paxitriol to paxilline and PC-M6 to 13-desoxypaxilline (Parker and Scott, 2004). The cloning and characterization of the paxilline gene cluster has made it possible to clone orthologous gene clusters from *N. lolii* and *E. festucae*.

MOLECULAR CLONING AND GENETIC ANALYSIS OF A GENE CLUSTER
FOR LOLITREM BIOSYNTHESIS

Using degenerate primers designed to conserve regions of *P. paxilli* and other fungal GGPP synthases, the *N. lolii* orthologue of *paxG*, which we have designated as *ltmG* (lolitrem biosynthesis), has been isolated (C. Young, G. Bryan, and B. Scott, unpublished results). Sequence analysis of a 25-kb genomic region around *ltmG* identified a *paxM* orthologue, *ltmM*, and a cytochrome P450 monooxygenase, *ltmK*. Analysis of RT-PCR showed the *ltm* genes are weakly expressed in culture but highly expressed in planta (C. Young, M. Bryant, G. Bryan, and B. Scott, unpublished results). Sequences adjacent to this *ltm* gene cluster have strong similarities to remnants of retroelements. It is likely that these retroelement platforms separate *ltmG*, *ltmM*, and *ltmK* from the remaining genes required for lolitrem biosynthesis. These retroelements are very abundant and highly dispersed within the genomes of both *N. lolii* and *E. festucae* (Scott and Young, 2003). A targeted deletion of *ltmM* has been constructed in *E. festucae* and associations established between perennial ryegrass and this mutant. These associations were shown to lack detectable levels of lolitrem B (C. Young, M. Christensen, B. Tapper, G. Bryan, and B. Scott, unpublished results). By contrast, the levels of ergovaline and peramine were similar to those found in wild-type associations. These results confirm that *ltmM* is required for lolitrem B biosynthesis.

The cloning of these indole-diterpene genes will now allow us to screen field isolates for the presence, distribution, and expression of these genes (Scott, 2001, 2004). Molecular screening will provide the certainty required for prediction of the toxin phenotype of naturally occurring endophytes, such as those that lack the ability to synthesize the mammalian toxins lolitrem B and ergovaline. Such isolates are desirable for use in pastoral agricultural systems to overcome animal toxicosis problems associated with endophyte synthesis of mammalian toxins, yet retain other bioprotective features such as the ability to resist insect herbivory (Fletcher, 1999; Popay et al., 1999). The availability of cloned toxin biosynthetic genes will also allow us to test how the expression of these genes in the symbiosis is regulated in response to both biotic and abiotic stress.

ACKNOWLEDGMENTS

This research was supported by grants MAU-X0127 and C10X0203 from the New Zealand Foundation for Research Science and Technology, and a grant (MAU103) from the Royal Society of New Zealand Marsden Fund. The authors would like to thank Andrea Bryant

(Massey) for technical assistance, and Wayne Simpson and Elizabeth Davies (AgResearch, Grasslands) for technical assistance and advice.

REFERENCES

Arachevaleta, M., C.W. Bacon, C.S. Hoveland, and D.E. Radcliffe. 1989. Effect of the tall fescue endophyte on plant response to environmental stress. Agron. J. 81:83–90.

Bush, L.P., H.H. Wilkinson, and C.L. Schardl. 1997. Bioprotective alkaloids of grass–fungal endophyte symbioses. Plant Physiol. 114:1–7.

Byrne, K.M., S.K. Smith, and J.G. Ondeyka. 2002. Biosynthesis of nodulisporic acid A: Precursor studies. J. Am. Chem. Soc. 124:7055–7060.

Christensen, M.J., A. Leuchtmann, D.D. Rowan, and B.A. Tapper. 1993. Taxonomy of *Acremonium* endophytes of tall fescue (*Festuca arundinacea*), meadow fescue (*F. pratensis*) and perennial rye-grass (*Lolium perenne*). Mycol. Res. 97:1083–1092.

Christensen, M.J., O.J.-P. Ball, R.J. Bennett, and C.L. Schardl. 1997. Fungal and host genotype effects on compatibility and vascular colonization by *Epichloë festucae*. Mycol. Res. 101:493–501.

Clay, K. 1990. Fungal endophytes of grasses. Ann. Rev. Ecol. Syst. 21:275–297.

Clay, K., and C. Schardl. 2002. Evolutionary origins and ecological consequences of endophyte symbiosis with grasses. Am. Nat. 160:S99–S127.

Craven, K.D., J.D. Blankenship, A. Leuchtmann, K. Hignight, and C.L. Schardl. 2001b. Hybrid fungal endophytes symbiotic with the grass *Lolium pratense*. Sydowia 53:44–73.

Craven, K.D., P.T.W. Hsiau, A. Leuchtmann, W. Hollin, and C.L. Schardl. 2001a. Multigene phylogeny of *Epichloë* species, fungal symbionts of grasses. Ann. Missouri Bot. Gard. 88:14–34.

Fletcher, L.R. 1999. "Non-toxic" endophytes in ryegrass and their effect on livestock health and production. p. 133–139. *In* D.R. Woodfield and C. Matthew (ed.) Ryegrass endophyte: An essential New Zealand symbiosis. New Zealand Grassland Association, Napier, New Zealand.

Fletcher, L.R., and I.C. Harvey. 1981. An association of a *Lolium* endophyte with ryegrass staggers. N. Z. Vet. J. 29:185–186.

Gallagher, R.T., A.D. Hawkes, P.S. Steyn, and R. Vleggaar. 1984. Tremorgenic neurotoxins from perennial ryegrass causing ryegrass staggers disorder of livestock: Structure elucidation of lolitrem B. J. Chem. Soc. Chem. Commun. 9:614–616.

Gwinn, K.D., and A.M. Gavin. 1992. Relationship between endophyte infestation level of tall fescue seed lots and *Rhizoctonia zeae* seedling disease. Plant Dis. 76:911–914.

Hill, N.S., W.C. Stringer, G.E. Rottinghaus, D.P. Belesky, W.A. Parrott, and D.D. Pope. 1990. Growth, morphological, and chemical component responses of tall fescue to *Acremonium coenophialum*. Crop Sci. 30:156–161.

Keating, T.A., D.E. Ehmann, R.M. Kohli, C.G. Marshall, J.W. Trauger, and C.T. Walsh. 2001. Chain termination steps in nonribosomal peptide synthetase assembly lines: Directed acyl-S-enzyme breakdown in antibiotic and siderophore biosynthesis. Chembiochem 2:99–107.

Kimmons, C.A., K.D. Gwinn, and E.C. Bernard. 1990. Nematode reproduction on endophyte-infected and endophyte-free tall fescue. Plant Dis. 74:757–761.

Kuldau, G.A., H.-F. Tsai, and C.L. Schardl. 1999. Genome sizes of *Epichloë* species and anamorphic hybrids. Mycologia 91:776–782.

Leuchtmann, A., C.L. Schardl, and M.R. Siegel. 1994. Sexual compatibility and taxonomy of a new species of *Epichloë* symbiotic with fine fescue grasses. Mycologia 86:802–812.

Malinowski, D.P., and D.P. Belesky. 2000. Adaptations of endophyte-infected cool-season grasses to environmental stresses: Mechanisms of drought and mineral stress tolerance. Crop Sci. 40:923–940.

Mantle, P.G., and C.M. Weedon. 1994. Biosynthesis and transformation of tremorgenic indole-diterpenoids by *Penicillium paxilli* and *Acremonium lolii*. Phytochemistry 36:1209–1217.

Marahiel, M.A., T. Stachelhaus, and H.D. Mootz. 1997. Modular peptide synthetases involved in nonribosomal peptide synthesis. Chem. Rev. 97:2651–2673.

McMillan, L.K., R.L. Carr, C.A. Young, J.W. Astin, R.G.T. Lowe, E.J. Parker, G.B. Jameson, S.C. Finch, C.O. Miles, O.B. McManus, W.A. Schmalhofer, M.L. Garcia, G.J. Kaczorowski, M.A. Goetz, J.S. Tkacz, and B. Scott. 2003. Molecular analysis of two cytochrome P450 monooxygenase genes required for paxilline biosynthesis in *Penicillium paxilli* and effects of paxilline intermediates on mammalian maxi-K ion channels. Mol. Gen. Genom. 270:9–23.

Moon, C.D., B. Scott, C.L. Schardl, and M.J. Christensen. 2000. The evolutionary origins of *Epichloë* endophytes from annual ryegrasses. Mycologia 92:1103–1118.

Munday-Finch, S.C., A.L. Wilkins, and C.O. Miles. 1996. Isolation of paspaline B, an indole-diterpenoid from *Penicillium paxilli*. Phytochemisty 41:327–332.

Panaccione, D.G. 1996. Multiple families of peptide synthetase genes from ergopeptine-producing fungi. Mycolog. Res. 100:429–436.

Panaccione, D.G., R.D. Johnson, J. Wang, C.A. Young, P. Damrongkool, B. Scott, and C.L. Schardl. 2001. Elimination of ergovaline from a grass–*Neotyphodium* endophyte symbiosis by genetic modification of the endophyte. Proc. Natl. Acad. Sci. USA 98:12 820–12 825.

Parker, E.J., and D.B. Scott. 2004. Indole-diterpene biosynthesis in ascomycetous fungi. p. 405–426. *In* Z. An (ed.) Handbook of industrial mycology. Marcel Dekker, New York.

Penn, J., I. Garthwaite, M.J. Christensen, C.M. Johnson, and N.R. Towers. 1993. The importance of paxilline in screening for potentially tremorgenic *Acremonium* isolates. p. 88–92. *In* D.E. Hume et al. (ed.) Proc. of the 2nd Int. Symp. on *Acremonium*/Grass Interactions, Palmerston North, New Zealand. 3–5 Feb. 1993. AgResearch, Grasslands Research Centre, Palmerston North.

Popay, A.J., D.E. Hume, J.G. Baltus, G.C.M. Latch, B.A. Tapper, T.B. Lyons, B.M. Cooper, C.G. Pennell, J.P.J. Eerens, and S.L. Marshall. 1999. Field performance of perennial ryegrass (*Lolium perenne*) infected with toxin-free fungal endophytes (*Neotyphodium* spp.). p. 113–122. *In* D.R. Woodfield and C. Matthew (ed.) Ryegrass endophyte: An essential New Zealand symbiosis. New Zealand Grassland Assoc., Napier, New Zealand.

Prestidge, R., D.R. Lauren, S.G. van der Zujpp, and M.E. di Menna. 1985. Isolation of feeding deterrents to Argentine stem weevil in cultures of endophytes of perennial ryegrass and tall fescue. N. Z. J. Agric. Res. 28:87–92.

Reinholz, J., and V.H. Paul. 2001. Toxin-free *Neotyphodium*-isolates achieved without genetic engineering—A possible strategy to avoid "ryegrass staggers." p. 261–271. *In* V.H. Paul and P.D. Dapprich (ed.) 4th Int. *Neotyphodium*/grass interactions symposium. 27–29 Sept. 2000. Univ. Paderborn, Soest, Germany.

Rowan, D.D. 1993. Lolitrems, peramine and paxilline: Mycotoxins of the ryegrass/endophyte interaction. Agric. Ecosystems Environ. 44:103–122.

Rowan, D.D., and D.L. Gaynor. 1986. Isolation of feeding deterrents against stem weevil from ryegrass infected with the endophyte *Acremonium loliae*. J. Chem. Ecol. 12:647–658.

Rowan, D.D., J.J. Dymock, and M.A. Brimble. 1990. Effect of fungal metabolite peramine and analogs on feeding development of Argentine stem weevil (*Listronotus bonariensis*). J. Chem. Ecol. 16:1683–1695.

Schardl, C.L. 2001. *Epichloë festucae* and related mutualistic symbionts of grasses. Fungal Gen. Biol. 33:69–82.

Schardl, C.L., and K. Clay. 1997. Evolution of mutalistic endophytes from plant pathogens. p. 221–238. *In* Carroll and P. Tudzynski (ed.) The Mycota V. Part B. Springer-Verlag, Berlin.

Schardl, C.L., A. Leuchtmann, H.-F. Tsai, M.A. Collett, D.M. Watt, and D.B. Scott. 1994. Origin of a fungal symbiont of perennial ryegrass by interspecific hybridization of a mutualist with the ryegrass choke pathogen, *Epichloë typhina*. Genetics 136:1307–1317.

Schardl, C.L., J. Wang, H.H. Wilkinson, and K.-R. Chung. 1999. Genetic analysis of biosynthesis and roles of anti-herbivore alkaloids produced by grass endophytes. p. 118–125. *In* G.T. Kumagai S. et al. (ed.) Proc. Int. Symp. Mycotoxicology 1999. Mycotoxin contamination: Health risk and prevention project. Keibundo Matsumoto Printing Co., Tokyo.

Schardl, C.L., and H.H. Wilkinson. 2000. Hybridization and cospeciation hypotheses for the evolution of grass endophytes. p. 63–83. *In* C.W. Bacon and J.F. White, Jr. (ed.) Microbial endophytes. Marcel Dekker, New York.

Scott, B. 2001. Molecular interactions between *Lolium* grasses and their fungal symbionts. p. 261–274. *In* G. Spangenberg (ed.) Molecular breeding of forage crops. Kluwer Academic Publ., Dordrecht, the Netherlands.

Scott, B. 2004. Functional analysis of the perennial ryegrass–*Epichloë* endophyte interaction. p. 133–144. *In* A. Hopkins et al. (ed.) Molecular Breeding of Forage and Turf. Proc. of the 3rd Int. Symp. on Molecular Breeding of Forage and Turf, Dallas, TX, and Ardmore, OK, USA. 18–22 May 2003. Vol. 11. Kluwer Academic Publ., Dordrecht, the Netherlands.

Scott, B., and C. Young. 2003. Genetic manipulation of clavicipitalean endophytes. p. 425–443. *In* J.F. White, Jr. et al. (ed.) Clavicipitalean fungi: Evolutionary biology, chemistry, biocontrol and cultural impacts. Marcel Dekker, New York.

Siegel, M.R., G.C.M. Latch, L.P. Bush, F.F. Fannin, D.D. Rowan, B.A. Tapper, C.W. Bacon, and M.C. Johnson. 1990. Fungal endophyte-infected grasses: Alkaloid accumulation and aphid response. J. Chem. Ecol. 16:3301–3315.

Steyn, P.S., and R. Vleggaar. 1985. Tremorgenic mycotoxins. Prog. Chem. Org. Nat. Prod. 48:1–80.

Tsai, H.-F., J.-S. Liu, C. Staben, M.J. Christensen, G.C.M. Latch, M.R. Siegel, and C.L. Schardl. 1994. Evolutionary diversification of fungal endophytes of tall fescue grass by hybridization with *Epichloë* species. Proc. Natl. Acad. Sci. USA 91:2542–2546.

von Döhren, H., U. Keller, J. Vater, and R. Zocher. 1997. Multifunctional peptide synthetases. Chem. Rev. 97:2675–2705.

West, C.P. 1994. Physiology and drought tolerance of endophyte-infected grasses. p. 87–99. *In* C.W. Bacon and J.F. White, Jr. (ed.) Biotechnology of endophytic fungi of grasses. CRC Press, Boca Raton, FL.

West, C.P., E. Izekor, D.M. Oosterhuis, and R.T. Robbins. 1988. The effect of *Acremonium coenophialum* on the growth and nematode infestation of tall fescue. Plant Soil 112:3–6.

Wiest, A., D. Grzegorski, B.-W. Xu, C. Goulard, S. Rebuffat, D.J. Ebbole, B. Bodo, and C. Kenerley. 2002. Identification of peptaibols from *Trichoderma virens* and cloning of a peptaibol synthetase. J. Biol. Chem. 277:20 862–20 868.

Young, C.A., L. McMillan, E. Telfer, and B. Scott. 2001. Molecular cloning and genetic analysis of an indole-diterpene gene cluster from *Penicillium paxilli*. Mol. Microbiol. 39:754–764.

GENE DISCOVERY AND MICROARRAY-BASED TRANSCRIPTOME ANALYSIS OF THE GRASS–ENDOPHYTE ASSOCIATION

German C. Spangenberg,[1,2,3,4] *Silvina A. Felitti,*[1,2] *Kate Shields,*[1,2] *Marc Ramsperger,*[1,2] *Pei Tian,*[1,2,3] *Eng Kok Ong,*[1,3] *Daniel Singh,*[1,4] *Erica Logan,*[1,4] *and David Edwards*[1,4]

BIOLOGY OF GRASS–ENDOPHYTE SYMBIOSES

*N*eotyphodium lolii, N. coenophialum, and *Epichloë festucae* are common symbiotic fungal endophytes of the temperate pasture grasses perennial ryegrass (*Lolium perenne*), tall fescue (*Festuca arundinacea*), and red fescue (*Festuca rubra*), respectively (Christensen et al., 1993). *Epichloë* taxa are ascomycete fungi (family Clavicipitaceae) that are ecologically obligate symbionts of grasses (Schardl et al., 1997). They comprise both the sexual *Epichloë* species and their asexual *Neotyphodium* derivatives (Glenn et al., 1996; Schardl, 1996). All establish asymptomatic associations with their host during the vegetative phase of growth, but the sexual species are capable of forming a stroma around the developing inflorescence that partially or completely blocks emergence of the floral meristem. Relative mutualism or antagonism of an *Epichloë*–grass symbiosis is largely related to the path of symbiont transmission (Schardl et al., 1997). Many of these fungi can propagate clonally in the floral meristems and consequently in the seed progeny of infected mother plants (vertical or matrilinear transmission). Alternatively, genotypes can transmit horizontally via sexual spores in a life cycle that also requires a third symbiont (the fly *Phorbia phrenione*) to mediate fungal mating (Schardl et al., 1997).

The benefits of the symbiosis for the host are increased seedling vigor and persistence and drought tolerance in marginal environments (Elbersen and West, 1996; Hill et al., 1990; Malinowski and Belesky, 1999). It also provides protec-

[1] Plant Biotechnology Centre, Primary Industries Research Victoria, Department of Primary Industries, La Trobe University, Bundoora, VIC 3086, Australia
[2] Molecular Plant Breeding Cooperative Research Centre, Australia
[3] Victorian Microarray Technology Consortium, Australia
[4] Victorian Bioinformatics Consortium, Australia

tion against some insect pests and nematodes (Breen, 1993; Elmi et al., 2000; Prestidge and Gallagher, 1988). Specific metabolites produced by the endophyte such as peramine and loline alkaloids provide protection from insect pests (Rowan and Gaynor, 1986; Wilkinson et al., 2000). However, other metabolites such as lolitrem B and ergovaline are toxic to grazing animals, causing conditions known as ryegrass staggers and fescue toxicosis, respectively (Gallagher et al., 1984; Yates et al., 1985). The most thoroughly studied compounds are alkaloids, including ergopeptine alkaloids, indole-isoprenoid lolitrems, pyrrolizidine alkaloids, and pyrrolopyrazine alkaloids (Bush et al., 1997; Scott, 2001).

Considerable variation exists among grass–endophyte associations for the production of different metabolites (Christensen et al., 1993, 1998; Moon et al., 2000; Siegel et al., 1990). Specific metabolite levels vary across endophyte strains, grass varieties, and environments. It is not currently known how much of this variation may be attributed to the endophyte genotype, the host genotype, or to environmental interactions. Endophytes also differ in their ability to confer herbivore resistance and in their insect-deterrent properties. This ability may be controlled by the host–endophyte interaction and thus may be subjected to selection, as has been demonstrated for the production of ergopeptine alkaloids in tall fescue (Panaccione et al., 2001).

In contrast to the information on alkaloids and animal toxicosis, the beneficial physiological aspects of the endophyte–grass interactions have not been well characterized in any system. Very little is known regarding the factors important in host colonization or nutrient exchange between plant and fungus. The physiological mechanisms which lead to increased plant vigor and enhanced tolerance to abiotic stresses unrelated to the reduction in pest damage to endophyte-infected grasses are unknown. Significant progress has been made in recent years in the molecular cloning and genetic analysis of *Epichloë–Neotyphodium* endophyte genes involved in the biosynthesis of lolines (Spiering et al., 2002), ergot alkaloids (Panaccione et al., 2001), and lolitrems (Scott, 2001, 2004). The isolation and characterization of genes involved in the grass–endophyte interaction would be critical for an effective manipulation of the mutualistic associations between grasses of the *Festuca-Lolium* complex and *Epichloë–Neotyphodium* endophytes.

FUNGAL GENOMICS

Genome technologies have revolutionized biology, and these advances have been driven by the ability to produce increasing amounts of sequence information at an ever-diminishing cost. High-throughput gene discovery by expressed sequence tag (EST) sequencing, initiated in 1995 (Adams et al., 1995), has expanded to become a powerful tool for biological analysis from organisms to ecosystems. As well as providing valuable information relating to the expressed portion of genomes, ESTs also provide a means to develop tools for further biological characterization through diversity and evolutionary analysis and gene expression studies (Rafalski et al., 1998). While EST sequencing is still the standard procedure for gene discovery in many organisms, a reduction in the cost of DNA sequencing has led to a move towards whole-genome sequencing.

The importance of fungal genomics was highlighted by the release of the *Saccharomyces cerevisiae* genome sequence in 1996 (Goffeau et al., 1996), the annotated genome sequence of the fission yeast *Schizosaccharomyces pombe* in 2002 (Wood et al., 2002) and completion of the genome sequencing of the filamentous fungus *Neurospora crassa* last year (Galagan et al., 2003). In addition to this, genome sequencing of other fungal species is nearing completion. These species include the human pathogens *Candida albicans* and *Cryptococcus neoformans*, and the phytopathogen *Magnaporthe grisea* (the causal agent of rice blast). Genome sequence information and EST collections from several other parasitic and symbiotic fungi are also being established (Tunlid and Talbot, 2002) (Table 4.1).

Table 4.1. Genome sequencing projects in fungal pathogens and mutualistic symbionts.

Species	Source URL and/or reference	Status (April 2004)
	Plant pathogens	
Blumeria graminis f. sp. *hordei* (barley powdery mildew)	http://cogeme.ex.ac.uk	3021 unique SQ†
Botrytis cinerae (gray mold)	http://cogeme.ex.ac.uk www.genoscope.cns.fr	2901 unique SQ
Cladosporium fulvum (tomato leaf mold)	www.ncbi.nlm.nih.gov/dbEST	1073 SQ
Fusarium graminearium (Fusarium head blight)	http://cogeme.ex.ac.uk www.broad.mit.edu/annotation/fungi/fusarium	4112 unique SQ
Fusarium sporotrichioides (Fusarium head blight)	http://www.genome.ou.edu/fsporo.html	7495 SQ
Magnaphorte griseae (rice blast)	http://cogeme.ex.ac.uk www.tigr.org/tdb/tgi/fungi.shtml www.fungalgenomics.ncsu.edu	8821 unique SQ 15355 unique SQ
Mycosphaerella graminicola (wheat leaf blotch)	http://cogeme.ex.ac.uk www.ncbi.nlm.nih.gov/dbEST	2926 unique SQ

(continued on next page.)

Table 4.1. (continued.)

Species	Source URL and/or reference	Status (April 2004)
Phytophthora infestans (late blight on potato and tomato)	http://cogeme.ex.ac.uk www.ncbi.nlm.nih.gov/dbEST	1414 unique SQ 38930 SQ
Phytophthora sojae (stem and root rot on soybean)	http://cogeme.ex.ac.uk www.ncbi.nlm.nih.gov/dbEST	5849 unique SQ 40187 SQ
	Other pathogens	
Metarhizium anisopliae (entomopathogen)	www.ncbi.nlm.nih.gov/dbEST	2946 SQ
Dactylaria candida (nematode-trapping fungus)		
	Symbiotic fungi	
Neotyphodium coenophialum (tall fescue endophyte)	http://hornbill.cspp.latrobe.edu.au/ endophyte.html	
Neotyphodium lolii (ryegrass endophyte)	http://hornbill.cspp.latrobe.edu.au/ endophyte.html	
Epichloë festucae (red fescue endophyte)	http://hornbill.cspp.latrobe.edu.au/ endophyte.html	
Paxillus involutus (ectomycorrhiza)		
Pisolithus (ectomycorrhiza)	www.ncbi.nlm.nih.gov/dbEST	2948 SQ
Glomus intraradices (AM, endomycorrhiza)	www.ncbi.nlm.nih.gov/dbEST	17363 SQ
	Human pathogens	
Candida albicans (opportunistic infections)	www.sanger.ac.uk/projects/C_albicans	
Aspergillus fumigatus (pulmonary pathogen)	www.sanger.ac.uk/projects/A_fumigatus www.tigr.org/tdb/tgi/fungi.shtml	28 Mb of SQ
Crytococcus neoformans (causing meningitis)	www.genome.ou.edu/cneo.html www.tigr.org/tdb/tgi/fungi.shtml	2333 SQ
Pneumocystis carinii (opportunistic infections, pneumonia)	www.sanger.ac.uk/projects/P_carinii	5674 unique SQ

† SQ, sequences.

The availability of whole-genome sequences from different fungi has emphasized the large diversity between different fungal species as well as provided an insight into common pathways, which have been conserved across evolutionary time. Genomic research into nonmodel fungi is still considered to be cost-effective through the application of EST sequencing and sequence annotation through comparison to model fungal genomes. Along with gene sequences, ESTs provide some measure of expression of genes through their representation in different cDNA libraries. The EST sequences also provide a valuable resource for the in silico identification of molecular markers such as simple sequence repeats (SSRs) (Robinson et al., 2004) or single nucleotide polymorphisms (SNPs) (Barker et al., 2003; Batley et al., 2003).

Fungal EST collections have been used to study gene expression underlying growth and differentiation processes in filamentous fungi. A collection of 24 000 ESTs representing some 2500 unique cDNAs from *N. crassa* have been obtained from different morphological and developmental stages by using cDNA libraries derived from germinating spores, branching hyphae or fruiting bodies (Nelson et al., 1997). In addition, changes in gene expression in *N. crassa* associated with circadian rhythm were assessed (Zhu et al., 2001). A large-scale comparison of sequence data from the filamentous fungus *N. crassa* with the complete genomic sequence of *S. cerevisiae* showed that *N. crassa* have a much higher proportion of genes without identifiable homologs. This suggests that the morphological complexity of *N. crassa* reflects the acquisition or maintenance of novel genes, consistent with its larger genome. One specific example is the identification of an additional *N. crassa* gene (NPH1 homolog) possibly involved in responses to light (Braun et al., 2000). In another study, digital gene expression profiles evaluated for asexually developing *Aspergillus nidulans* tissues and compared with *N. crassa* vegetative tissue led to the identification of novel genes in *A. nidulans* asexual development and stress response (Prade et al., 2000).

GENE DISCOVERY IN *EPICHLOË–NEOTYPHODIUM* ENDOPHYTES

No comprehensive genomic resource for *Neotyphodium* and *Epichloë* species has been generated until recently (Spangenberg et al., 2001; Felitti et al., 2004). Only 700 gene sequences, including three *N. uncinatum* cDNAs encoding candidate genes putatively involved in loline biosynthesis, three *N. coenophialum* genes (two peptide synthase genes and one in planta-expressed gene), highly conserved gene sequences such as actin, β-tubulin, elongation factor-1, and ribosomal RNA genes, as well as several cDNA and genomic clones from *Neotyphodium* and *Epichloë*, were previously deposited in GenBank.

We have established an endophyte EST resource comprising 9507 sequences from *Epichloë* and *Neotyphodium* endophytes (Spangenberg et al., 2001; Felitti et al., 2004). Redundancy of these sequences was assessed by clustering and sequence assembly using cap3 (Huang and Madan, 1999). This identified a total of 3942 unique genes represented by 2414 unique ESTs and 1528 assembled contigs (Table 4.2).

Table 4.2. Discovery of expressed sequence tags (ESTs) and EST-derived simple sequence repeats (SSRs) from five cDNA libraries generated from in vitro grown *Epichloë festucae, Neotyphodium coenophialum*, and *N. lolii*.

Number of sequences	9507
Number of contigs	1528
Number of singletons	2414
Number of SSRs	613

This endophyte EST collection has been annotated by comparison with the public nucleotide and protein sequence databases GenBank and Swiss-Prot. The ESTs were then manually assigned into functional categories based on their similarity to previously characterized genes (Fig. 4.1). The recent adoption of structured gene ontologies (GO) within the genomics community enables the direct comparison of genes and their expressed products between divergent organisms. Gene-ontology annotation may either be primary, through manual curation, or secondary, where GO annotation is derived through sequence identity to a previously annotated sequence. We have developed tools for automated GO annotation of ESTs based on comparison with a database of GO-annotated sequences. Application of this tool to the nonredundant endophyte EST data set enabled the structured GO annotation of 25% of 3942 endophyte genes from five different cDNA libraries (Fig. 4.1).

Comparative analysis between the ESTs from *N. coenophialum* and those from *Epichloë festucae* and *N. lolii* identified common genes among these fungal species. Similarities of ESTs between the two *Neotyphodium* endophytes and *E. festucae* illustrate conserved genes among these fungi, identifying 237 common genes between *N. coenophialum* and *E. festucae*, and 538 common genes between *N. coenophialum* and *N. lolii*. The majority of common genes showed identity to predicted genes of unknown function, followed by genes involved in protein synthesis, genes encoding hypothetical proteins, and conserved genes involved in energy production. All these endophyte sequences and respective annotation data are maintained within a custom MySQL database using the ASTRA database schema (Love et al., 2003), allowing for integrated searching of sequences or sequence annotation using a user-friendly web interface (http://hornbill.cspp.latrobe.edu .au/).

DISCOVERY OF EST-DERIVED SSR AND SNP MARKERS

The EST-derived SSRs (EST-SSRs) for both *N. coenophialum* and *N. lolii* were identified from the established endophyte EST resource using SSR Primer (Robinson et al., 2004). The SSR loci were identified within 9.7% of the *N. coenophialum* sequences and 6.3% of *N. lolii* sequences. This is similar to the EST-SSR frequency identified within other eukaryotic organisms (Gao et al., 2003). A variety of SSR motifs were identified, with trinucleotide repeat arrays being the most common. The characterization of SSR loci from the *Neotyphodium* and *Epichloë* EST resource and the use of SSR markers for the assessment of genetic variation within and between endophyte species has been previously described (van Zijll de Jong et al., 2003, 2004, this volume).

Increasingly, SNPs are becoming the marker of choice in genetic analysis. They are used in agriculture as markers in breeding programs and have many uses in human genetics, such as the detection of alleles associated with genetic diseases and the identification of individuals. Offering the potential for generating very high-density genetic maps, SNPs are valuable for genome mapping (Rafalski, 2002). The relatively low mutation rate of SNPs also makes them excellent markers for studying complex genetic traits and as a tool for the understanding of genome evolution (Syvanen, 2001). The ability to efficiently mine DNA sequence datasets for SNPs using in silico analysis tools has greatly expanded the utilization of SNPs in genome research. We have applied the AutoSNP discovery tool (Barker et al., 2003; Batley et al., 2003) to the *Neotyphodium* and *Epichloë* endophyte EST resource, leading to the identification of a total of 1979 polymorphisms or one SNP for every 250 bp of aligned sequence. This is similar to the frequency of SNPs observed in some plant studies (Germano and Klein, 1999; Coryell et al., 1999), but greater than that observed for human studies where SNP frequency can be as low as one or two SNPs per thousand bases (Clifford et al., 2000; Deutsch et al., 2001). The SNPs identified in the endophyte EST dataset correspond to 1136 transitions, 525 transversions, and 333 insertion/deletions (indels). This high frequency of transitions has been observed in previous SNP discovery programs (Deutsch et al., 2001; Garg et al., 1999; Picoult-Newberg et al., 1999) and reflects the high frequency of C to T mutation due to deamination of *S*-methylcytosine to thymine over evolutionary time (Coulondre et al., 1978).

FUNGAL TRANSCRIPTOMICS

The measurement of gene expression has undergone rapid advancement with the development of microarray technology and sequence-based methods for expression profiling. Sequence-based methods have the potential to more accurately determine the quantitative levels of gene expression due to their extended linear dynamics. Furthermore, they do not require prior sequence information, so have the advantage of identifying novel genes or assessing gene expression for previously uncharacterized organisms. The predominant methods for sequence-based expression analysis are serial analysis of gene expression (SAGE) (Velculescu et al., 1995) and massively parallel signature sequencing (MPSS) (Brenner et al., 2000a, 2000b). Of these, only SAGE has been broadly adopted. While MPSS provides several major benefits over SAGE, the high costs involved with the process have led to its limited adoption. However, the use of MPSS for annotating genomes with expressed genes is likely to lead to the wider adoption of MPSS. Hybridization-based microarrays have become the tool of choice for transcriptomic analysis, primarily because of their capacity to analyze multiple samples simultaneously.

The availability of fungal genomic and EST sequences has made it possible to construct DNA microarrays for the simultaneous analysis of the expression levels of large sets of genes in several species of fungal pathogens and symbionts. The mRNA levels of 6000 yeast genes were analyzed simultaneously under specific genetic and physiological conditions (Goffeau, 2000). These genome-wide transcriptome analyses involved assessments for cell cycle, meiosis, sporulation, ploidy, rich vs. minimal medium, osmotic stress, oxidative and chemical stresses, deletion, overexpression, or activation of transcription factors. Recently, up to 300 complete expression profiles were generated in which transcript levels of 287 mutants and 13 compound-treated yeast cultures were analyzed (Hughes et al., 2000). Using these expression profiles, a reference database was constructed and demonstrated that the cellular pathways, altered under different environmental conditions or through modification of the genotype, could be determined by pattern-matching expression profiles (Hughes et al., 2000). Examination of profiles caused by deletions of uncharacterized genes validated the utility of this approach and experimentally confirmed that eight previously uncharacterized open reading frames encode proteins required for sterol metabolism, cell wall function, mitochondrial respiration, or protein synthesis (Hughes et al., 2000). Clock-controlled genes have been studied in *N. crassa* using two different cDNA microarrays containing more than 1000 genes

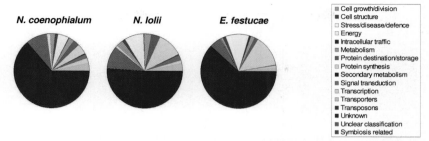

Fig. 4.1. *Functional annotation of expressed sequence tags from Epichloë–Neotyphodium endophytes.*

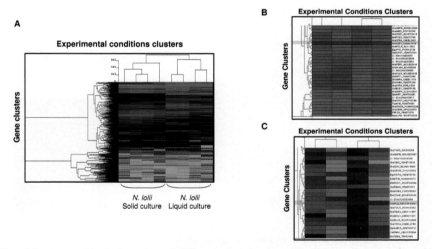

Fig. 4.2. *Hierarchical clustering of different signal values (Euclidean distance) from N. lolii grown in liquid culture vs. N. lolii grown in solid culture using GeneSight software (BioDiscovery, Inc., El Segundo, CA). (A) Hierarchical clustering of mean signal values. (B) Hierarchical clustering of ratio of median signal values showing N. lolii genes up-regulated in liquid culture. (C) Hierarchical clustering of ratio of median signal values showing N. lolii genes with similar expression levels in both culture conditions. Adapted from Felitti et al., 2004.*

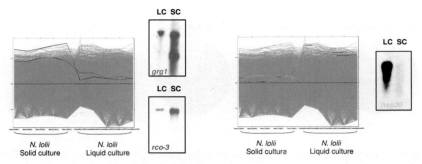

Fig. 4.3. *Neotyphodium lolii transcriptome analysis. Time series plots of mean signal values: grg1, homologue to a glucose-repressible gene; rco-3, homologue to a glucose transporter gene; and hsp30, homologue to a 30-kDa heat shock protein from Neurospora crassa. The N. lolii cDNAs were used as probes in Northern hybridization analysis. This analysis allowed for validation of microarray-based gene expression data. Adapted from Felitti et al., 2004.*

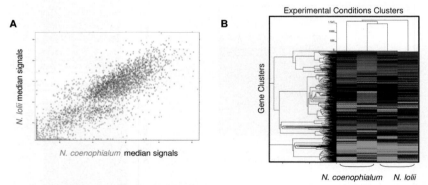

Fig. 4.4. *Comparative transcriptome analysis of N. coenophialum and N. lolii. (A) Scatter plot of median signal values showing N. coenophialum and N. lolii prevalent gene expression. (B) Hierarchical clustering of ratio median signal values showing species-prevalent gene expression. Adapted from Felitti et al., 2004.*

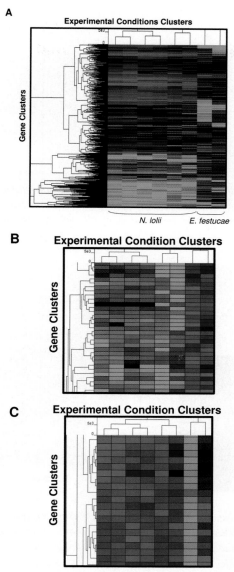

Fig. 4.5. *Hierarchical clustering of mean signal values (Euclidean distance) from N. lolii and E. festucae grown in liquid culture using GeneSight software (BioDiscovery, Inc., El Segundo, CA). The genes that are shown in red are highly expressed, while the ones that are shown in green have low expression levels. (A) Neotyphodium lolii and E. festucae-prevalent gene expression. (B) Epichloë festucae-prevalent gene expression showing predominantly amino acid metabolism genes. (C) Neotyphodium lolii-prevalent gene expression showing numerous gene classes. Adapted from Felitti et al., 2004.*

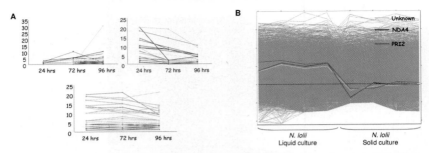

Fig. 4.6. (A) *Time series plots of median signal values showing numerous N. lolii gene classes expression levels.* (B) *Time series plot of median signal values showing a gene of unknown function coordinately expressed with cell cycle genes. Adapted from Felitti et al., 2004.*

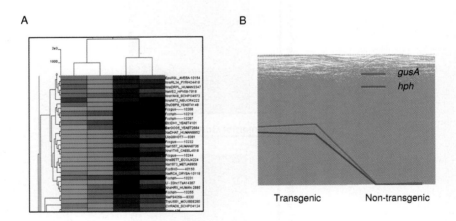

Fig. 4.7. *Analysis of Neotyphodium gene expression using the EndoChip-containing gusA and hph genes and RNA prepared from in vitro grown transgenic Neothyphodium strain FM13 (Lp1/pNOM and pAN7-1) expressing chimeric hph and gus genes (Murray et al., 1992).* (A) *Hierarchical clustering of mean signal values showing differential gene expression in different endophyte genotypes at target loci.* (B) *Time series plot of mean signal values showing the expression levels of both transgenes in the transgenic and the wild-type strains. Adapted from Felitti et al., 2004.*

(Nowrousian et al., 2003; Correa et al., 2003). Nutrient-dependent transcript variations in *N. crassa* were studied using a microarray comprising 4700 cDNAs (Aign and Hoheisel, 2003). A microarray based on available *Candida albicans* genome sequence data (Tzung et al., 2001) has been used to identify new cellular targets of several transcriptional regulators that play a central role in the control of metabolism in yeast and hyphal morphogenesis in *C. albicans* (Murad et al., 2001). In one of the few published studies of microarray-based transcriptome analysis in plant–fungal mutualistic associations, novel symbiosis-related genes in the *Eucalyptus globulus–Pisolithus tinctorius* ectomycorrhiza association were identified (Voiblet et al., 2001).

One bottleneck in fungal transcriptomics is the lack of integrated bioinformatics tools for the analysis of genome sequence expression data (Soanes et al., 2002). The utility of the yeast genome has been aided by custom-designed databases, such as the yeast proteome database (www.incyte.com/sequence/proteome/) and the Stanford genome database (http://genome-www.stanford.edu/Saccharomyces). A database for comparative analyses of EST sequences of phytopathogenic fungi (http://cogene.man.ac.uk and www.cs.man.ac.uk/norm/glims) has been developed (Paton et al., 2000; Soanes et al., 2002). The DNA sequence information that is becoming available for human, animal, and plant fungal pathogens and mutualistic species, as well as for several of their hosts (Tunlid and Talbot, 2002), now enable detailed genomic analysis of the corresponding fungal–host interactions. Recent advances in bioinformatics suggest a move towards the adoption of generic open-source databases designed to maximize the integration and interrogation of diverse data sources. The expansion in use of the Ensembl database schema (Hubbard et al., 2002) is an example of this convergence in bioinformatics databases and its facility for viewing related data from several different organisms makes it an ideal model for comparative genomics in fungi.

MICROARRY-BASED TRANSCRIPTOME ANALYSIS

The availability of partially sequenced endophyte cDNA clones enables the construction of specific endophyte microarrays. We have assembled a *Neotyphodium* and *Epichloë* unigene set from ESTs generated from the partial sequencing of randomly selected cDNA clones from six cDNA libraries prepared from *Neotyphodium* and *Epichloë* endophytes grown in vitro and in planta (Felitti et al., 2004; Spangenberg et al., 2001). This unigene set, representing a total of 4662 unique genes, was used to produce two generations of *Epichloë–*

Neotyphodium endophyte cDNA microarrays: the Nchip and the EndoChip. The Nchip consists of 3806 genes derived from four cDNA libraries from in vitro grown *Epichloë–Neotyphodium* endophytes, while the EndoChip contains 18 264 features interrogating 5325 *Epichloë–Neotyphodium* genes derived from six cDNA libraries including in planta expressed endophyte sequences. These microarrays have been applied to examine changes in *Epichloë–Neotyphodium* endophyte gene expression in response to growth in different culture conditions (Felitti et al., 2004). Hierarchical cluster analysis (Fig. 4.2) and time series plots (Fig. 4.3) demonstrated differential gene expression patterns for *N. lolii* grown under different culture conditions. For example, *grg1* and *rco-3* orthologues showed higher expression levels in *N. lolii* endophyte grown on solid culture, whereas the *N. lolii hsp30* orthologue showed higher expression levels in *N. lolii* endophyte grown in liquid culture. Northern hybridization analysis using *N. lolii* cDNAs as probes allowed for the validation of microarray-based gene expression results (Fig. 4.3).

Since microarray-based gene expression analysis is dependent on hybridization between complementary strands of DNA, microarrays may be used to study gene expression in related species that share DNA sequence homology. Microarray-based gene expression analysis with RNA from in vitro grown *N. coenophialum* and *N. lolii* demonstrated its utility in comparative transcriptome analysis of both endophyte species (Fig. 4.4). Furthermore, microarray-based transcriptome analysis with total RNA from *N. lolii* and *E. festucae* grown in vitro has enabled the study of genome-specific gene expression in asexual and sexual forms of grass endophytes (Fig. 4.5). These studies demonstrate the power of the *Epichloë–Neotyphodium* endophyte cDNA microarrays in genome-wide transcriptome analysis for different endophyte species, as well as gene expression responses associated with specific developmental stages of the life cycle of asexual and sexual forms of grass endophytes.

The *Epichloë–Neotyphodium* endophyte cDNA microarrays are also useful tools for the analysis of unannotated genes through expression profiling using template genes of known function. For example, expression profiling experiments allowed identification of novel unannotated *N. lolii* genes which are coordinately expressed with two endophyte genes encoding orthologues from cell cycle genes in yeast and *N. crassa* (Fig. 4.6). This experimental approach can thus be used to identify novel unannotated *Epichloë–Neotyphodium* endophyte genes putatively involved in specific metabolic pathways and other cellular processes as candidates for detailed functional analysis through gene disruption or gene silencing. Furthermore, microarray analysis using EndoChip provides a tool in endophyte functional genomics to assess global gene expression changes in response to the introduction of novel genes or following disruption through homologous recombination or silencing of target genes. To demonstrate this capability, RNA from in vitro grown *Neotyphodium* strain FM13 (Lp1/pNOM-101 and pAN7-1) transformed with chimeric *hph* and *gusA* genes (Murray et al., 1992) was used in microarray analysis demonstrating *hph* and *gusA* transgene expression in transformed endophyte (Fig. 4.7).

The use of the EndoChip, together with a 15K unigene microarray arising from the generation and analysis of more than 44 000 ESTs in perennial ryegrass (Sawbridge et al., 2003; Spangenberg et al., 2001), allows for the concomitant genome-wide transcriptome analysis of *Epichloë–Neotyphodium* endophytes and their hosts (Felitti et al., 2004; Spangenberg et al., 2001).

SUMMARY

The first genomics and transcriptomics resources and tools for *Epichloë–Neotyphodium* endophytes have been developed. They provide the basis for large-scale gene and molecular marker discovery, enabling detailed molecular genetics and functional genomics studies of both endophyte biology and the interactions between endophytes and their grass hosts. The development and proof-of-concept applications of the *Epichloë–Neotyphodium* endophyte unigene microarray, EndoChip, complemented by similar developments with the design and generation of ryegrass host unigene microarrays, have opened the way for the molecular dissection of the grass–endophyte association at the transcriptome level. The incorporation of these tools and data resulting from their application to be expanded with proteomic and metabolomic data and analysis tools within an integrated endophyte bioinformatics database provides a general functional genomics resource for *Epichloë–Neotyphodium* grass endophytes.

ACKNOWLEDGMENTS

This work was supported by the Department of Primary Industries, Victoria, Australia; the Molecular Plant Breeding Cooperative Research Centre, Australia; the Victorian Microarray Technology Consortium, Australia; the Victorian Bioinformatics Consortium, Australia; and the Argentinian Council for Scientific and Technical Research. The *Neotyphodium* strain FM13 (Lp1/pNOM-101 and pAN7-1) was kindly provided by Prof. Barry Scott, Institute of Molecular BioSciences, Massey University, Palmerston North, New Zealand.

REFERENCES

Adams, M.D., A.R. Kerlavage, R.D. Fleischmann, R.A. Fuldner, C.J. Bult, N.H. Lee, E.F. Kirkness, K.G. Weinstock, et al. 1995. Initial assessment of human gene diversity and expression patterns based upon 83-million nucleotides of cDNA sequence. Nature (London) 377:3–17.

Aign, V., and J.D. Hoheisel. 2003. Analysis of nutrient-dependent transcript variations in *Neurospora crassa*. Fungal Genet. Biol. 40:225–233.

Barker, G., J. Batley, H. O'Sullivan, K.J. Edwards, and D. Edwards. 2003. Redundancy based detection of sequence polymorphisms in expressed sequence tag data using autoSNP. Bioinformatics 19:421–422.

Batley, J., G. Barker, H. O'Sullivan, K.J. Edwards, and D. Edwards. 2003. Mining for single nucleotide polymorphisms and insertions/deletions in maize expressed sequence tag data. Plant Physiol. 132:84–91.

Braun, E.L., A.L. Halpern, M.A. Nelson, and D.O. Natvig. 2000. Large scale comparison of fungal sequence information: Mechanism of innovation in *Neurospora crassa* and gene loss in *Saccharomyces cerevisiae*. Genomics Res. 10:416–430.

Breen, J.P. 1993. Enhanced resistance to three species of aphids (Homoptera, Aphididae) in *Acremonium* endophyte-infected turf-grasses. J. Econ. Entomol. 86:1279–1286.

Brenner, S., M. Johnson, J. Bridgham, G. Golda, D.H. Lloyd, D. Johnson, S.J. Luo, S. McCurdy, et al. 2000a. Gene expression analysis by massively parallel signal sequencing (MPSS) on microbead arrays. Nat. Biotechnol. 18:630–634.

Brenner, S., S.R. Williams, E.H. Vermaas, T. Storck, K. Moon, C. McCollum, J.I. Mao, S.J Luo, et al. 2000b. In vitro cloning of complex mixtures of DNA on microbeads: Physical separation of differentially expressed cDNAs. Proc. Natl. Acad. Sci. USA 97:1665–1670.

Bush, L.P., H.H. Wilkinson, and C.L. Schardl. 1997. Bioprotective alkaloids of grass–fungal endophyte symbioses. Plant Physiol. 114:1–7.

Christensen, M.J., H.S. Easton, W.R. Simpson, and B.A. Tapper. 1998. Occurrence of the fungal endophyte *Neotyphodium coenophialum* in leaf blades of tall fescue and implications for stock health. N. Z. J. Agric. Res. 41:595–602.

Christensen, M.J., A. Leuchtmann, D.D. Rowan, and B.A. Tapper. 1993. Taxonomy of *Acremonium* endophytes of tall fescue (*Festuca arundinacea*), meadow fescue (*F. pratensis*) and perennial ryegrass (*Lolium perenne*). Mycol. Res. 97:1083–1092.

Clifford, R., M. Edmonson, Y. Hu, C. Nguyen, T. Scherpbier, and K.H. Buetow. 2000. Expression-based genetic/physical maps of single nucleotide polymorphisms identified by the cancer genome anatomy project. Genome Res. 10:1259–1265.

Correa, A., Z.A. Lewis, A.V. Greene, I.J. March, R.H. Gomer, and D. Bell-Pedersen. 2003. Multiple oscillators regulate circadian gene expression in *Neurospora*. Proc. Natl. Acad. Sci. USA 100:13597–13602.

Coryell, V.H., H. Jessen, J.M. Schupp, D. Webb, and P. Keim. 1999. Allele-specific hybridisation markers for soybean. Theor. Appl. Genet. 101:1291–1298.

Coulondre, C., J.H. Miller, P.J. Farabaugh, and W. Gilbert. 1978. Molecular basis of base substitution hot spots in *Escherichia coli*. Nature (London) 274:775–780.

Deutsch, S., C. Iseli, P. Bucher, S.E. Antonarakis, and H.S. Scott. 2001. A cSNP map and database for human chromosome 21. Genome Res. 11:300–307.

Elbersen, H.W., and C.P. West. 1996. Growth and water relations of field-grown tall fescue as influenced by drought and endophyte. Grass Forage Sci. 51:333–342.

Elmi, A.A., C.P. West, R.T. Robbins, and T.L. Kirkpatrick. 2000. Endophyte effects on reproduction of a root-knot nematode (*Meloidogyne marylandi*) and osmotic adjustment in tall fescue. Grass Forage Sci. 55:166–172.

Felitti, S.A., K. Shields, M. Ramsperger, P. Tian, T. Webster, E.K. Ong, T. Sawbridge, and G. Spangenberg. 2004. Gene discovery and microarray-based transcriptome analysis in grass endophytes. p. 145–153. *In* A. Hopkins et al. (ed.) Proc. of the 3rd Int. Symp., Molecular Breeding of Forage and Turf. Kluwer Academic Publ., Dordrecht, the Netherlands.

Galagan, J.E., S.E. Calvo, K.A. Borkovich, E.U. Selker, N.D. Read, D. Jaffe, W. Fitzhugh, L.-J. Ma, et al. 2003. The genome sequence of the filamentous fungus *Neurospora crassa*. Nature (London) 422:859–868.

Gallagher, R.T., A.D. Hawkes, P.S. Steyn, and R. Vleggaar. 1984. Tremorgenic neurotoxins from perennial ryegrass causing ryegrass staggers disorder of livestock: Structure elucidation of lolitrem B. J. Chem. Soc. Chem. Comm. 9:614–616.

Gao, L.F., J.F. Tang, H.W. Li, and J.Z. Jia. 2003. Analysis of microsatellites in major crops assessed by computational and experimental approaches. Molec. Breed. 12:245–261.

Garg, K., P. Green, and D.A. Nickerson. 1999. Identification of candidate coding region single nucleotide polymorphisms in 165 human genes using assembled expressed sequence tags. Genome Res. 9:1087–1092.

Germano, J., and A.S. Klein. 1999. Species specific nuclear and chloroplast single nucleotide polymorphisms to distinguish *Picea glauca*, *P. mariana* and *P. rubens*. Theor. Appl. Genet. 99:37–49.

Glenn, A.E., C.W. Bacon, R. Price, and R.T. Hanlin. 1996. Molecular phylogeny of *Acremonium* and its taxonomic implications. Mycologia 88:369–383.

Goffeau, A. 2000. Four years of post-genomic life with 6000 yeast genes. FEBS Lett. 480:37–41.

Goffeau, A., B.G. Barrell, H. Bussey, R.W. Davis, B. Dujon, H. Feldmann, F. Galibert, J.D. Hoheisel, et al. 1996. Life with 6000 genes. Science (Washington, DC) 274:546–567.

Hill, N.S., W.C. Stringer, G.E. Rottinghaus, D.P. Belesky, W.A. Parrott, and D.D. Pope. 1990. Growth, morphological, and chemical component responses of tall fescue to *Acremonium coenophialum*. Crop Sci. 30:156–161.

Huang, X., and A. Madan. 1999. CAP3: A DNA sequence assembly program. Genome Res. 9:868–877.

Hubbard, T., D. Barker, E. Birney, G. Cameron, Y. Chen, L. Clark, T. Cox, J. Cuff, et al. 2002. The Ensembl genome database project. Nucleic Acids Res. 30:38–41.

Hughes, T.R., M.J. Marton, A.R. Jones, C.J. Roberts, R. Stoughton, C.D. Armour, H.A. Bennett, E. Coffey, et al. 2000. Functional discovery via a compendium of expression profiles. Cell 102:109–126.

Love, C.G., J. Batley, and D. Edwards. 2003. Applied computational tools for crop genome research. J. Plant Biotechnol. 5:193–195.

Malinowski, D.P., and D.P. Belesky. 1999. *Neotyphodium coenophialum*-endophyte infection affects the ability of tall fescue to use sparingly available phosphorus. J. Plant. Nutr. 22:835–853.

Moon, C.D., B. Scott, C.L. Schardl, and M.J. Christensen. 2000. The evolutionary origins of *Epichloë* endophytes from annual ryegrasses. Mycologia 92:1103–1118.

Murad, A.M.A., C. D'Enfert, C. Gaillardin, H. Tournu, F. Tekaia, D. Talibi, D. Marechal, V. Marchais, et al. 2001. Transcript profiling in *Candida albicans* reveals new cellular functions for the transcriptional repressors CaTup1, CaMig1 and CaNrg1. Mol. Microbiol. 42:981–993.

Murray, F.R., G.C.M. Latch, and D.B. Scott. 1992. Surrogate transformation of perennial ryegrass, *Lolium perenne*, using genetically modified *Acremonium* endophyte. Mol. Gen. Genet. 233:1–9.

Nelson, M.A., S. Kang, E.L. Braun, M.E. Crawford, P.L. Dolan, P.M. Leonard, J. Mitchell, A.M. Armijo, et al. 1997. Expressed sequences from conidial, mycelial, and sexual stages of *Neurospora crassa*. Fungal Genet. Biol. 21:348–363.

Nowrousian, M., G.E. Duffield, J.J. Loros, and J.C. Dunlap. 2003. The frequency gene is required for temperature-dependent regulation of many clock-controlled genes in *Neurospora crassa*. Genetics 164:923–33.

Panaccione, D.G., R.D. Johnson, J. Wang, C.A. Young, P. Damrongkool, B. Scott, and C.L. Schardl. 2001. Elimination of ergovaline from a grass–*Neotyphodium* endophyte symbiosis by genetic modification of the endophyte. Proc. Natl. Acad. Sci. USA 98:12 820–12 825.

Paton, N.W., S.A. Khan, A. Haynes, F. Moussouni, A. Brass, K. Elibeck, C.A. Goble, S. Hubbard, and S.G. Oliver. 2000. Conceptual modelling of genomic information. Bioinformatics 16:548–558.

Picoult-Newberg, L., T.E. Ideker, M.G. Pohl, S.L. Taylor, M.A. Donaldson, D.A. Nickerson, and M. Boyce-Jacino. 1999. Mining SNPs from EST databases. Genome Res. 9:167–174.

Prade, R.A., P. Ayoubi, S. Krishnan, S. Macwana, and H. Russell. 2000. Accumulation of stress and inducer-dependent plant-cell-wall-degrading enzymes during asexual development in *Aspergillus nidulans*. Genetics 157:957–967.

Prestidge, R.A., and R.T. Gallagher. 1988. Endophyte fungus confers resistance to ryegrass: Argentine stem weevil larval studies. Ecol. Entomol. 13:429–435.

Rafalski, A. 2002. Applications of single nucleotide polymorphisms in crop genetics. Curr. Opin. Plant Biol. 5:94–100.

Rafalski, J.A., M. Hanafey, G-H. Miao, A. Ching, J-M. Lee, M. Dolan, and S. Tingey. 1998. New experimental and computational approaches to the analysis of gene expression. Acta. Biochim. Pol. 45:929–934.

Robinson, A.J., C.G. Love, J. Batley, G. Barker, and D. Edwards. 2004. Simple sequence repeat marker loci discovery using SSR primer. Bioinformatics 20:1–2.

Rowan, D.D., and D.L. Gaynor. 1986. Isolation of feeding deterrents against Argentine stem weevil from ryegrass infected with the endophyte *Acremonium loliae*. J. Chem. Ecol. 12:647–658.

Sawbridge, T., E.K. Ong, C. Binnion, M. Emmerling, R. McInnes, K. Meath, N. Nguyen, K. Nunan, et al. 2003. Generation and analysis of expressed sequence tags in perennial ryegrass (*Lolium perenne* L.). Plant Sci. 165:1089–1100.

Schardl, C.L. 1996. *Epichloë* species: Fungal symbionts of grasses. Ann. Rev. Phytopathol. 34:109–130.

Schardl, C.L., A. Leuchtmann, K-R.Chung, D. Penny, and M.R. Siegel. 1997. Coevolution by common descent fungal symbionts (*Epichloë* spp.) and grass hosts. Mol. Biol. Evol. 14:133–143.

Scott, B. 2001. Molecular interactions between *Lolium* grasses and their fungal symbionts. p. 261–274. *In* G. Spangenberg (ed.) Proc. of the 2nd Int. Symp., Molecular Breeding of Forage Crops. Kluwer Academic Publ., Dordrecht, the Netherlands.

Scott, B. 2004. Functional analysis of the perennial ryegrass–*Epichloë* endophyte interaction. p. 145–153. *In* A. Hopkins et al. (ed.) Proc. of the 3rd Int. Symp., Molecular Breeding of Forage and Turf. Kluwer Academic Publ., Dordrecht, the Netherlands.

Siegel, M.R., G.C.M. Latch, L.P. Bush, F.F. Fannin, D.D. Rowan, B.A. Tapper, C.W. Bacon, and M.C. Johnson. 1990. Fungal endophyte infected grass: Alkaloid accumulation and aphid response. J. Chem. Ecol. 16:3301–3316.

Soanes, D.M., W. Skinner, J. Keon, J. Hargreaves, and N.J. Talbot. 2002. Genomics of phytopathogenic fungi and the development of bioinformatic resources. Mol. Plant–Microb. Interact. 15:421–427.

Spangenberg, G., R. Kalla, A. Lidgett, T. Sawbridge, E.K. Ong, and U. John. 2001. Breeding forage plants in the genome era. p. 1–39. *In* G. Spangenberg (ed.) Proc. of the 2nd Int. Symp., Molecular Breeding of Forage Crops. Kluwer Academic Publ., Dordrecht, the Netherlands.

Spiering, M.J., H.H. Wilkinson, J.D. Blandenship, and C.L. Schardl. 2002. Expressed sequence tags and genes associated with loline alkaloid expression by the fungal endophyte *Neotyphodium uncinatum*. Fungal Genet. Biol. 36:242–254.

Syvanen, A.C. 2001. Genotyping single nucleotide polymorphisms. Nat. Rev. Genet. 2:930–942.

Tunlid, A., and N.J. Talbot. 2002. Genomics of parasitic and symbiotic fungi. Curr. Opin. Microbiol. 5:513–519.

Tzung, K-W., R.M. Williams, S. Scherer, N. Federspiel, T. Jones, N. Hansen, V. Bivolarevic, L. Huizar. 2001. Genomic evidence for a complete sexual cycle in *Candida albicans*. Proc. Natl. Acad. Sci. USA 98:3249–3253.

Van Zijll de Jong, E., N.R. Bannan, J. Batley, K.M. Guthridge, G.C. Spangenberg, K.F. Smith, and J.W. Forster. 2004. Genetic diversity in the perennial ryegrass fungal endophyte *Neotyphodium lolii*. p. 155–164. *In* A. Hopkins et al. (ed.) Proc. of the 3rd Int. Symp., Molecular Breeding of Forage and Turf. Kluwer Academic Publ., Dordrecht, the Netherlands.

Van Zijll de Jong, E., K.M. Guthridge, G.C. Spangenberg, and J.W. Forster. 2003. Development and characterisation of EST-derived simple sequence repeat (SSR) markers for pasture grass endophytes. Genome 46:277–290.

Velculescu, V.E., L. Zhang, B. Vogelstein, and K.W. Kinzler. 1995. Serial analysis of gene expression. Science (Washington, DC) 270:484–487.

Voiblet, C., S. Duplessis, N. Encelot, and F. Martin. 2001. Identification of symbiosis-regulated genes in *Eucalyptus globulus–Pisolithus tinctorius* ectomycorrhiza by differential hybridization of arrayed cDNAs. Plant J. 25:181–191.

Wilkinson, H.H., M.R. Siegel, J.D. Blankenship, A.C. Mallory, L.P. Bush, and C.L. Schardl. 2000. Contribution of fungal loline alkaloids to protection from aphids in a grass–endophyte mutualism. Mol. Plant–Microbe. Interact. 13:1027–1033.

Wood, V., R. Gwillan, M-A. Rajandream, M. Lyne, R. Lyne, A. Stewart, J. Sgouros, N. Peat, et al. 2002. The genome sequence of *Schizosaccharomyces pombe*. Nature (London) 415:871–880.

Yates, S.G., R.D. Plattner, and G.B. Garner. 1985. Detection of ergopeptine alkaloids in endophyte infected, toxic Ky-31 tall fescue by mass spectrometry/mass spectrometry. J. Agric. Food Chem. 33:719–722.

Zhu, H., M. Nowrousian, D. Kupfer. H.V. Colot, G. Berrocaltito, H. Lai, D. Bell-Pedersen, B.A. Roe, et al. 2001. Analysis of expressed sequence tags from two starvation, time-of-day-specific libraries *of Neurospora crassa* reveals novel clock-controlled genes. Genetics 157:1057–1065.

MOLECULAR GENETIC MARKER-BASED ANALYSIS OF THE GRASS–ENDOPHYTE SYMBIOSIS

Eline van Zijll de Jong,[1,3] Kevin F. Smith,[2,3] German C. Spangenberg,[1,3] and John W. Forster[1,3]

Early studies of different grass–endophyte associations demonstrated that considerable variation for the expression of beneficial and deleterious agronomic characters could be observed between strains (Bacon, 1988; Bacon and Siegel, 1988), and that potential existed for the manipulation of these characters through artificial inoculations (Clay, 1989; Siegel et al., 1987). However, corresponding data on levels of genetic variation among endophytes was lacking to support the selection of such strains. As a consequence, first-generation molecular genetic markers based on isoenzyme polymorphism were developed. Isoenzymes do not detect genetic variation at the level of DNA directly, but instead assay protein primary-structure variation arising from mutational changes. Nonetheless, isoenzymes were used to study the genetic variation of a large number of *Neotyphodium* and *Epichloë* endophyte genotypes (Leuchtmann and Clay, 1990). Low levels of genetic variation were detected within *N. coenophialum* (from tall fescue, *Festuca arundincea*) and *N. lolii* (from perennial ryegrass, *Lolium perenne*). Although *N. lolii* isolates from only a single cultivar of perennial ryegrass were analyzed, a larger number of *N. coenophialum* isolates from wild populations and cultivated varieties of tall fescue were included, suggesting that minimal intraspecific diversity is a general property. However, due to codon redundancy, conservative amino acid substitutions, and positive selection of functional characteristics, isoenzymes may reveal only a subset of the possible genetic variation, providing a relatively conservative measure of genetic variability in endophytes. In addition, isoenzyme markers are not suitable for the analysis of endophytes in planta, due to commonality of enzyme systems with the plant host. Further studies have, however, demonstrated

[1] Plant Biotechnology Centre, Primary Industries Research Victoria, Department of Primary Industries, La Trobe University, Bundoora, VIC 3086, Australia
[2] Primary Industries Research Victoria, Department of Primary Industries, Private Bag 105, Hamilton, VIC 3300, Australia
[3] Molecular Plant Breeding Cooperative Research Centre, Australia

that the grouping of endophyte isolates based on isoenzyme genotype corresponded closely to distinctive morphological and cultural characters (Christensen et al., 1993; Leuchtmann, 1994). This observation led to the widespread use of isoenzyme markers to confirm the taxonomic classification of *Neotyphodium* and *Epichloë* isolates (Cabral et al., 1999; Craven et al., 2001a; Leuchtmann and Schardl, 1998; Leuchtmann et al., 1994; Naffaa et al., 1998).

DNA-based markers have been developed to overcome some of the limitations of isoenzyme markers. This process has, however, been limited by the availability of DNA sequence information from endophytes. Consequently, nonspecific genetic markers such as random amplified polymorphic DNAs (RAPDs) (Groppe et al., 1995; Liu et al., 1995) and amplified fragment length polymorphisms (AFLPs) (Arroyo Garcia et al., 2002; Brem and Leuchtmann, 2003; Tredway et al., 1999; Wilkinson et al., 2000) that do not require any prior DNA sequence knowledge have been used. These systems allow many loci to be analyzed simultaneously, providing a more accurate measure of the genetic diversity of endophytes. The nonspecificity of these markers, however, limits their use to the analysis of endophytes to fungal cultures, and the RAPD technique is seriously limited by reproducibility problems. RAPD markers have been used for the identification of *N. coenophialum* and *N. lolii* (Liu et al., 1995) and to study the diversity of *E. bromicola* isolates in *Bromus erectus* (Groppe et al., 1995). AFLPs have been used to analyze genetic variation in *E. festucae* and *E. bromicola* (Arroyo Garcia et al., 2002; Brem and Leuchtmann, 2003; Tredway et al., 1999), to determine phylogenetic relationships between *Neotyphodium* and *Epichloë* endophytes (Brem and Leuchtmann, 2003; Tredway et al., 1999) and for segregation analysis of *E. festucae* to identify the genetic locus controlling loline alkaloid production (Wilkinson et al., 2000).

Until recently, only a small number of specific genetic markers for anonymous or known genes (sequence tagged site or STS loci) and microsatellite or simple sequence repeat (SSR) loci had been developed (Doss and Welty, 1995; Doss et al., 1998; Groppe et al., 1995; Moon et al., 1999). These systems are sufficiently specific to allow in planta identification and assessment of endophyte genetic diversity. However, the STS markers only permitted diagnostic detection of endophytes in planta, without evidence for genetic polymorphism (Doss and Welty, 1995; Doss et al., 1998). It was not known if the SSR markers which detected genetic polymorphism provided an accurate measure of the genetic diversity of endophytes because of the small number available (Groppe et al., 1995; Moon et al., 1999).

Here we describe the development of a large set of expressed sequence tag (EST)-derived SSR markers and the use of these markers for the identification and assessment of genetic diversity arising within and between endophyte taxa. Comparisons are made with preexisting molecular genetic marker systems. Preliminary findings on intraspecific genetic variation in the perennial ryegrass endophyte *N. lolii* are described, along with the application of molecular genetic marker systems for phylogenetic analysis of endophytes. The expression of endophyte-related characters may be influenced not only by endophyte genotype but also by interactions between the genotypes of host and symbiont. Recent developments in molecular genetic marker systems for co-genotyping grass–endophyte associations to allow detailed genetic analysis are discussed.

DEVELOPMENT AND CHARACTERIZATION OF SSR SYSTEMS

Earlier studies had generated only a small number of SSR markers for the genetic analysis of endophytes. Most of these SSR loci were identified in nonenriched genomic DNA libraries constructed from strains of LpTG-2 and *E. typhina* (Moon et al., 1999). The hybridization-based screening of more than 200 000 phage plaques from these libraries led to the isolation of nine unique SSR loci. Other SSR loci were discovered in the 3'-untranslated region (3'-UTR) of the *N. lolii* 3-hydroxy-3-methylglutaryl coenzyme A reductase gene (Moon et al., 1999) and by sequence analysis of a polymorphic band from RAPD fingerprinting of *E. bromicola* (Groppe et al., 1995).

More recently, a large number of SSR loci have been identified from clones of *N. coenophialum* and *N. lolii* cDNA libraries (van Zijll de Jong et al., 2003). The isolation of SSRs associated with expressed sequences has proved to be a more efficient process than genomic cloning. As the genic component accounts for a larger proportion of the total DNA in organisms with small genome sizes, a higher proportion of the total SSR population may potentially be recovered compared with organisms with large genome sizes. Unique SSR loci were detected in 9.7% of *N. coenophialum* ESTs (78/801) and 6.3% of *N. lolii* ESTs (122/1951). The SSR arrays formerly discovered in genomic DNA sequences were most commonly composed of mono-, di-, or trinucleotide repeats with perfect or complex repeat structures and relatively high repeat unit numbers (Moon et al., 1999). By contrast, arrays with trinucleotide repeat units, perfect repeat structures and low repeat unit numbers were most common in ESTs (van

Zijll de Jong et al., 2003). These features are common in EST-SSRs from plants (Cardle et al., 2000) and probably reflect the need to preserve codon register.

Analysis of amplification efficiency for genomic DNA-derived SSR markers was performed using a panel of 30 isolates representing six *Neotyphodium* taxa and two *Epichloë* taxa, with 82% producing efficient amplification from the majority of isolates (Moon et al., 1999). Similarly, the majority of the EST-derived SSR markers (79%) showed efficient amplification from the majority of isolates in tests performed across a panel of 20 isolates representing four *Neotyphodium* taxa and six *Epichloë* taxa (van Zijll de Jong et al., 2003). The level of transfer across species was higher for EST-derived SSR markers than genomic DNA-derived SSR markers (65% compared with 36%), as expected from the more conserved nature of expressed DNA sequences. This inference is also supported by the common occurrence of null alleles with genomic DNA-derived SSR loci in studies of additional *Neotyphodium* and *Epichloë* taxa (Moon et al., 2000).

The genomic DNA-derived SSR markers, however, revealed more polymorphism than EST-derived SSR markers (Moon et al., 1999; van Zijll de Jong et al., 2003). Unlike some EST-derived SSR markers, none of the genomic DNA-derived SSR markers detected monomorphic loci in the genotypes screened. Additionally, the genomic DNA-derived SSR markers detected genetic variation within and between *Neotyphodium* and *Epichloë* species, while although all of the EST-derived SSR markers detected variation between *Neotyphodium* and *Epichloë* species, only half that number detected variation within *Neotyphodium* and *Epichloë* species. However, AFLP analysis of the same isolates also detected limited levels of polymorphism within *Neotyphodium* and *Epichloë* species, suggesting that higher levels of conservation of EST-derived SSR loci was not the limiting factor for the detection of intraspecific genetic variation in this study (van Zijll de Jong et al., 2003).

Genomic DNA-derived SSR markers used in combination were able to resolve the majority of isolates to the level of known isoenzyme groupings, and some markers detected variation within these groupings (Moon et al., 1999). The polymorphism detected with EST-SSR markers also permitted the assignment of isolates to known taxonomic groupings (van Zijll de Jong et al., 2003). Like the genomic DNA-derived SSR markers, the majority of EST-derived SSR markers did not unequivocally discriminate one taxon from all other taxa, but different taxa could be distinguished using appropriate marker combinations.

The specificity of SSR markers permits in planta endophyte detection. Genomic DNA-derived SSR markers were shown to be specific for *Neotyphodium* and *Epichloë* endophytes, such that amplification products were not obtained from fungal species belonging to the same family or from unrelated fungal

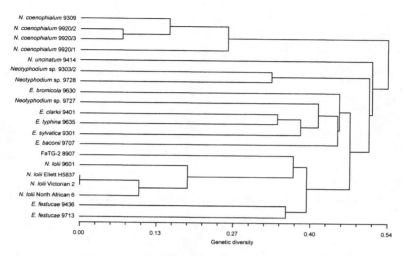

Fig. 5.1. *UPGMA (unweighted pair group method of arithmetic averages) phenogram for 20 Neotyphodium and Epichloë isolates using measurements of average taxonomic distance based on the genetic diversity of 77 EST-derived SSR loci.*

species (Groppe and Boller, 1997; Moon et al., 1999). The conserved nature of coding sequences resulted in some EST-derived SSR markers (3/20) showing-cross-amplification from unrelated fungal species, but the signals detected from these fungi were weak compared with those from the endophyte (van Zijll de Jong et al., 2004). Both genomic DNA-derived SSR markers and EST-derived SSR markers have been shown to detect endophytes in a number of grass species including *N. lolii* in perennial ryegrass, *N. occultans* in annual ryegrasses *L. multiflorum* and *L. rigidum*, *N. coenophialum* in tall fescue, and *N. uncinatum* in meadow fescue (*F. pratensis*). The markers were shown to detect endophytes in the leaf sheaths and blades and in seeds of infected grasses.

The EST-derived SSR markers appear to be generally more sensitive for in planta detection than genomic DNA-derived SSR markers. Tests performed with in vitro composed admixtures of genomic DNA isolated from a cultured endo-phyte and an endophyte-free grass genotype revealed that some EST-derived SSR markers were able to detect an endophyte at DNA mass ratios as low as 1:5000 (van Zijll de Jong et al., 2004). The sensitivity, however, was found to vary between EST-derived SSR markers, possibly reflecting differences in the kinetic efficiency of PCR amplification. Genomic DNA-derived SSR markers were reported to often yield very low amounts of amplification products from

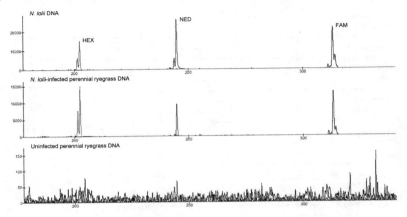

Fig. 5.2. *Detection of N. lolii in the leaf sheath of perennial ryegrass with three fluorochrome-labeled EST-derived SSR markers (NLESTA1TA10-FAM, NLESTA1CC10-HEX and NCESTA1FH03-NED). The results of in planta detection are compared with those obtained from cultured N. lolii and infected plant tissue.*

endophytes in planta (Moon et al., 1999), and a second round of amplification was often necessary (Moon et al., 2000). These differences once again probably arise in part from differences in the levels of sequence conservation of priming sites in coding sequences and noncoding genomic DNA. An SSR marker has been developed not only for detection, but also quantification of the level of endophyte infection in grass tissue, based on competitive PCR (Groppe and Boller, 1997). Implementation of this marker has been used to detect correlations between fungal concentrations and vegetative vigor of plants (Groppe et al., 1999) and compatibility between genetically modified endophyte strains and their host plants (Panaccione et al., 2001).

ANALYSIS OF INTRASPECIFIC GENETIC DIVERSITY

Further studies of genetic diversity in perennial ryegrass, tall fescue, and meadow fescue endophytes from diverse origins with isoenzyme markers (Christensen et al., 1993) detected greater levels of intraspecific genetic diversity than in the initial studies (Leuchtmann and Clay, 1990). Six different isoenzyme groupings were identified among 21 *N. lolii* isolates, three different isoenzyme groupings identified among 10 *N. coenophialum* isolates, while among six *N. uncinatum* isolates two different isoenzyme groupings were identified. Isoenzyme markers may, however, provide conservative measures of intraspecific variation. A study of selected isolates from representative isoenzyme groupings with genomic DNA-derived SSR markers led to the detection of genetic variation within some isoenzyme groupings (Moon et al., 1999).

Initial studies with EST-derived SSR markers detected low levels of genetic variation within *N. coenophialum* and *N. lolii* (van Zijll de Jong et al., 2003). However, only a small number of isolates were studied, and many were obtained from the same grass variety or grass varieties with putative common ancestry. The low levels of genetic variation detected within *Neotyphodium* taxa are not unexpected, as these endophytes show an asexual reproductive mode. Higher levels of genetic variation are detected within *Epichloë* taxa (Leuchtmann and Schardl, 1998; Moon et al., 1999; van Zijll de Jong et al., 2003). Diversity in *Epichloë* is correlated with mode of transmission. Isoenzyme studies detected higher levels of intraspecific variation in species such as *E. typhina*, which is predominately horizontally or sexually transmitted, than in species such as *E. festucae*, which is predominately vertically or asexually transmitted (Leuchtmann and Schardl, 1998). A study of two natural populations of *E. festucae* with AFLPs detected evidence of clonal lineages in these populations (Arroyo Garcia et al., 2002).

Further SSR marker studies in larger populations containing *Neotyphodium* endophytes are necessary to confirm the results obtained with isoenzyme markers. The development of SSR markers for in planta endophyte detection allows the efficient assessment of the genetic diversity of endophytes in large grass populations. Genetic diversity of *N. lolii* was investigated in a subsample of plants from a globally distributed perennial ryegrass germplasm collection of wild and cultivated accessions (van Zijll de Jong et al., 2004). Endophyte viability in this collection may have been affected by suboptimal seed storage conditions (Welty et al., 1987). Although only a single genotype was examined for most accessions, only 21 accessions from a total of 132 were found to harbor endophytes. These accessions were subsequently resampled to detect larger numbers of endophyte-positive plants. The incidence of endophyte within these accessions was variable, ranging from 2 to 100%.

Assessment of the genetic diversity of *N. lolii* with 18 EST-derived SSR markers detected low levels of genetic variation. Most of the variation was within the range of variation detected in previous studies. The more diverse endophytes detected in some genotypes from accessions from Tunisia, Morocco, Spain, Sardinia, and Sweden appear to belong to different endophyte taxa. Perennial ryegrass is known to be host to two different endophyte taxa: *N. lolii* and *Lolium*

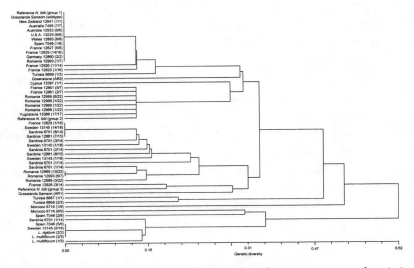

Fig. 5.3. *UPGMA phenogram for endophytes in Lolium accessions and varieties using measurements of average taxonomic distance based on the genetic diversity of 18 SSR loci. Indicated in parentheses following the accession name are the number of genotypes containing this endophyte strain and the total number of genotypes analyzed from this accession. Reference N. lolii group 1 are isolates from perennial ryegrass varieties Aries, Banks, Ellett, Fitzroy, Vedette, and Victorian, and an ecotype from North Africa. Reference N. lolii group 2 is an isolate from the perennial ryegrass variety Aries. Reference N. lolii group 3 is an isolate from an ecotype from Belgium. The low-toxin-producing endophytes AR1 (in cultivar Grasslands Samson) and AR5 (in cultivar Greenstone) are included for comparison, along with the resident wild-type endophyte in Grasslands Samson.*

perenne taxonomic group 2 (LpTG-2) (Christensen et al., 1993). Some of the diverse endophytes may belong to the second group of endophytes in perennial ryegrass. The LpTG-2 endophytes have a heteroploid genome containing up to two gene sequence variants that are similar to the single gene sequence variants found in *E. festucae* and *E. typhina* (Schardl et al., 1994). Two amplification products were often detected in these diverse endophytes, suggestive of heteroploid genomes. Other diverse endophytes were genetically similar to endophytes in annual ryegrass and Italian ryegrass, which are host to the taxon *N. occultans* (Moon et al., 2000).

The geographic pattern of genetic diversity in *N. lolii* was not randomly distributed (van Zijll de Jong et al., 2004). Endophytes in accessions from western Europe (and the New World), eastern Europe, the Mediterranean basin, and North Africa clustered to form four distinct groups, although endophytes in accessions from Romania were separated into two of these groups; the eastern Europe group and the Mediterranean basin group. The geographical pattern of endophyte diversity shows similarities with both ancient and more recent routes of dispersion of perennial ryegrass. A study of chloroplast DNA variation in *Lolium* species, which like the endophyte is maternally inherited, identified three clusters of populations which appear to have originated in the Middle East and to have spread following the migration of the first farmers along three routes: a northwestern route (Danubian movement), a southwestern route (Mediterranean movement which includes the British Isles), and a North African continental route (Balfourier et al., 2000). Perennial ryegrass populations are associated with the first two migration routes, while annual ryegrass populations are associated with the third route. *N. lolii* was also predominant in accessions along the first two migration routes, and genetic differentiation of *N. lolii* along these two migration routes was evident.

Genetically similar endophytes occurred along the northwestern route in accessions from the former Yugoslavia and from Romania, which were distinct from genetically similar endophytes that occurred along the southwestern movement in accessions from Corsica and Wales. Endophytes in the New World accessions from the USA, Australia, and New Zealand, which are thought to be derived from British germplasm sources, were genetically identical to the endophytes in the accession from Wales. This observation supports a correlation between endophyte diversity and dispersal routes that are more recent in historical terms. Endophytes in accessions from France, which appears from the distribution of *Lolium* populations to be located at the crossroads of these two movements (Balfourier et al., 2000), were also found to cluster with endophytes in accessions from the two European movements. There were, however, some differences in the geographic pattern of diversity in *N. lolii* and the chloroplast genome of perennial ryegrass. For example, the endophytes in accessions from Sardinia were genetically distant from other endophytes along the southwestern movement in accessions from France (including Corisca), Wales, and the New World (USA, Australia, and New Zealand). Such anomalies are consistent with an isoenzyme study which showed genetic differentiation between *N. lolii* isolates from France (western Europe) and New Zealand (New World) from *N. lolii* isolates from Spain (Mediterranean basin) (Christensen et al., 1993).

Additional accessions along the three routes of migration as well as from the center of origin must be studied in order to confirm the origins of the perennial ryegrass–*N. lolii* symbiosis. The low levels of genetic variation detected in *N. lolii*, however, support the model of a monophyletic origin.

PHYLOGENETIC ANALYSIS

To date, only limited applications of molecular markers to endophyte phylogenetic analysis have been made. The majority of phylogenetic studies have used DNA sequence analysis of specific loci such as the β-tubulin gene or translation elongation factor 1-α gene (Craven et al., 2001a, 2001b; Moon et al., 2000, 2002; Schardl et al., 1997, 1994; Tsai et al., 1994). However, AFLP markers have been used to estimate phylogenetic relationships within and between *Epichloë* and *Neotyphodium* endophytes (Brem and Leuchtmann, 2003; Tredway et al., 1999). *Epichloë festucae* was found to be monophyletic, with isolates forming distinct clades according to their host grass species (Tredway et al., 1999). These results provided some support for endophyte–host coevolution (Schardl et al., 1997), but the phylogenetic relationships among *E. festucae* isolates did not correspond to the known taxonomic classification of host grasses. The AFLP analysis of *E. bromicola* also found genetic differentiation of isolates according to their host grass species (Brem and Leuchtmann, 2003). In both of these studies, phylogenetic trees based on AFLP had high bootstrap support, but differences were detected between these trees and phylogenetic trees based on gene sequences (Brem and Leuchtmann, 2003; Tredway et al., 1999).

The use of AFLP and SSR polymorphism data for phylogenetic analysis is limited by the occurrence of homoplasy due to size identity of distinct alleles (Grimaldi and Crouau-Roy, 1997; Vekemans et al., 2002). Phylogenetic analysis of SSR polymorphism data is also hindered by the irregular rate of molecular evolution of SSR arrays (Colson and Goldstein, 1999). In addition, the majority of *Neotyphodium* species are heteroploid (Craven et al., 2001a; Moon et al., 2000, 2002; Schardl et al., 1994; Tsai et al., 1994). This property renders prediction of phylogenetic relationships from AFLP or SSR polymorphism data problematic, as the multiple products that occur in these species tend to exaggerate estimates of evolutionary divergence.

DNA sequence analysis of SSR flanking regions in EST-derived SSR loci, however, has proven informative for studying the genetic relationships among *Neotyphodium* and *Epichloë* species (E. van Zijll de Jong, unpublished data). Phylogenetic analysis of the flanking regions of five loci has provided support for the theory of a common origin of *Neotyphodium* and *Epichloë* endophytes. Both intron, exon, and UTR sequences were surveyed, and although locus-specific differences were detected, consistent patterns of relationship based on putative genome sharing were observed. These relationships were consistent with other studies of phylogenetic affinity between specific *Neotyphodium* and *Epichloë* species.

CO-GENOTYPING OF GRASS–ENDOPHYTE SYMBIOSES

Expressed sequence tag-derived SSR markers from endophytes and genomic-DNA derived SSR markers from perennial ryegrass (Jones et al., 2001, 2002) were used to assess the genetic diversity of endophytes and host grasses collected from Australian farms with variable incidence of endophyte-related livestock toxicosis and varieties (Grasslands Samson, Greenstone and Tolosa) containing endophyte strains (AR1, AR5, and NEA2, respectively) with reduced toxicity effects (van Zijll de Jong et al., 2004). A total of 20 endophyte-specific loci and 22 host-specific loci were used in this study. The endophyte strain AR1, which is unable to produce the animal toxins lolitrem B and ergovaline, and endophyte strains AR5 and NEA2, which are unable to produce lolitrem B, were genetically distinct from the wild-type endophytes in grasses collected from farms with variable incidence of livestock toxicosis (data not shown). The genetic diversity of endophytes collected from the farms was low. Endophytes from paddocks with no or minor incidence of livestock toxicosis were genetically indistinguishable from endophytes from paddocks with a major incidence of livestock toxicosis. The genetic diversity among host grasses was higher, suggesting that quantitative variability in endophyte-related effects may result from variation in host genotype, while qualitative variation associated with endophyte strains such as AR1, AR5, and NEA2 may be attributable to the endophyte genotype.

Other studies support the concept of host-mediated genetic control of endophyte phenotypic traits. Artificial inoculations of different tall fescue cultivars with different *N. coenophialum* isolates revealed certain isolates and certain cultivars that had consistently higher levels of the toxin ergovaline (Christensen et al., 1998). Host genetic control of endophyte toxin levels has been detected in perennial ryegrass, as demonstrated by significant general combining effects and smaller specific combining effects in a partial diallel crossing scheme between different endophyte-containing genotypes (Easton et al., 2002).

Preliminary results from quantitative trait locus (QTL) analysis of a perennial ryegrass mapping family also support the role of host genetic factors in the control of endophyte-specific traits. The F_1 (Northern African$_6$ × Aurora$_6$) population is based on a two-way pseudo-testcross structure, and has been used to develop a functionally associated marker-based genetic map populated by EST-RFLP and EST-SSR loci (Forster et al., 2004). The parental genotypes, derived from a Moroccan ecotype and a British cultivar bred from a Swiss ecotype, are both genetically and phenotypically diverse, and the progeny set has been extensively used for QTL mapping. The Northern African$_6$ (NA$_6$) parent is endophyte-positive, while the Aurora$_6$ (AU$_6$) parent is endophyte-negative. A semiquantitative enzyme-linked immunosorbent assay (ELISA) test was used to detect progeny of maternal origin from NA$_6$, and the variation in ELISA scores in these genotypes, containing a single resident endophyte, was used to detect several genomic regions on the maps of both parents that significantly affect the trait. The ELISA score may be potentially correlated with endophyte incidence, efficiency of colonization and associated metabolic traits. In the study of Easton et al. (2002), as much as 65% of the genetically controlled variation in toxin concentration was a function of mycelial mass.

Co-genotyping of endophytes and host grasses in mapping families with SSR markers provides the tools for identification of qualitative variation arising from the endophyte and quantitative variation arising from the host grass. However, specific SSR markers are not currently available for the genotyping of all host grass species, such as the brome grasses (*Bromus* spp.) and cocksfoot (*Dactylis glomerata*). Diversity array technology (DArT^{TM}) provides a sequence-independent method for the simultaneous co-genotyping of the endophyte and host grass. This technology is a hybridization-based method that uses genomic DNA fragments immobilized on a solid-state surface to assay for the presence of complementary genomic DNA fragments in the test genotype, generating a specific hybridization pattern or genetic fingerprint (Jaccoud et al., 2001).

DArT slides composed of two genomic representations from a number of endophyte-free perennial ryegrass plants and isolated *N. lolii* cultures have been developed. Tests demonstrated specific hybridization of labeled probe DNA from isolated *N. lolii* to endophyte-derived array features and labeled probe DNA from uninfected perennial ryegrass to plant-derived features, and hybridization to both endophyte and plant features by *N. lolii*-infected perennial ryegrass.

Differences were observed in the hybridization patterns obtained with probe DNA derived from different endophyte-infected perennial ryegrass genotypes.

However, DNA sequence analysis of the individual array features detected high levels of redundancy. Further analysis is required to increase the number of

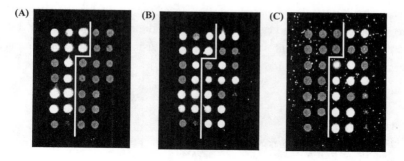

Fig. 5.4. *Hybridization to a subgrid of a DArT slide with probe DNA from (A) uninfected perennial ryegrass, (B) N. lolii-infected perennial ryegrass, and (C) N. lolii. Sample DNA was labeled with Cy-3 and the empty vector control was labeled with Cy-5. A white-colored spot indicates hybridization of the sample DNA, and a gray-colored spot indicates hybridization of the empty vector control only. Plant features are located on the left section of the subgrid and endophyte features are on the right section of the subgrid.*

informative characters on the array for accurate measurement of genetic variation in endophytes and host grasses using this technology.

SUMMARY

The development of efficient molecular marker systems based on EST-SSR variation has permitted significant advances in endophyte taxonomy, phylogenetics, and population genetics. The SSRs associated with genic sequences are generally less variable than genomic DNA-derived SSRs. However, the relative ease of isolation of these sequences, along with the potential for correlation with functional variation, compensates for any disadvantages. Apart from the current studies of intraspecific variation in *N. lolii*, equivalent studies may be performed in other endophyte taxa, defining the extent and nature of genetic diversity. The EST-SSR markers also provide important diagnostic tools to monitor the incidence, identity, and stability of endophytes in cultivated varieties.

In the future, major advances in the study of grass–endophyte interactions are likely to be made through co-genotyping and the quantitative analysis of both biparental cross-derived and linkage disequilibrium populations (Forster et al.,

2004). The identification of host-specific genomic regions that specify general and endophyte genotype-specific control of phenotypic traits will permit the marker-assisted breeding of improved symbioses. Further advances in molecular marker technology will be based on the development of single nucleotide polymorphism markers from candidate genes of both host and endophyte. These markers are capable of high-throughput analysis using automated technology platforms.

ACKNOWLEDGMENTS

The authors thank Nathaniel Bannan, Nola McFarlane, Dr. Kevin Reed (DPI), Dr. Adrian Leuchtmann (ETH, Zürich, Switzerland), and Dr. Alan Stewart (Pyne Gould Guinness Seeds, Christchurch, New Zealand) for genetic resource contributions. The QTL analysis was performed by Kathryn Guthridge, Dr. Noel Cogan, and Anita Vecchies. The valuable advice of Dr. Alan Stewart and Prof. Michael Hayward (Rhydgoch Genetics, Aberystwyth, UK) is gratefully acknowledged. This work was supported by the Australian Molecular Plant Breeding Cooperative Research Centre.

REFERENCES

Arroyo Garcia, R., J.M. Martinez Zapater, B. Garcia Criado, and I. Zabalgogeazcoa. 2002. Genetic structure of natural populations of the grass endophyte *Epichloë festucae* in semiarid grasslands. Mol. Ecol. 11:355–364.

Bacon, C.W. 1988. Procedure for isolating the endophyte from tall fescue and screening isolates for ergot alkaloids. Appl. Environ. Microb. 54:2615–2618.

Bacon, C.W., and M.R. Siegel. 1988. Endophyte parasitism of tall fescue. J. Prod. Agric. 1:45–55.

Balfourier, F., C. Imbert, and G. Charmet. 2000. Evidence for phylogeographic structure in *Lolium* species related to the spread of agriculture in Europe. A cpDNA study. Theor. Appl. Genet. 101:131–138.

Brem, D., and A. Leuchtmann. 2003. Molecular evidence for host-adapted races of the fungal endophyte *Epichloë bromicola* after presumed host shifts. Evolution 57:37–51.

Cabral, D., M.J. Cafaro, B. Saidman, M. Lugo, P.V. Reddy, and J.F. White. 1999. Evidence supporting the occurrence of a new species of endophyte in some South American grasses. Mycologia 91:315–325.

Cardle L., L. Ramsay, D. Milbourne, M. Macaulay, D. Marshall, and R. Waugh. 2000. Computational and experimental characterization of physically clustered simple sequence repeats in plants. Genetics 156:847–854.

Christensen, M.J., A. Leuchtmann, D.D. Rowan, and B.A. Tapper. 1993. Taxonomy of *Acremonium* endophytes of tall fescue (*Festuca arundinacea*), meadow fescue (*F. pratensis*) and perennial ryegrass (*Lolium perenne*). Mycol. Res. 97:1083–1092.

Christensen, M.J., H.S. Easton, W.R. Simpson, and B.A. Tapper. 1998. Occurrence of the fungal endophyte *Neotyphodium coenophialum* in leaf blades of tall fescue and implications for stock health. N. Z. J. Agric. Res. 41:595–602.

Clay, K. 1989. Clavicipitaceous endophytes of grasses: Their potential as biocontrol agents. Mycol. Res. 92:1–12.

Colson, I., and D.B. Goldstein. 1999. Evidence for complex mutations at microsatellite loci in *Drosophila*. Genetics 152:617–627.

Craven, K.D., J.D. Blankenship, A. Leuchtmann, K. Hignight, and C.L. Schardl. 2001a. Hybrid fungal endophytes symbiotic with the grass *Lolium pratense*. Sydowia 53:44–73.

Craven, K.D., P.T.W. Hsiau, A. Leuchtmann, W. Hollin, and C.L. Schardl. 2001b. Multigene phylogeny of *Epichloë* species, fungal symbionts of grasses. Ann. Missouri Bot. Garden 88:14–34.

Doss, R.P., and R.E. Welty. 1995. A polymerase chain reaction-based procedure for detection of *Acremonium coenophialum* in tall fescue. Phytopathology 85:913–917.

Doss, R.P., S.L. Clement, S.R. Kuy, and R.E. Welty. 1998. A PCR-based technique for detection of *Neotyphodium* endophytes in diverse accessions of tall fescue. Plant Dis. 82:738–740.

Easton, H.S., G.C.M. Latch, B.A. Tapper, and O.J.P. Ball. 2002. Ryegrass host genetic control of concentrations of endophyte-derived alkaloids. Crop Sci. 42:51–57.

Forster, J.W., E.S. Jones, J. Batley, and K.F. Smith. 2004. Molecular marker-based genetic analysis of pasture and turf grasses. p. 197–239. *In* A. Hopkins et al. (ed.) Molecular breeding of forage and turf. Kluwer Academic Publ., Dordrecht, the Netherlands.

Grimaldi, M.C., and B. Crouau-Roy. 1997. Microsatellite allelic homoplasy due to variable flanking sequences. J. Mol. Evol. 44:336–340.

Groppe, K., and T. Boller. 1997. PCR assay based on a microsatellite-containing locus for detection and quantification of *Epichloë* endophytes in grass tissue. Appl. Environ. Microb. 63:1543–1550.

Groppe, K., I. Sanders, A. Wiemken, and T. Boller. 1995. A microsatellite marker for studying the ecology and diversity of fungal endophytes (*Epichloë* spp.) in grasses. Appl. Environ. Microb. 61:3943–3949.

Groppe, K., T. Steinger, I. Sanders, B. Schmid, A. Wiemken, and T. Boller. 1999. Interaction between the endophytic fungus *Epichloë bromicola* and the grass *Bromus erectus*: Effects of endophyte infection, fungal concentration and environment on grass growth and flowering. Mol. Ecol. 8:1827–1835.

Jaccoud, D., K. Peng, D. Feinstein, and A. Kilian. 2001. Diversity arrays: A solid state technology for sequence information independent genotyping. Nucleic Acids Res. 29(4):e25.

Jones, E.S., M.P. Dupal, R. Kölliker, M.C. Drayton, and J.W. Forster. 2001. Development and characterisation of simple sequence repeat (SSR) markers for perennial ryegrass (*Lolium perenne* L.). Theor. Appl. Genet. 102:405–415.

Jones, E.S., M.P. Dupal, J.L. Dumsday, L.J. Hughes, and J.W. Forster. 2002. An SSR-based genetic linkage map for perennial ryegrass (*Lolium perenne* L.). Theor. Appl. Genet. 105:577–584.

Leuchtmann, A. 1994. Isozyme relationships of *Acremonium* endophytes from 12 *Festuca* species. Mycol. Res. 98:25–33.

Leuchtmann, A., and K. Clay. 1990. Isozyme variation in the *Acremonium/Epichloë* fungal endophyte complex. Phytopathology 80:1133–1139.

Leuchtmann, A., and C.L. Schardl. 1998. Mating compatibility and phylogenetic relationships among two new species of *Epichloë* and other congeneric European species. Mycol. Res. 102:1169–1182.

Leuchtmann, A., C.L. Schardl, and M.R. Siegel. 1994. Sexual compatibility and taxonomy of a new species of *Epichloë* symbiotic with fine fescue grasses. Mycologia 86:802–812.

Liu, D., R. van Heeswijck, G. Latch, T. Leonforte, M. Panaccio, C. Langford, P. Cunningham, and K. Reed. 1995. Rapid identification of *Acremonium lolii* and *Acremonium coenophialum* endophytes through arbitrarily primed PCR. FEMS Microbiol. Lett. 133:95–98.

Moon, C.D., B.A. Tapper, and B. Scott. 1999. Identification of *Epichloë* endophytes in planta by a microsatellite-based PCR fingerprinting assay with automated analysis. Appl. Environ. Microb. 65:1268–1279.

Moon, C.D., B. Scott, C.L. Schardl, and M.J. Christensen. 2000. The evolutionary origins of *Epichloë* endophytes from annual ryegrasses. Mycologia 92:1103–1118.

Moon, C.D., C.O. Miles, U. Jarlfors, and C.L. Schardl. 2002. The evolutionary origins of three new *Neotyphodium* endophyte species from grasses indigenous to the Southern Hemisphere. Mycologia 94:694–711.

Naffaa, W., C. Ravel, and J.J. Guillaumin. 1998. A new group of endophytes in European grasses. Ann. Appl. Biol. 132:211–226.

Panaccione, D.G., R.D. Johnson, J.H. Wang, C.A. Young, P. Damrongkool, B. Scott, and C.L. Schardl. 2001. Elimination of ergovaline from a grass–*Neotyphodium* endophyte symbiosis by genetic modification of the endophyte. Proc. Natl. Acad. Sci. USA 98:12 820–12 825.

Schardl, C.L., A. Leuchtmann, K.R. Chung, D. Penny, and M.R. Siegel. 1997. Coevolution by common descent of fungal symbionts (*Epichloë* spp.) and grass hosts. Mol. Biol. Evol. 14:133–143.

Schardl, C.L., A. Leuchtmann, H.F. Tsai, M.A. Collett, D.M. Watt, and D.B. Scott. 1994. Origin of a fungal symbiont of perennial ryegrass by interspecific hybridization of a mutualist with the ryegrass choke pathogen, *Epichloë typhina*. Genetics 136:1307–1317.

Siegel, M.R., G.C.M. Latch, and M.C. Johnson. 1987. Fungal endophytes of grasses. Annu. Rev. Phytopathol. 25:293–315.

Tredway, L.P., J.F. White, B.S. Gaut, P.V. Reddy, M.D. Richardson, and B.B. Clarke. 1999. Phylogenetic relationships within and between *Epichloë* and *Neotyphodium* endophytes as estimated by AFLP markers and rDNA sequences. Mycol. Res. 103:1593–1603.

Tsai, H.F., J.S. Liu, C. Staben, M.J. Christensen, G.C.M. Latch, M.R. Siegel, and C.L. Schardl. 1994. Evolutionary diversification of fungal endophytes of tall fescue grass by hybridization with *Epichloë* species. P. Natl. Acad. Sci. USA 91:2542–2546.

van Zijll de Jong, E., K.M. Guthridge, G.C. Spangenberg, and J.W. Forster. 2003. Development and characterization of EST-derived simple sequence repeat (SSR) markers for pasture grass endophytes. Genome 46:277–290.

van Zijll de Jong, E., N.R. Bannan, J. Batley, K.M. Guthridge, G.C. Spangenberg, K.F. Smith, and J.W. Forster. 2004. Genetic diversity in the perennial ryegrass fungal endophyte *Neotyphodium lolii*. p. 155–164. *In* A. Hopkins et al. (ed.) Molecular breeding of forage and turf. Kluwer Academic Publ., Dordrecht, the Netherlands.

Vekemans, X., T. Beauwens, M. Lemaire, and I. Roldan-Ruiz. 2002. Data from amplified fragment length polymorphism (AFLP) markers show indication of size homoplasy and of a relationship between degree of homoplasy and fragment size. Mol. Ecol. 11:139–151.

Welty, R.E., M.D. Azevedo, and T.M. Cooper. 1987. Influence of moisture content, temperature and length of storage on seed germination and survival of endophytic fungi in seeds of tall fescue and perennial ryegrass. Phytopathology 77:893–900.

Wilkinson, H.H., M.R. Siegel, J.D. Blankenship, A.C. Mallory, L.P. Bush, and C.L. Schardl. 2000. Contribution of fungal loline alkaloids to protection from aphids in a grass–endophyte mutualism. Mol. Plant–Microbe Interact. 13:1027–1033.

Section III

Ecology and Agronomy

A HIERARCHICAL FRAMEWORK FOR UNDERSTANDING THE ECOSYSTEM CONSEQUENCES OF ENDOPHYTE–GRASS SYMBIOSES

Marina Omacini, Enrique J. Chaneton, and Claudio M. Ghersa[1]

Understanding the linkages between endophyte–grass symbioses, community dynamics, and ecosystem function sets a new challenge to current knowledge on the ecological role of microbial symbionts. Microorganisms associated with higher plants, either mutualistic or parasitic, are integral components of terrestrial ecosystems that may fundamentally affect transfers of energy and matter through food webs. So far, work on the impact of plant symbionts on ecosystem functioning has been concerned mainly with nitrogen-fixing bacteria and arbuscular mycorrhizal (AM) fungi (van der Heijden and Cornelissen, 2002). In contrast, the consequences of grass endophyte infections on ecosystem-level processes have received little empirical attention (Clay, 1994, 1997). Here we argue that, through their primary effects on individual hosts and neighborhood interactions, *Neotyphodium*-like endophytes can significantly influence ecosystem dynamics. Furthermore, endophytic fungi offer an ideal model system to examine how plant symbionts may alter feedbacks between above- and belowground ecosystem compartments (van der Putten et al., 2001; Wardle, 2002).

Fungal endophytes are widespread symbionts of cool-season grasses in temperate regions of the globe (Clay, 1990; Bacon and Hill, 1997). *Neotyphodium* endophytes systemically infect grass shoots, forming asymptomatic symbioses with no signs of damage in the host tissues. However, fungal endophytes induce physiological and biochemical changes in their host grasses that may in turn affect other community members, including neighboring plants, consumers at various trophic levels, decomposers, and microbial mutualists (Clay 1990, 1994, 1997; Bernard et al., 1997; Malinowski and Belesky, 2000; Omacini et al., 2001, 2004; Clay and Schardl, 2002). As shown by other chapters in this volume,

[1] IFEVA–CONICET, Facultad de Agronomía, Universidad de Buenos Aires, Av. San Martín 4453, 1417 Buenos Aires, Argentina

major progress has been made during the last decade or so into the biology of grass–endophyte associations. Indeed, since Clay's (1994, 1997) first evaluation of the possible roles of grass endophytes in biodiversity and community dynamics, several studies have revealed endophyte impacts on attributes relevant to ecosystem function (e.g., Clay and Holah, 1999; Franzluebbers et al., 1999; Omacini et al., 2001, 2004; Rudgers et al., 2004).

In this chapter we propose a hierarchical framework to synthesize current evidence on effects of *Neotyphodium* endophytes on various ecosystem attributes associated with energy flow and nutrient cycling. We start by defining a hierarchy of ecological domains that span different spatiotemporal scales, ranging from individual host plants through local plant neighborhoods and up to whole biotic communities, which allows us to trace the many consequences of endophyte infection. We then discuss evidence on how endophytes may influence ecological processes operating in each scale domain, and how effects perceived at low hierarchical levels may propagate up through higher ones. Our aim here is not to review the literature exhaustively, but to concentrate on recent studies illustrating the potential significance of endophytes for ecosystem dynamics. We use this framework to identify gaps in existing knowledge of grass–endophyte interactions that may impinge upon biodiversity and ecosystem functioning.

A HIERARCHICAL APPROACH TO ENDOPHYTE IMPACTS ON ECOSYSTEM FUNCTIONING

Ecosystems are complex biological entities whose structure and dynamic behavior can be observed across a range of spatial and temporal scales. Such complexity may be conveniently organized into various levels or *scale domains*, forming a nested hierarchy in both space and time (Allen et al., 1984; Wiens, 1989). Relevant domains are more or less arbitrarily defined by the observer with a given objective in mind (O'Neill et al., 1986). Nonetheless, in any hierarchically organized system, slow changing, higher-level entities and processes generate the context in which small-scale, low-level entities dynamically interact at relatively fast time scales. On the other hand, integration of mechanisms operating at small scales among lower-level components determines the emergent properties of higher-level entities perceived at larger spatial scales (O'Neill et al., 1986; Wiens, 1989). To assess the diversity of interactions whereby *Neotyphodium* endophytes may be linked to ecosystem function, we define a three-level hierarchy comprising the individual host plant, local neighborhood, and whole biotic community as recognizable structural entities. These entities roughly correspond with different spatiotemporal domains and ecological processes, which might be directly or indirectly influenced by endophyte infections (Fig. 6.1).

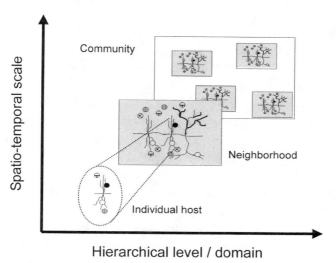

Fig. 6.1. *Hierarchical levels or scale domains in which the impacts of grass endophytes on attributes and processes relevant to ecosystem functioning may be observed. Circles represent above- and belowground organisms which are directly or indirectly affected by endophyte presence (black: Neotyphodium; white: other microbial symbionts; patterned: consumers and decomposers).*

At the lower hierarchical level, phenotypic traits of individual hosts are under direct influence from the fungus metabolism and mycelium growth patterns. Endophyte-induced changes at this level may not only affect resource acquisition and utilization by the host plant, but also the host's immediate relations with primary consumers (herbivores, pathogens) and other mutualistic symbionts (e.g., mycorrhizal fungi). At a higher level, the neighborhood is recognized as a distinct entity, comprising a diverse assemblage of species that belong in different trophic levels and co-occur within a relatively small area, sharing a common resource pool (Wiens, 1989). Most plant interactions with their abiotic environment and with other community components take place at the neighborhood scale (Huston and DeAngelis, 1994; Reynolds et al., 2003). Hence, at this intermediate-level, endophyte presence may alter microhabitat conditions, plant coexistence, tritrophic interactions, rhizosphere composition, litter quality and persistence, detrital food webs, and nutrient supply rates. Finally, the whole-community level is made up of a collection of interacting neighborhood patches (Fig. 6.1) dynamically connected by horizontal movement of organisms and abiotic resources (Wiens, 1989). At this level, emergent ecosystem attributes such as primary productivity, soil carbon/nutrient pools, food-web structure, and invasion resistance result from integrating variation in structural components and process rates at the neighborhood scale. It is suggested that endophyte–grass symbioses will have immediate effects on ecosystem function at this higher level as host species become a dominant community component (Grime, 1998).

The heuristic value of this approach lies in that it allows one to focus on patterns observed on a given scale domain, while recognizing causal mechanisms from lower levels and dynamic constraints imposed by higher-level processes (O'Neill et al., 1986; Wiens, 1989). For instance, endophyte control of energy transfer from host plants to upper trophic levels of insect food webs (Omacini et al., 2001) depends on species-specific responses of herbivorous insects to various fungal alkaloids (Faeth and Bultman, 2002; Bultman et al., 2003), but will additionally be constrained by movement patterns and density-dependent aggregate behavior of searching parasitoids. In the following sections, we review evidence for endophyte impacts in each hierarchical level and highlight interactions considered to be of particular importance to community dynamics and ecosystem function.

ENDOPHYTE EFFECTS AT THE HOST PLANT LEVEL

To assess endophyte impacts on ecosystem function, a comprehensive understanding of interactions at the individual host level is needed. Terrestrial plants acquire, allocate, and release essential resources in ways that strongly influence patterns of carbon and mineral nutrients cycling at various scales (Aerts and Chapin, 2000). Grass endophytes have been shown to alter many such individual-based processes, although as yet underlying mechanisms for effects observed on host ecophysiology are not fully understood (for a review, see Malinowski and Belesky, 2000). Also, substantial work has revealed the influence of endophyte infection on host plant tolerance to biotic and abiotic stress, and how genetic variation affects the outcome of *Neotyphodium*–grass associations (see Chapters 7, 8). These issues will not be discussed further here. We briefly discuss endophyte-induced changes in plant morphology, physiology, and biochemistry that are key to carbon and nutrient accrual by the *Neotypho-*

dium–grass symbiotum, and therefore are likely to affect ecosystem processes at larger scales.

Plant Growth, Resource Use, and Biochemistry

Physiological effects associated with endophyte presence may result in either enhanced or reduced biomass growth of endophyte-infected (E+) plants relative to uninfected ones (E-). Endophyte-induced changes in leaf gas exchange parameters have been reported for some grass species (Malinowsky and Belesky, 2000; Morse et al., 2002). For example, Morse et al. (2002) found that *Neotyphodium*-infected Arizona fescue (*Festuca arizonica*) had lower stomatal conductance and net photosynthesis rates than noninfected conspecifics when grown under high water supply. However, this pattern was reversed under severe water shortage, which accounted for the higher growth rate and water-use efficiency of E+ plants in stressful environments (Morse et al., 2002). Similar differences were found between E+ and E- plants in other experiments manipulating water availability (Malinowsky and Belesky, 2000).

Increased tillering, shoot biomass, and/or reproductive output of E+ plants were reported by several studies with *Festuca* spp. and *Lolium* spp. (West, 1994). In general, however, effects of endophyte infection on host plant growth varied drastically with soil nutrient availability. Endophytes appear to enhance shoot biomass production on nitrogen-rich soils, but may have less clear-cut effects under low N supply (Malinowski and Belesky, 2000). Detailed work on the N economy of tall fescue (*Festuca arundinacea*) suggested that E+ plants may be more efficient at using available N than E- plants (Arachevaleta et al., 1989). Malinowski and Belesky (2000) reported that E+ tall fescue genotypes were consistently outperformed by E- genotypes, except when grown at low levels of soil P availability. Most recent work on Arizona fescue by Faeth et al. (2004) showed that E+ plants produced consistently less biomass than E- plants, irrespective of soil water and nutrient levels.

If mineral nutrient uptake rates and tissue concentrations are influenced by endophyte presence (Arachevaleta et al., 1989; Belesky and Fedders, 1996), this could have far-reaching consequences on recycling processes through the quality of aboveground litter shed by host plants (Aerts and Chapin, 2000). Yet, data available so far on N and P contents of E+ vs. E- lines show inconsistent patterns. Omacini et al. (2001) found an endophyte-mediated increase in total shoot N content of Italian ryegrass (*Lolium multiflorum*) monocultures. In contrast, Bultman et al. (2004) reported lower leaf protein N levels in infected tall fescue. Not surprisingly, relative changes in N and P tissue concentrations of E+ vs. E– plants also depend strongly on soil nutrient levels (Malinowski and Belesky, 2000).

Biomass allocation in host grasses, particularly root-to-shoot ratios, can be under substantial control from endophyte infection (Ahlholm et al., 2002; Belesky and Fedders, 1996; Morse et al., 2002; West, 1994). Endophytes can modify underground patterns of root growth in ways that may affect not only resource uptake (Malinowski and Belesky, 2000), but also soil carbon inputs and microbial processes at the neighborhood scale (Omacini et al., 2004). In some cases, E+ plants were found to produce more root biomass than their E– counterparts (West, 1994), whereas in others the reverse was true (e.g., Faeth et al., 2004). Nevertheless, some recent studies failed to detect consistent changes in root biomass associated with fungal infection in various *Neotyphodium*–grass symbioses (Ahlholm et al., 2002; Müller, 2003; Vila Aiub et al., 2004).

Endophyte-infected grasses exhibit fundamental differences in their secondary chemistry. Secondary metabolites that are greater or are exclusively present in E+ plants include a range of alkaloid (Bush et al., 1993; Siegel et al., 1990; see Chapter 2) and phenolic compounds (Malinowski and Beleski, 2000; Malinowski et al., 1998). The role of fungal alkaloids in reducing herbivory levels has been well established (Clay and Schardl, 2002; see Chapter 7). However, concentrations of different alkaloids vary widely across host grasses and with endophyte strain (Bush et al., 1993; Siegel et al., 1990; see Section 2, this volume). Moreover, alkaloid production in endophytic grasses may drastically change with soil nitrogen and water availability (Arachevaleta et al., 1989; Bultman and Bell, 2003; Malinowski and Belesky, 2000). Although less well studied, phenolics produced by endophytic grasses can be released through root exudates, thus having important effects on rhizosphere chemistry leading, for example, to increased P availability (Malinowski et al., 1998).

Host Plant–Consumer Interactions

By modifying the quantity of biomass produced, and the nutritional quality of host tissues, fungal endophytes may alter energy flows to above- and below-ground herbivores. A majority of studies indicate that endophytes associated with tall fescue and perennial ryegrass tend to decrease aboveground herbivory (Clay and Schardl, 2002; Faeth and Bultman, 2002). The same result has been reported for other host grasses (e.g., Italian ryegrass, Omacini et al., 2001). Fungal alkaloids of E+ plants may alter insect life-history traits and demographic parameters, causing a decline in the intrinsic growth rates of herbivore populations (Faeth and Bultman, 2002; Siegel et al., 1990). Endophyte-mediated protection against herbivorous consumers signifies that host grasses may, in turn, accumulate greater biomass and nutrient stocks than their more susceptible neighbors (Clay, 1994). However, caution is necessary in scaling up results obtained at the individual level because resistance to herbivory varies with prevailing habitat conditions, as well as with fungal, grass, and insect species (Bultman and Bell, 2003; Bultman et al., 2004; Chapter 7).

Endophytes may also reduce damage by root herbivores (Clay, 1990; Bernard et al., 1997). Pedersen et al. (1988) found densities of phytophagous nematodes to be higher in the roots and surrounding soil of E– tall fescue plants than in E+ plant rhizospheres. Chu-Chou et al. (1992) observed that abundance of parasitic nematodes (*Helcotylencus* spp. and *Tylenchus* spp.) was suppressed in field plots planted with E+ tall fescue, whereas nonparasitic nematodes were unaffected. Recent work on meadow fescue (*Festuca pratensis*) showed that grass grub (*Costelytra zealandica*) larval densities and damage to roots were lower in a field sowed with an E+ genotype than in an adjacent endophyte-free pasture (Popay et al., 2003). The precise mechanisms for endophyte effects on soilborne herbivores are still to be determined (see Malinowski and Belesky, 2000).

The AM fungi are important root symbionts in grasslands, where they occur on a broad range of host plant species (van der Heijden and Sanders, 2002), including endophytic grasses. The AM fungi are regarded as plant mutualists that increase nutrient supply while obtaining carbohydrates from the host. It has been shown that endophyte infection reduced AM colonization in tall fescue (Guo et al., 1992), perennial ryegrass (Müller, 2003), and Italian ryegrass (Omacini et al., unpublished data). However, declines in AM fungal colonization may not consistently change host plant productivity or biomass allocation (e.g., Müller, 2003), suggesting that endophyte suppression of root symbionts may have no immediate cost for the plant. Malinowski and Belesky (2000) argued that increased water and P uptake by the roots of E+ tall fescue could compensate for decreases in AM fungi. On the other hand, mycorrhizal infections may influence plant–herbivore interactions above- and belowground (van der Heijden and Sanders, 2002). To our knowledge, only one glasshouse study (Vicari et al., 2002) examined such complex indirect interactions on endophyte-associated grasses.

To summarize, positive (mutualistic) effects of endophytes on individual hosts depend on the specific grass–fungus association and on environmental conditions, especially resource supply and loss rates to herbivory. Ecological differences between E+ and E− biotypes may even be amplified under more stressful conditions (Malinowski and Belesky, 2000). In our hierarchical framework (Fig. 6.1), extrinsic factors such as soil nutrient availability reflect the balance of input (e.g., mineralization) and output (e.g., uptake/leaching) flows at the neighborhood scale (Huston and DeAngelis, 1994; Reynolds et al., 2003). These higher-level processes therefore constrain the outcome of host–endophyte interactions.

ENDOPHYTE-DRIVEN DYNAMICS AT THE NEIGHBORHOOD LEVEL

Given the above endophyte effects on individual hosts, it is reasonable to assume that infected grasses elicit distinct physical and/or chemical signals, influencing the behavior of other organisms in their neighborhoods. Here, the neighborhood domain is defined by the physical proximity to a focal, endophyte-infected plant. Grass biotypes differing in structural or chemical traits may create different microenvironments and spatial relationships within the community by affecting a variety of processes such as resource competition, litter accumulation/decomposition, seedling recruitment, tritrophic interactions, and nutrient mineralization.

As we expand the scale of observation, ecosystem complexity should be expected to increase as well. Thus, while a large number of species and ecosystem functions might be affected at the neighborhood scale, this complexity might also act to filter out the influence of endophyte–grass associations on ecosystem dynamics. For example, in species-rich plant communities, E+ plants would be less likely to exert a large impact on certain processes such as decomposition because litter produced by host grasses would be mixed with a greater diversity of materials derived from co-occurring plant species (Wardle, 2002). We suggest that dominance by E+ genotypes at the neighborhood scale may be a necessary prerequisite for endophyte-driven ecosystem changes to become apparent (Clay 1994, 1997; Grime, 1998).

In examining endophyte modulation of neighborhood-scale dynamics, we focus on three generalized subsystems (Swift and Anderson, 1993; Wardle, 2002), which comprise plants and their microbial symbionts, herbivore–consumer trophic webs, and decomposer soil webs, respectively (Fig. 6.2). We first assess endophyte effects on multispecies assemblages within each subsystem by treating them as relatively independent entities; second, we discuss how feedbacks among subsystems may be mediated by endophyte presence.

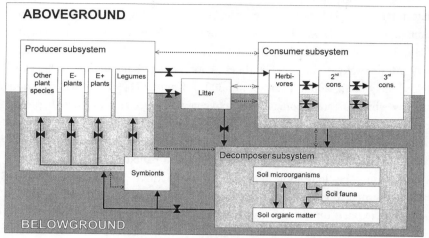

Fig. 6.2. *Generalized model for the control of ecosystem functioning by grass endophytes. The ecosystem structure (large box) is subdivided into three main subsystems, each constituted of various aboveground and/or belowground components. Different taxa are lumped into broad functional groups defined by their roles in ecosystem function. Solid arrows show energy and/or material flows between various compartments; broken arrows represent information flows (feedbacks) among subsystems. Focal points of endophyte modulation are indicated by valve symbols.*

Producer Subsystem

Plant competition for resources at the neighborhood scale is a major factor driving potential effects of endophyte–grass associations on biodiversity and ecosystem function (Grime, 1998). Grassland patches containing several plant species constitute the arena in which the demographic consequences of the symbiosis will be perceived. For example, enhanced seed germination rates and/or seedling survival associated with endophyte infection (Clay, 1987; Vila Aiub et al., 2004) may facilitate local dominance by host grasses. A number of greenhouse and field studies have demonstrated that plant–plant interactions can be strongly modified by endophytic fungi. In domesticated grasses, endophyte-infected plants generally outcompete uninfected conspecific or heterospecific neighbors, either with or without herbivores present (Clay, 1994, 1997; Clay and Schardl, 2002). However, studies conducted on native grasses show that competitive dominance by infected genotypes may not be the rule (Faeth and Sullivan, 2003; Faeth et al., 2004).

There exist several mechanisms whereby endophytes could conceivably mediate interactions between co-occurring plants, including exploitative resource competition (Clay, 1994; West, 1994), recruitment limitation by litter deposition (Clay, 1994), release of allelochemicals (Bush et al., 1993; Sutherland et al., 1999; Malinowski et al., 1999), and herbivore-mediated interactions (Clay et al., 1993; Popay et al., 2003; Vicari et al., 2002). Connell (1990) discussed various ways by which apparent competition between plant species may arise due to the presence of a third species from the same or a different trophic level. We are not aware of any study testing for such indirect effects in the context of *Neotyphodium*–grass associations. Moreover, there is still a paucity of field experiments demonstrating the long-term outcome of competitive interactions involving endophyte-infected grasses (Clay and Holah, 1999; Clay and Schardl, 2002; see *Endophyte Impacts on Community-Level Dynamics*, below).

Interactions between endophyte-infected pasture grasses and clover (*Trifolium* spp.) are a good example of how endophyte–grass symbiosis may alter nitrogen cycling, at least on a neighborhood scale. Sutherland and Hoglund (1989) found that infected perennial ryegrass suppressed white clover (*T. repens*) in sown pastures. In a greenhouse study, Malinowski et al. (1999) observed that competitive suppression of red clover (*T. pratense*) by E+ tall fescue depended on the particular grass genotype. The genotype that most strongly reduced clover yield had twice the concentration of loline alkaloids in the roots; hence, allelopathy was invoked as a possible mechanism for the observed effect (Malinowski et al., 1999; Sutherland et al., 1999). More importantly, Snell and Quigley (1993) detected a reduction in root length and nodulation of subterranean clover (*T. subterraneum*), which translated into shoot and root biomass reductions, in pots amended with litter from E+ perennial ryegrass but not in those with litter from E– plants. These studies suggest that endophytes–grass associations may indirectly decrease N inputs to local soil patches through their negative impact on the abundance and root-symbiotic activity of nitrogen-fixing leguminous plants.

Consumer Subsystem

A growing body of evidence suggests that endophyte infections may have substantial impacts on energy transfers beyond the herbivore level through effects on individual growth and population size of higher level consumers (predators, parasitoids, pathogens). In general, studies have been limited to endophyte symbioses with tall fescue and perennial ryegrass, and have typically involved three-species food chains (Grewal et al., 1995; Bultman et al., 2003). Faeth and Bultman (2002) provide a review of examples for negative and positive effects from endophytes on development and survival rates of parasitoids attacking herbivorous insects.

In a field study on aphid–parasitoid food webs naturally established on Italian ryegrass monocultures, *Neotyphodium*-infected plants supported significantly lower aphid and leaf-miner densities than adjacent endophyte-free plants (Omacini et al., 2001). Further, presence of fungal endophytes caused an eight-fold decrease in parasitized aphids of the dominant species, *Rhopalosiphum padi*. This resulted from disproportionally higher parasitism rates for aphids colonizing E– patches. Interaction strengths calculated for various primary and secondary parasitoid species attacking aphid species suggested that energy flows to upper trophic levels were more evenly distributed in endophyte-controlled insect food webs (Omacini et al., 2001). Endophyte-driven effects on food-web dynamics occurred in the absence of significant changes in host plant productivity and would have been generated by deterrence of *R. padi* by loline-type alkaloids (Clay and Schardl, 2002; Omacini and Bush, unpublished data).

In a follow-up experiment, no density responses to endophyte presence were observed in the aphid *Sipha maydis*, which displaced *R. padi* from ryegrass patches. Interestingly, however, the number of ants attending aphid colonies on E+ grass patches dropped markedly relative to those recorded in E– patches (E. Chaneton et al., unpublished data). These results help to expand our appreciation of the consequences of endophyte infection in natural systems by showing that indirect effects need not be driven by density changes of intermediary herbivore species (see also Bultman et al., 2003), and that endophytes may have ramifying effects on other mutualistic interactions.

Indirect effects of *Neotyphodium* on multitrophic interactions may be more difficult to pinpoint in soil food webs. Changes in root productivity and defensive secondary chemicals produced by host–fungal symbioses can alter the performance of root herbivores, and may thus propagate to third- and fourth-level consumers in soil communities (Wardle, 2002). In addition, dominance by endophytic grasses may increase plant cover, affecting microhabitat conditions for various ground-dwelling invertebrates. Prestidge and Marshall (1997) reported greater activity (pitfall trapping) of predatory invertebrates in paddocks sown with E+ than in those with E– perennial ryegrass. Grewal et al. (1995) found that litter-dwelling beetles feeding on E+ tall fescue plants were more susceptible to soil-borne entomopathogenic nematodes than were those fed E– plants. As it is the case for aboveground food webs, endophyte effects on trophic dynamics of soil biota appear to be transmitted mainly by subtle changes in host plant chemistry, rather than by changes in primary productivity. However, we emphasize that more experimental work is needed in this area.

Decomposer Subsystem

Litter decomposition is a key process in terrestrial ecosystem functioning (Swift and Anderson, 1993). Clay (1994, 1997) suggested that litter produced by endophyte-infected grasses could break down more slowly, and that subsequent buildup of a thick litter layer could affect soil communities and retard nutrient cycling. Interestingly, Franzluebbers et al. (1999) found small changes in the structure of soil microbial communities beneath tall fescue patches with contrasting levels of endophyte infection. Nonetheless, they also found a substantial decrease in microbial biomass and soil respiration rates, which were consistent with greater concentrations of soil organic C under highly infected fescue. Recently, Omacini et al. (2004) hypothesized that endophyte may alter litter decomposition rates by changing the substrate quality/quantity to decomposers and/or the microenvironment for decomposition.

In microcosm/litterbag experiments, it was found that the litter from E+ plants (Italian ryegrass) decomposed at a lower rate than the litter from E- plants (Omacini et al., 2004). Moreover, covering the soil with a thick litter layer from E+ plants reduced decomposition of litter produced by *Bromus unioloides*, an endophyte-free species. These results are consistent with the idea that endophyte byproducts can persist in dead plant material (Bush et al., 1993), decreasing host-litter quality and hence energy inputs to soil food webs. This study also suggested that endophyte impacts on decomposition were mediated by a decline in soil microbial activity (Omacini et al., 2004; see also Franzluebbers et al., 1999). However, potential consequences for higher-trophic levels in the decomposer subsystem (Fig. 6.2; Wardle, 2002) are unknown. The lower numbers of microarthropods (collembola and oribatid mites) observed in E+ vs. E- fescue pastures (Bernard et al., 1997) suggest that bottom-up cascades in soil food webs elicited by endophyte-infected plants are certainly possible.

Feedback between Subsystems

Empirical evidence suggests that feedback mechanisms between above- and belowground compartments play a major role in the regulation of terrestrial ecosystems (Wardle, 2002). In our neighborhood-scale model (Fig. 6.2), feedbacks are portrayed as flows of information connecting the three principal subsystems (Swift and Anderson, 1993). In particular, plant interactions with soil microbes can influence local diversity through positive or negative feedbacks on plant growth resulting from changes in soil community composition (Bever, 2003; Reynolds et al., 2003). Such interactions would be crucial to patch dynamic processes shaping grassland structure at the community level (see Fig. 6.1).

Through effects on the host plant's root growth and biochemistry, endophytic fungi may alter the soil biota in ways that will feed back on the future population performance of the endophyte–grass association and its competitive relations with neighboring plants (Matthews and Clay, 2001). Yet, little attention has been devoted to possible endophyte mediation of plant–soil microbe feedbacks. Matthews and Clay (2001) found lower biomass of E+ fescue plants when cultured in soil previously occupied by its own conspecifics than when cultured in soil previously dominated by other (noninfected) grasses. They argued that various mechanisms could explain such a negative feedback (Bever, 2003), including a reduction of soil nutrient availability, an increase of endophyte-resistant soil pathogens, and the accumulation of secondary compounds in the root environment (Matthews and Clay, 2001). On the other hand, if endophytes reduce mycelial growth (Guo et al., 1992; Müller, 2003) and spore production (Chu-Chou et al., 1992) from AM fungi in the root zone, then mycorrhiza-dependent species growing nearby could be negatively affected. This would generate a positive feedback for infected grass species. But to our knowledge, such an indirect modification of plant–mycorrhiza relations by endophyte-infected plants, and its consequences for nutrient utilization patterns in grasslands, has never been examined.

Plant–soil feedbacks can also be influenced by herbivorous consumers that change the quantity or quality of dead organic matter returned to the soil (Bardgett et al., 1998; Wardle, 2002). For example, grazing animals may induce changes in plant litter chemistry that will affect decomposition and nutrient mineralization rates. It has been shown that defoliation of endophyte-infected plants often increases foliar concentrations of fungal alkaloids during regrowth (Bultman et al., 2004). If herbivore-induced changes in tissue chemistry of E+ plants persist upon the host death, we would expect a decrease in litter decomposition rate (Omacini et al., 2004) and a likely decrease in soil nutrient availability, at least in the short term (Bardgett et al., 1998). It is as yet unclear what are the long-term consequences of such feedback loops on the species composition of endophyte-infected grass patches.

ENDOPHYTE IMPACTS ON COMMUNITY-LEVEL DYNAMICS

Endophyte impacts on neighborhood processes may propagate horizontally to affect community-level attributes at larger spatiotemporal scales. The influence of endophyte–grass associations on ecosystem function will scale up depending on patch dynamic processes at the community level (Fig. 6.1). Two aspects are relevant here; that is, the sign of endophyte-driven feedback interactions at the neighborhood level, and the ability of E+ genotypes to invade nearby patches occupied by other plant species. The net outcome of positive and negative feedback effects on plant growth will determine the probability for infected host grasses to dominate a local patch (Matthews and Clay, 2001; Bever, 2003). Whichever the mechanism, a positive feedback on host plant fitness will allow E+ genotypes to expand across community space by increasing the host's propagule pressure on neighboring patches. Crucially, the invasion capacity of E+ genotypes will depend on seed/seedling traits, allowing the plant to become established in dense canopies (Clay, 1987), and on spatial heterogeneity in environmental conditions that modify the relative performance of host–fungus associations (Malinowski and Belesky, 2000; Bultman et al., 2004). Frequent reports of infected grass species increasing in abundance across successional time (Clay and Schardl, 2002) suggest both these ecological mechanisms operate in concert within natural and agricultural systems.

Ecological theory suggests two general ways through which endophytic grasses may contribute to community functioning. The *mass ratio* hypothesis indicates that the extent to which a plant species affects ecosystem attributes would be associated with its contribution to total community biomass (Grime, 1998). This highly intuitive idea was at the core of our hierarchal argument so far. In contrast, the *keystone species* hypothesis suggests that nondominant species which account for only a small fraction of total biomass may still exert major effects on ecosystem processes (Power et al., 1996). For instance, endophyte infection in pastures and natural grasslands, even if grass hosts are present in low frequency, may force managers to remove grazing animals to avoid potential toxicosis. It follows that presence of infected genotypes may thus initiate wholesale, cascading effects on community structure and ecosystem function (see also Grime, 1998, for a different perspective on subordinate species). We now discuss recent findings for community-level effects of *Neotyphodium*–grass associations and explore their consequences for various ecosystem attributes. The latter discussion is recognizably more speculative given the lack of direct empirical evidence.

Plant Community Dynamics

Few published studies directly compared plant community dynamics in similar sites differing only in endophyte infection levels. Clay (1997) hypothesized that plant composition and diversity should be altered through the enhanced vigor and stress tolerance ability of infected grasses. He also suggested that indirect mechanisms involving endophyte effects on microhabitat conditions and microbial soil communities should be taken into account as potential drivers of plant dynamics (Matthews and Clay, 2001). These ideas were broadly consistent with the results of a recent field experiment on tall fescue pastures (Clay and Holah, 1999). Sowing of either E+ or E– fescue seeds on large, replicated plots containing a naturally diverse seed bank revealed the effects of *Neotyphodium* on host plant dominance and whole-community dynamics. After 4 yr of succession, plant species diversity was substantially lower on E+ plots, where infected tall fescue accounted for 90% of total biomass (vs. 60% of uninfected tall fescue in E– plots).

In another study conducted on a 3-yr-old field dominated by tall fescue, Spyreas et al. (2001) found that plant richness decreased as endophyte infection frequency increased but only in unmowed vegetation plots, whereas species richness was positively correlated with infection frequency in mowed plots. Interestingly, infection frequencies in tall fescue were increased by heavy mowing. The results of this experiment indicated that management can modify the predicted impact of endophytic grasses on community dynamics by altering environmental conditions that favor competitive dominance by E+ genotypes (Spyreas et al., 2001).

If endophyte-infected grasses are capable of displacing other plant species that colonize during early succession, we might also expect later-successional, undisturbed communities to become less invasible when dominated by E+ plants. Clay and Schardl (2002) reported that, after 7 yr of succession, the abundance of forbs and woody plants in plots sowed with E+ tall fescue (see Clay and Holah, 1999) was remarkably lower than in E– control plots. However, the mechanism behind the observed decrease in community invasibility has not been revealed. On the other hand, the literature on *Neotyphodium–* infected tall fescue and perennial ryegrass, and our own experience with Italian ryegrass in Argentinean grasslands (e.g., Vila Aiub et al., 2004) indicate that infected genotypes may be more aggressive invaders than their endophyte-free conspecifics. It is intriguing that research done to date on the role of mutualistic symbionts in plant invasion processes (Richardson et al., 2000), except for the work of Rudgers et al. (2004), has ignored this dual role of *Neotyphodium* endophytes.

Endophyte presence may affect the spatial heterogeneity of plant community. The experiments discussed above underline the importance of endophyte infection in controlling overall plant diversity through host plant dominance. Yet, community dominance by E+ plants does not necessarily imply extinction of patches dominated by E– conspecifics. This is because of the fact that *Neotyphodium* transmission through the host's seeds is imperfect, and thus infected plants produce both E+ and E– progeny (Faeth and Sullivan, 2003). We do not know what the consequences of maintaining such intraspecific diversity are for ecosystem dynamics, but theory suggests that small-scale heterogeneity in processes such as litter decomposition (see Omacini et al., 2004) may help to sustain ecosystem productivity in the long term (Grime, 1998).

Ecosystem Functioning

Individual plant species may strongly influence ecosystem productivity (Grime, 1998; Wardle, 2002). However, the enhanced biomass production of infected host grasses may not be accompanied by an overall increase in aboveground primary productivity (e.g., Clay and Holah, 1999). Endophyte-driven effect on plant productivity at the neighborhood level will scale up to the whole community depending on factors such as the functional identity of the species being displaced by E+ plants, compensatory responses from co-occurring plant species, and niche complementarity between cool-season E+ genotypes and other plants from different functional types (Wardle, 2002). Rudgers et al. (2004) demonstrated that plant species diversity was more strongly negatively correlated with the biomass of E– fescue than with that of E+ fescue, which suggested that endophyte presence altered the relationship between diversity and community productivity.

In addition, community-level reductions in herbivory and secondary productivity are possible given that fungal alkaloids can decrease biomass consumption by livestock, game, and phytophagous insects (Clay, 1994; Bacon and Hill, 1997; Clay and Schardl, 2002). The importance of tall fescue endophytes for livestock production has been documented in the USA where high endophyte incidence in tall fescue pastures can diminish livestock daily weight gains (see Section IV, this volume). Endophyte-induced changes in biomass consumption by wild vertebrates are less well known (Clay, 1994). In contrast, susceptibility to invertebrate pests (e.g., the Argentine stem weevil, *Listronotus bonariensis*) in perennial ryegrass lacking infection with *Neotyphodium lolii* suggests that endophyte presence may strongly regulate energy flows to herbivores in some agronomic systems.

Litter accumulation by endophyte-infected genotypes may strongly affect carbon flows to soil food webs. It has been shown that reduced quality of the litter deposited by E+ plants can alter soil microbial activity (Franzluebbers et al., 1999; Omacini et al., 2004), but effects on mineral nutrient availability due to decreased litter breakdown have not been demonstrated. Moreover, we are not aware of any study showing an increase in total litter biomass derived from the community dominance by E+ grasses. Only one study to date characterized the size of soil organic C and N pools beneath E+ vs. E– fescue pastures (Franzluebbers et al., 1999). The lower bulk density and greater soil organic C found in E+ plots was tentatively assigned to a decrease in forage intake by domestic herbivores, and thus a greater quantity of plant detritus entering the soil. Overall, the evidence so far suggests that dominance by endophytic grasses might enhance the net ecosystem productivity (a measure of total carbon balance) as a consequence of the combined effects of increased carbon uptake by plants and reduced consumption by heterotrophic communities, resulting in soil organic matter accumulation.

FUTURE CHALLENGES

We are just beginning to understand how fungal endophytes may alter ecosystem processes. The hierarchical framework introduced here makes it clear that available knowledge on the ecosystem function consequences of endophyte–grass interactions diminishes as we scale up from the host plant through the whole community domain. Overall, most evidence on the functional roles of endophytes comes from over-simplified agricultural systems and small-scale experiments in controlled environments. These biases compromise the ecological relevance of such findings for complex natural communities (Faeth and Sullivan, 2003). In addition, most experimental studies on endophyte–grass interactions were too short-term to tell something about the population dynamics of the fungus–grass symbiosis in natural settings (Clay, 1994; Clay and Schardl, 2002).

We identify three major areas for conceptual development and future research on the ecosystem roles of endophytic fungi. Firstly, we suggest that these symbiotic interactions should be incorporated into theoretical arguments about the *rules of assembly* in ecological communities (Grime, 1998; van der Putten et al., 2001; Wardle, 2002). The work of Clay and Holah (1999), Spyreas et al. (2001), and Rudgers et al. (2004) strongly supports this contention. Although ecosystems vary greatly in complexity and diversity, it is tempting to predict that the functional significance of a given species increases with its relative contribution to community biomass (Grime, 1998). Yet, some evidence reviewed in this chapter suggests that this may not be always the case when looking at the effects of endophyte–grass symbioses on ecosystem function. Nevertheless, we stress that demonstrating whether endophytic fungi may act as keystone symbionts will require carefully designed experiments in which infection is manipulated and various ecosystem attributes are measured at relevant scales (see Power et al., 1996).

Secondly, we believe more experimental work is need to reveal the potentially large impacts of endophyte–grass associations on food-web structure and function. In particular, we still know little about interactions between *Neotyphodium* endophytes and soil organisms as mediated by host plant growth and biochemistry. Endophyte-induced changes in components of belowground communities not only may be reflected in the regulation of various ecosystem processes, but may additionally provide feedback on the population dynamics of infected grasses and their immediate neighbors.

Therefore, we finally propose that the so-called *feedback approach* (Bever, 2003) may be applied to tackle plant–soil and plant–plant interactions in the presence of fungal endophytes in a more realistic, multispecific context. Interactions at this neighborhood level may help us understand why some endophyte-infected plant species are becoming so invasive in seminatural ecosystems (Clay, 1994, 1997). Ultimately, future ecosystem responses to ongoing environmental changes (increased temperature, elevated atmospheric CO_2, nitrogen deposition, etc.) in many temperate grasslands may be shaped by functional roles of endophytic grasses.

SUMMARY

Neotyphodium endophytes are integral components of terrestrial ecosystems that fundamentally affect transfers of energy and matter through food webs. Here we propose a three-level hierarchical framework to synthesize current evidence on *Neotyphodium* effects on ecosystem functioning at various spatiotemporal scales. We focus on the individual host plant, local plant neighborhood, and whole-community levels to discuss how and to what extent grass endophytes influence key ecological processes. This framework allows one to identify gaps in existing knowledge of grass–endophyte interactions in natural and human-made ecosystems.

ACKNOWLEDGMENTS

We thank O. Sala for previous discussion on hierarchy theory, and A. Mella, P. Roset, and M. Rabadan for help with manuscript preparation. Our work is partly funded by Fundación Antorchas, Consejo Nacional de Investigaciones Científicas y Técnicas (CONICET) y Agencia Nacional de Promoción Científica (FONCYT). M.O. was supported by a doctoral fellowship from the University of Buenos Aires.

REFERENCES

Aerts, R., and F.S. Chapin III. 2000. The mineral nutrition of wild plants revisited: A re-evaluation of processes and patterns. Adv. Ecol. Res. 30:1–67.

Ahlhom, J.U., M. Helander, S. Lehtimaki, P. Wäli, and K. Saikkonen. 2002. Vertically transmitted fungal endophytes: Different responses of host-parasite systems to environmental conditions. Oikos 99:177–183.

Allen, T.F.H., R.V. O'Neill, and T.W. Hoekstra. 1984. Interlevel relations in ecological research and management: Some working principles from hierarchy theory. Forest Service Gen. Tech. Rep. RM-110. Rocky Mountain Forest and Range Exp. Stn., Fort Collins, CO.

Arachevaleta, M., C.W. Bacon, C.S. Hoveland, and D.E. Radcliffe. 1989. Effect of the tall fescue endophyte on plant response to environmental stress. Agron. J. 81:83–90.

Bacon, C.W., and N.S. Hill (ed.) 1997. Neotyphodium/Grass Interactions. Proc. Int. Symp. Acremonium/Grass Interactions, 3rd, Athens, GA, USA. 28–31 May 1997. Plenum Press, New York.

Bardgett, R., D. Wardle, and G. Yeates. 1998. Linking above-ground and below-ground interactions: How plant responses to foliar herbivory influence soil organisms. Soil Biol. Biochnol. 14:1867–1978.

Belesky, D.P., and J.M. Fedders. 1996. Does endophyte influence regrowth of tall fescue? Ann. Bot. (London) 78:499–505.

Bernard, E.C., K.D. Gwinn, C.D. Pless, and C.D. Williver. 1997. Soil invertebrate species diversity and abundance in endophyte infected tall fescue pastures. p. 383–388. In C.W. Bacon and N. S. Hill (ed.) Neotyphodium/Grass Interactions. Proc. Int. Symp. Acremonium/Grass Interactions, 3rd, Athens, GA, USA. 28–31 May 1997. Plenum Press, New York.

Bever, J.D. 2003. Soil community feedback and the coexistence of competitors: Conceptual frameworks and empirical tests. New Phytol. 157:465–473.

Bultman, T.L., and G.D. Bell. 2003. Interaction between fungal endophytes and environmental stressors influences plant resistance to insects. Oikos 103:182–190.

Bultman, T.L., G. Bell, and W. Martin. 2004. A fungal endophyte mediated reversal of wound-induced resistance and constrains tolerance in a grass. Ecology 85:679–685.

Bultman, T.L., M.R. McNeill, and S.L. Goldson. 2003. Isolate-dependent impacts of fungal endophytes in a multitrophic interaction. Oikos 102:491–496.

Bush, J.K., F. Fannin, M. Siegel, D.L. Dahlman, and H. Burton. 1993. Chemistry, occurrence and biological effects of saturated pyrrolizidine alkaloids associated with endophyte-grass interactions. Agric. Ecosvst. Environ. 44:81–102.

Clay, K. 1987. Effects of fungal endophytes on the seed and seedling biology of *Lolium perenne* and *Festuca arundinacea*. Oecologia 73:358–362.

Clay, K. 1990. Fungal endophytes of grasses. Annu. Rev. Ecol. Syst. 21:275–295.

Clay, K. 1994. The potential role of endophytes in ecosystems. p. 73–86. *In* C.W. Bacon and J. White (ed.) Biotechnology of endophytic fungi of grasses. CRC Press, Boca Raton, FL.

Clay, K. 1997. Consequences of endophyte-infected grasses on plant diversity. p. 93–108. *In* C. W. Bacon and N. S. Hill (ed.) *Neotyphodium/*Grass Interactions. Proc. Int. Symp. *Acremonium/* Grass Interactions, 3rd, Athens, GA, USA. 28–31 May 1997. Plenum Press, New York.

Clay, K., and J. Holah. 1999. Fungal endophyte symbiosis and plant diversity in successional fields. Science (Washington, DC) 285:1742–1744.

Clay, K., S. Marks, and G.P. Cheplick. 1993. Effects of insect herbivory and fungal endophyte infection on competitive interactions among grasses. Ecology 74:1767–1777.

Clay, K., and C.L. Schardl. 2002. Evolutionary origins and ecological consequences of endophyte symbiosis with grasses. Am. Midl. Nat. 160:99–127.

Connell, J.H. 1990. Apparent vs. real competition in plants. p. 9–26. *In* J.B. Grace and D.A. Tilman (ed.) Perspectives on plant competition. Academic Press, San Diego, CA.

Chu-Chou, M., B. Guo, Z.Q. An, J.W. Hendrix, R. Ferriss, D.I. Siegel, C. Dougherty, and P. Burrus. 1992. Suppression of mycorrhizal fungi in fescue by the *Acremonium coenophialum* endophyte. Soil Biol. Bioch. 24:633–637.

Faeth, S., and T.L. Bultman. 2002. Endophytic fungi and interactions among host plants, herbivores, and natural enemies. p. 89–123. *In* T. Tscharntke and B.A. Hawkins (ed.) Multitrophic interactions. Cambridge Univ. Press, Cambridge.

Faeth, S., M. Helander, and K. Saikkonen. 2004. Asexual *Neotyphodium* endophytes in a native grass reduce competitive abilities. Ecol. Lett. 7:304–313.

Faeth, S., and T. Sullivan. 2003. Mutualistic asexual endophytes in a native grass are usually parasitic. Am. Midl. Nat. 161:310–325.

Franzluebbers, A.J., N. Nazih, J.A. Stuedemann, J.J. Fuhrmann, H.H. Schomberg, and P.G. Hartel. 1999. Soil carbon and nitrogen pools under low and high endophyte infected tall fescue. Soil Sci. Soc. Am. J. 63:1687–1694.

Grewal, S., P. Grewal, and R. Gaugler. 1995. Endophytes of fescue grasses enhance susceptibility of *Popillia japonica* larvae to an entomophagous nematode. Entomol. Exp. Appl. 74:219–224.

Grime, J.P. 1998. Benefits of plant diversity to ecosystems: Immediate, filter and founder effects. J. Ecol. 86:902–910.

Guo, B., J. Hendrix, Z. An, and R. Ferriss. 1992. Role of *Acremonium* endophyte of fescue on inhibition of colonization and reproduction of mycorrhizal fungi. Mycologia 84:882–885.

Huston, M.A., and D.L. DeAngelis. 1994. Competition and coexistence: The effects of resource transport and supply rates. Am. Nat. 144:954–977.

Malinowski, D.A., G.A. Alloush, and D.P. Belesky. 1998. Evidence for chemical changes on the root surface of tall fescue in response to infection with the fungal endophyte *Neotyphodium coenophialum*. Plant Soil 205:1–12.

Malinowski, D.P., and D.P. Belesky. 2000. Adaptations of endophyte-infected cool-season grasses to environmental stresses: Mechanisms of drought and mineral stress tolerance. Crop Sci. 40:923–940.

Malinowski, D.A., D.P. Belesky, and J.M. Fedders. 1999. Endophyte infection may affect the competitive ability of tall fescue growth with red clover. J. Agron. Crop Sci. 183:91–102.

Matthews, J.W., and K. Clay. 2001. Influence of fungal endophyte infection on plant/soil feedback and community interactions. Ecology 82:500–509.

Morse, L., T. Day, and S. Faeth. 2002. Effect of *Neotyphodium* endophyte infection on growth and leaf gas exchange of Arizona fescue under contrasting watering availability regimes. Environ. Exp. Bot. 48:257–268.

Müller, J. 2003. Artificial infection by endophytes affects growth and mycorrhizal colonization of *Lolium perenne*. Funct. Plant Biol. 30:419–424.

Omacini, M., E. Chaneton, C.M. Ghersa, and C. Muller. 2001. Symbiotic fungal endophytes control insect host-parasite interaction webs. Nature (London) 409:78–81.

Omacini, M., E. Chaneton, C.M. Ghersa, and P. Otero. 2004. Do foliar endophytes affect grass litter decomposition? A microcosm approach using *Lolium multiflorum*. Oikos 104:581–590.

O'Neill, R.V., D.L. DeAngelis, J.B. Waide, and T.F.H. Allen. 1986. A hierarchical concept of ecosystems. Princeton Univ. Press, Princeton, NJ.

Pedersen, J., R. Rodriguez-kabana, and R. Shelby. 1988. Ryegrass cultivars and endophyte in tall fescue affect nematodes in grass and succeeding soybean. Agron. J. 80:811–814.

Popay, A.I., R.J. Townsend, and L. Fletcher. 2003. The effect of endophyte (*Neotyphodium uncinatum*) in meadow fescue on grass grub larvae. N. Z. Plant Prot. 56:123–128.

Power, M., D. Tilman, J.A. Estes, B.A. Menge, W. Bond, L.S. Mills, G.C. Daily, J.C. Castilla, J. Lubchenco, and R.T. Paine. 1996. Challenges in the quest for keystones. BioScience 46:609–620.

Prestidge, R.A., and S.L. Marshall. 1997. The effects of *Neotyphodium*-infected perennial ryegrass on the abundance of invertebrate predators. p. 195–198. *In* C.W. Bacon and N.S. Hill (ed.) *Neotyphodium*/Grass Interactions. Proc. Int. Symp. *Acremonium*/Grass Interactions, 3rd, Athens, GA, USA. 28–31 May 1997. Plenum Press, New York.

Reynolds, H., A. Packer, J.D. Bever, and K. Clay. 2003. Grassroots ecology: Plant-microbe-soil interactions as drivers of plant community structure and dynamics. Ecology 84:2281–2291.

Richardson, D.M., N. Allsopp, C. D'Antonio, S. Milton, and M. Rejmanek. 2000. Plant invasions—The role of mutualisms. Biol. Rev. 75:65–93.

Rudgers, J.A., J.M. Koslow, and K. Clay. 2004. Endophytic fungi alter relationships between diversity and ecosystem properties. Ecol. Lett. 7:42–51.

Siegel, M.R., G.C.M. Latch, L.P. Bush, F.F. Fannin, D.D. Rowan, B.A. Tapper, C.W. Bacon, and M.C. Johnson. 1990. Fungal endophyte-infected grasses: Alkaloid accumulation and aphid response. J. Chem. Ecol. 16:3301–3315.

Snell, F.J., and P.E. Quigley. 1993. Allelopthic effects of endophyte in perennial ryegrass residues on young subterranean clover plants. p. 343–344. *In* A.M. Baker, J.R. Crush, and L.M. Humpries (ed.) Proc. XVII Int. grassland congress, Palmerston North, NZ.

Spyreas, G., D.J. Gibson, and B.A. Middleton. 2001. Effects of endophyte infection in tall fescue (*Festuca arundinacea*, Poaceae) on community diversity. Int. J. Plant. Sci. 162:1237–1245.

Sutherland, B.L., and J.H. Hoglund. 1989. Effect of ryegrass containing the endophyte *Acremonium lolii*, on the performance of associated white clover and subsequent crops. Proc. N. Z. Grassland Assoc. 50:265–269.

Sutherland, B.L., D.E. Hume, and B.A. Tapper. 1999. Allelopathic effects of endophyte-infected perennial ryegrass extracts on white clover seedlings. N. Z. J. Agric. Res. 42:19–26.

Swift, M.J., and J.M. Anderson. 1993. Biodiversity and ecosystem function in agricultural systems. p. 523. *In* E.D. Shulze and H.H. Mooney (ed.) Biodiversity and ecosystem function. Springer-Verlag, Berlin.

Van der Heijden, M.G.A., and J.H.C. Cornelissen. 2002. The critical role of plant–microbe interactions for biodiversity and ecosystem functioning: Arbuscular mycorrhizal associations as an example. *In* M. Loreau and S. Naeem (ed.) Biodiversity and ecosystem functioning: Synthesis and perspectives. Oxford Univ. Press, Oxford.

Van der Heijden, M.G.A., and I.R. Sanders (ed.) 2002. Mycorrhizal ecology. Springer-Verlag, Berlin.

van der Putten, W.H., L.E.M. Vet, J.H. Harvey, and F.L. Wackers. 2001. Linking above and belowground multitrophic interactions of plants, herbivores, pathogens, and their antagonists. Trends Ecol. Evol. 16:547–554.

Vicari, M., P. Hatcher, and A. Ayres. 2002. Combined effect of foliar and mycorrhizal endophytes on an insect herbivore. Ecology 83:2452–2464.

Vila Aiub, M.M., P. Gundel, and C.M. Ghersa. 2004. Fungal endophyte infection changes growth attributes in *Lolium multiflorum* Lam. Aust. Ecol. In press.

Wardle, D. 2002. Communities and ecosystems: Linking the aboveground and belowground components. Princeton Univ. Press, Princeton, NJ.

West, C.P. 1994. Physiology and drought tolerance of endophyte-infected grasses. p. 87–102. *In* C.W. Bacon and J.J. White, Jr. (ed.) Biotechnology of endophytic fungi of grasses. CRC Press, Boca Raton, FL.

Wiens, J.A. 1989. Spatial scaling in ecology. Funct. Ecol. 3:385–397.

BIOTIC RESPONSES IN ENDOPHYTIC GRASSES

Alison J. Popay[1] and Stacy A. Bonos[2]

In grasses, mutualism between host plants and the endophytic clavicipitaceous fungi manifests itself as improved growth and survival of individual plants and a predominance of infected plants in natural and managed grassland communities. Enhancements in host fitness derived from infection are at least partly attributable to protection from biotic stresses, such as herbivory and disease, mediated by the production of secondary metabolites. Endophyte infection, however, provides no guarantee of resistance to herbivory, and there are instances where there is not only no effect on herbivores, but also where insect performance is enhanced on endophyte-infected plants. These effects are strain specific and influenced substantially by host plant genotype.

Biotic responses can be defined as those that occur when the presence of the endophyte alters the interaction between a host plant and the organisms associated with it. In the first report of such a response, Prestidge et al. (1982) demonstrated a negative association between Argentine stem weevil (*Listronotus bonariensis*) populations and the incidence of *Neotyphodium lolii* infection in perennial ryegrass. By 1994, more than 40 species of insects spanning several different orders had been found to be adversely affected by endophytes (Breen, 1994; Clement et al., 1994; Popay and Rowan, 1994). These effects were shown mainly in the economically important grasses tall fescue (*Festuca arundinacea*) and perennial ryegrass (*Lolium perenne*), or by bioassaying the alkaloids produced by endophyte infection. In the 10 yr since these reviews were published, relatively few insects have been added to this already impressive list. Instead, emphasis has shifted to exploring the biotic responses in the diversity of endophyte–host plant associations that exist worldwide. Some endophytes have been deliberately sought to resolve the mammalian toxicity associated with the wild-type endophytes in tall fescue and perennial ryegrass widely used for agriculture

[1] AgResearch, Ruakura Research Centre, Private Bag 3123, Hamilton, New Zealand
[2] Department of Plant Biology and Pathology, Cook College, Rutgers University, New Brunswick, NJ USA

in the USA, Australia, and New Zealand. In addition, more research attention has also been given to the bioactivity of endophytes in native grasses.

It is our intention in this review to concentrate on developments in our understanding of biotic responses in endophytic grasses during the last 10 years rather than to revisit information already covered in the extensive 1994 reviews.

BIOTIC RESPONSES IN WILD GRASSES

Wild and native grasses are frequently infected with species of *Epichloë* or asexual *Neotyphodium* endophytes. In natural populations of grasses, the proportion of plants infected ranges from very low to 100%, suggesting that no consistent increase in host fitness can be attributed to endophyte in these situations, although few have been extensively studied. Nevertheless, there are examples of high infection frequency in native habitats, which are most likely maintained by infected plants having a selective advantage over uninfected ones. The role that endophyte-mediated bioactivity has in providing that selective advantage for infection is not clear.

Insect Resistance

Epichloë endophytes are, in evolutionary terms, the forerunners of the asexual *Neotyphodium* endophytes. They colonize leaves and culms of vegetative parts of infected plants and form sexual structures (stromata) around developing inflorescences, which often partly or completely sterilize the host. In a survey of 14 species of grasses containing six species of *Epichloë* from natural sites in Switzerland, only one (*F. gigantea* infected with *E. festucae*) contained *N*-formylloline and *N*-acetylloline alkaloids, another (*F. rubra* infected with *E. festucae*) contained ergovaline, while six plant–endophyte associations contained low levels of peramine (Leuchtmann et al., 2000). Those plants that formed stroma were less likely to contain alkaloids compared with those with asymptomatic fungal infections.

Despite the apparent lack of bioactive alkaloids, there is some evidence that *Epichloë* endophytes may influence herbivory. Leaves sampled in the vicinity of stroma on *Agrostis heimalis* infected with *Epichloë* deterred feeding by fall armyworm (*Spodoptera frugiperda*), whereas nonstromal leaves had no such effect (White et al., 1993). These authors suggested that compounds produced by the stroma mycelium may have been responsible for the deterrence. In contrast to these results, tillers with stromata showed more leaf damage by unspecified microherbivores than vegetative or flowering tillers in wild populations of *Brachypodium sylvatica*, almost all of which were infected with the endophyte *E. sylvatica* (Brem and Leuchtmann, 2001). Similarly, field observations of a native North American grass, *Panicum agrostoides*, with a high frequency of infection by the endophytic fungus *Balansia henningsiana*, showed no differences between infected and uninfected plants in leaf area damaged by herbivores (Clay et al., 1989). Leaves of naturally and artificially infected *B. sylvatica* plants which are not known to produce any of the common alkaloids associated with endophyte infection did, however, reduce growth, development, and survival of fall armyworm with little evidence of feeding deterrency (Brem and Leuchtmann, 2001). The absence of alkaloids in stroma-forming infections and apparent attractiveness of tillers with stromata may relate to the dependency of *Epichloë* on a symbiotic fly for sexual reproduction (Leuchtmann et al., 2000), whereas the results with fall armyworm illustrate the potential of these endophytes to reduce herbivory in their natural habitats.

The endophyte *Neotyphodium starrii* is the most extensively studied of the asexual endophytes infecting native grasses, particularly in Arizona tall fescue (*F. arizonica*). The high incidence of infection of *F. arizonica* in its native habitat suggests there is a selective advantage for the infected state, although it is seldom that all tillers on infected plants contain endophyte. No biotic stress tolerance has been observed which could account for the high frequency of infection. Endophyte infection did not reduce consumption of clipped leaves by the redlegged grasshopper (*Melanoplus femurrubrum*) (Lopez et al., 1995) or by another native grasshopper, *Xanthippus corallipes* (Saikkonen et al., 1999), and had no effect on leaf cutting ant (*Acromyrmex versicolor*) queen survival, worker production, or size of fungal gardens (Tibbets and Faeth, 1999). In contrast to these studies, however, *N. starrii* in *Bromus anomalous* reduces feeding and oviposition by adult Argentine stem weevil (Bell and Prestidge, 1991) and growth, development, and survival of the larvae in laboratory studies (Bell and Prestidge, 1999). *Neotyphodium starrii* in *B. anomalous* produces up to 50 μg g^{-1} of peramine, but negligible amounts of lolitrem B and no ergovaline or lolines (Bell and Prestidge, 1999; Siegel et al., 1990). Only low concentrations of peramine (<1.9 μg g^{-1}) have been found in *F. arizonica* infected with *N. starrii* (Saikkonen et al., 1999).

In other examples of native grass endophytes affecting insect herbivores, an Australasian grass (*Echinopogon ovatus*) infected with an endophytic fungus serologically related to *N. lolii* also deters feeding by Argentine stem weevil

(Miles et al., 1998). This endophyte does not synthesize peramine but produces the alkaloids paxilline and *N*-formylloline. Another *Bromus* species, *B. setifolius*, in its native South American habitat is infected by the fungus *N. tembladerae*, which produces both peramine and ergovaline but not lolitrem B (White et al., 2001). These authors found a high frequency of infection in plant communities inhabited by leaf cutting ants and demonstrated a feeding preference by fall armyworm for endophyte-free (E–) over endophyte-infected (E+) plants. Bazely et al. (1997) reported that leaves from *F. rubra* plants containing an endophyte (probably an anamorph of *E. festucae*), which produces ergovaline and peramine, deterred feeding and reduced survival of locusts (*Locusta migratoria*). In studies of endophyte associations with wild barley (*Hordeum* spp.), populations of Russian wheat aphid (*Diuraphis noxia*) declined on two of four accessions containing endophytes (Clement et al., 1997). The two accessions showing resistance to aphids produced *N*-formylloline and ergovaline, while the two accessions that had no effect have not been found to produce these alkaloids (TePaske et al., 1993).

Disease Resistance

Studies on antifungal properties of endophytes in native grasses are relatively scarce but suggest, as for insects, that the presence of antifungal compounds may be a common feature of biotic responses of these endophyte/grass associations. In separate studies, for instance, two *Balansia* spp. have been found to have antifungal properties (Clay et al., 1989; Stovall and Clay, 1991) and timothy (*Phleum pratense*) infected with *E. typhinum* had greater resistance to *Cercospora* leaf spot than E– timothy (Koshino et al., 1987). Recently, Yue et al. (2000a) published an extensive study of 23 endophyte–plant combinations in a range of grass species, including *Agrostis*, *Festuca*, *Lolium*, and *Poa*. Of these, 22 showed some antifungal activity against *Cryphonectria parasitica*, with the greatest activity shown by *E. festucae* and *N. tembladerae*. Fungal inhibitors identified from cultures of *E. festucae* included indole derivatives, a sesquiterpene, and a diacetamide.

In a study of *Neotyphodium* spp. infecting native grasses in China, the percentage of fungi isolated from leaves of infected *Agropyron cristatum*, *Elymus cylindricus*, and *F. rubra* was lower than that isolated from equivalent E– plants (Nan and Li, 2000). Fungi isolated were mainly species of *Alternaria*, *Cladosporium*, and *Fusarium*. These authors also showed that spore suspensions of *A. alternata* and several *Fusarium* species applied to leaves of *E. cylindricus* caused lesions regardless of the endophyte status of the plants but, with the exception of the pathogen *F. equiseti*, significantly fewer and smaller lesions developed on E+ than on E− leaves.

INVERTEBRATE RESPONSES IN FORAGE AND TURF GRASSES

Invertebrate Resistance and Wild-Type Endophytes

Wild-type endophytes in tall fescue grown for forage and turf uses in the USA produce the alkaloids peramine, ergovaline, and three loline derivatives, while those in perennial ryegrass in New Zealand, Australia, and the USA produce peramine, ergovaline, and lolitrem B. Previous reviews have outlined the bioactivity that endophytes in these grasses show against a diverse range of invertebrate herbivores (Bacon et al., 1997; Breen, 1994; Clement et al., 1994; Popay and Rowan, 1994). Included in that list are economically important foliar-chewing lepidopteran pests such as fall armyworm and sod webworm (*Parapediasia* spp.); stem-boring larvae of the bluegrass billbug (*Sphenophorus parvulus*) and Argentine stem weevil; pests that suck plant sap near the base of the plant such as hairy chinch bug (*Blissus leucopterus hirtus*) and pasture mealybug (*Balanococcus poae*); and other plant suckers such as leafhoppers and aphids (Homoptera: Aphididae). Some insect pests of ryegrass such as frit fly (*Oscinella frit*) and leatherjackets (*Tipula* spp.) are unaffected by endophyte (Lewis and Clements, 1986; Lewis and Vaughan, 1997).

The response of insects to endophyte infection in tall fescue and ryegrass depends, to a degree, on the part of the plant on which they feed. Those insects which defoliate grasses seldom severely damage their hosts and are not consistently affected by endophyte infection. Weight gain and development rate of fall armyworm can be reduced by E+ tall fescue (Breen, 1994; Clement et al., 1994; Popay and Rowan, 1994), with the magnitude of negative effects varying with particular grass–endophyte associations (Braman et al., 2002). Other research, however, has shown either no major effects of *N. coenophialum* infection on this insect (e.g., Breen, 1993a) or, in some cases, increased larval mass and accelerated development on E+ plants (Bultman and Conard, 1998; Bultman and Bell, 2003). Effects of endophyte infection on fall armyworm may be age related, with neonates exhibiting a preference for uninfected leaves whereas fourth-instars consumed equal amounts of infected and uninfected material. Larval age also influences the response of the armyworm species *Pseudaletia unipunctata*

(Eichenseer and Dahlman, 1993) and *P. convecta* (McDonald et al., 1993) to endophyte. Some lepidopteran insects such as black cutworm (*Agrotis ipsilon*) (Williamson and Potter, 1997) and common cutworm (*A. infusa*) (McDonald et al., 1993) are either not affected or only mildly affected by endophyte. Similarly, two leaf-feeding Orthoptera in Australia, the black field cricket (*Teleogryllus commodus*) and the Queensland field cricket (*T. oceanicus*) show some feeding preference for E− perennial ryegrass in early instars but not in later instars (McDonald et al., 1993). Notably, E+ tall fescue deters feeding by bluegrass webworm (*P. teterella*) larvae and reduces their survival (Koga et al., 1997), although activity appears to be confined to the leaf sheaths (Kanda et al., 1994). Endophyte infection of turf perennial ryegrass cultivars, however, may alter behavior of this insect as emigration increased with the proportion of E+ perennial ryegrass in greenhouse pots (Richmond and Shetlar, 1999).

Insects that feed near the base of the plant in the region of the meristem inflict more severe damage on plants than leaf grazers and appear to be more sensitive to the presence of endophyte. Damage to perennial ryegrass by Argentine stem weevil larvae (Popay et al., 1999) and populations of pasture mealybug in tall fescue (Pennell and Ball, 1999) and perennial ryegrass (Popay et al., 2000a; C. Pennell, A.J. Popay, O.J.-P. Ball, and D.E. Hume et al., 2004, unpublished data) are dramatically reduced by the presence of endophyte in the field. Bluegrass billbug damage can be severe in Kentucky bluegrass (*Poa pratensis*) and E− tall fescue turf. Overseeding E+ perennial ryegrass into Kentucky bluegrass may be a useful biocontrol option to reduce damage by this insect (Richmond et al., 2000). Black beetle adults, which shred the base of tillers, are also highly sensitive to endophyte infection in both tall fescue (Popay, 2004) and perennial ryegrass (Ball et al., 1994, 1997a). Population densities of chinch bugs that feed near the crown of plants are also much lower in E+ than in E− tall fescue and ryegrass. The highly mobile third-instar chinch bug larvae avoid toxic E+ turf grass cultivars (Carriere et al., 1998) and emigrate more quickly from 100% E+ perennial ryegrass stands than from those with lower infection levels (Richmond and Shetlar, 2000).

The homopteran plant-sucking insects, such as aphids and leafhoppers, are similar to the leaf grazers in that they exhibit diverse responses to endophyte infection in tall fescue and ryegrass, depending on the insect species as well as the endophyte–plant association (Breen, 1993b; Clement et al., 1994; Davidson and Potter, 1995; Popay and Rowan, 1994). Survival of the bird cherry-oat aphid (*Rhopalosiphum padi*) is reduced by *N. coenophialum* in tall fescue but not by *N. lolii* in perennial ryegrass, whereas other aphid species such as greenbug (*Schizaphis graminum*) are affected by both these endophyte–grass associations (Breen, 1993b). Likewise, endophyte-mediated resistance to the Russian wheat aphid has been demonstrated in a wide range of grass cultivars in both perennial and annual ryegrass (*L. multiflorum*) and tall fescue (Clement et al., 1996). Leaf age and senescence and concentration of endophyte are other factors affecting aphid performance on E+ plants (Breen, 1993b).

Although fungal growth does not typically extend into the roots, root-feeding invertebrates are affected by endophyte infection in fescues and also in ryegrass, though generally to a lesser extent in the latter. Mode of feeding is an important determinant of the strength of the effects. Survival of root-feeding aphids such as *Aploneura lentisci*, like their foliar counterparts, is severely impaired by endophyte infection in tall fescue and meadow fescue (*F. pratensis*) (Schmidt and Guy, 1997). For root-chewing scarab larvae (white grubs), however, evidence for adverse effects of endophyte infection are more equivocal. Endophyte infection has reduced survival and weight gain of white grubs in greenhouse studies (Potter et al., 1992) and has increased tolerance to herbivory by *Cyclocephala lurida* in rooting boxes (Crutchfield and Potter, 1995). Survival of *Exomala orientalis* was reduced in E+ tall fescue, but there was no effect of endophyte in strong creeping red fescue (*F. rubra* L. *rubra*) on survival and weight gain of either this insect or of *P. japonica* in greenhouse pots (Koppenhoffer et al., 2003). In the field, clear evidence for effects of endophyte on scarab larvae has been difficult to obtain. Murphy et al. (1993) found fewer white grubs in E+ than in E– tall fescue plots in one of two field trials but in several other trials, E+ tall fescue exerted no measurable effect on larval populations of Japanese beetle (*Popillia japonica*) (Davidson and Potter, 1995; Potter et al., 1992). Koppenhoffer et al. (2003) found higher populations of white grubs in E+ than in E– plots of tall fescue. In contrast with these studies, meadow fescue infected with its natural endophyte, *N. uncinatum*, was strongly deterrent to the native New Zealand grass grub, *Costelytra zealandica*, in laboratory and pot trials. This was reflected in marked differences in herbage yield and visible damage between E+ and E– swards in the field (Popay et al., 2003a). In perennial ryegrass, Prestidge and Ball (1993) demonstrated significant reductions in weight gain of second-instar larvae by *N. lolii* infection of ryegrass in laboratory experiments, but found no effects of infection on growth and survival of third-instar grubs in pot trials or on population densities in the field.

Endophyte infection of tall fescue has also been associated with reductions in populations of some endo- and ecto- plant-parasitic nematode species in the field and in pot trials, while other species have been unaffected (Elmi et al.,

2000). The wild-type endophyte in New Zealand ryegrass can also reduce populations of plant-parasitic nematodes (Stewart et al., 1993; Eerens et al., 1998) albeit not consistently (Watson et al., 1995).

Seed Feeding

Protection of seed is an important but often overlooked component of biotic responses of endophytic grasses. These seeds contain high concentrations of both hyphae and alkaloids which could be expected to deter feeding by seed predators and inhibit fungal infection of the seed. The presence of *N. coenophialum* in tall fescue seed reduces survival and population growth of two stored grain pests, the confused flour beetle (*Tribolium confusum*) (Yoshimatsu et al., 1998) and the sawtoothed grain beetle (*Oryzaephilus surinamensis*) (Yoshimatsu et al., 1999). In a field test, E– ryegrass seed disappeared at a faster rate than infected seeds, but the seed predator was not identified (Popay et al., 2000b). Birds were not responsible, although they are able to discriminate between E+ and E– seed (Madej and Clay, 1991; Pennell and Rolston, 2003). Consumption of seed by mice (Barger and Tannenbaum, 1998) and various earthworm species (Baltus, 2001) is not affected by endophyte.

Endophyte Diversity and Biotic Response

The key to exploiting endophytes for biotic stress protection in turf and forage grasses without negative impacts on grazing animals lies in the diversity of endophytes that exist in *Lolium* and *Festuca* species in natural habitats of these grasses. Seed from plants of these species were collected mainly in Europe and screened for the presence of the known classes of compounds in order to identify strains that lack the mammalian toxins, ergot alkaloids in tall fescue and ryegrass and lolitrem B in the latter (Tapper and Latch, 1999). The majority of collections screened in New Zealand had chemical profiles similar to the wild-type endophytes, suggesting that these strains of endophyte predominate in naturalized and wild populations of tall fescue and perennial ryegrass. Nevertheless, there were strains which differed from wild-type endophytes in producing different combinations of the known alkaloids. Comparative assays conducted with insects on some of these strains in planta have identified potentially useful strains.

In tall fescue, the introduction of endophytes that do not synthesize ergot alkaloids and produce a different spectrum of loline derivatives appears to have consequences for the spectrum of invertebrate herbivore resistance. Such is the case for a nontoxic endophyte AR542 (MaxQ) introduced into tall fescue cultivars in the USA. Tall fescue infected with AR542, like wild-type-infected tall fescue, is strongly resistant to pasture mealybug (Pennell and Ball, 1999) and deterrent to black beetle adults (Popay, 2004). Damage by larval black beetle is greatly reduced in the field by the presence of AR542 in tall fescue (Wheatley et al., 2003), despite the absence of ergovaline, the only compound shown to have bioactivity against this insect (Ball et al., 1997a). On the other hand, AR542 does not confer resistance to lesion nematodes (*Pratylenchus* spp.) (Timper and Bouton, 2004) and shows less resistance to the bird cherry-oat aphid than wild-type (Hunt and Newman, 2004).

In New Zealand, endophytes deficient in lolitrem B but producing ergovaline and peramine gave strong resistance to Argentine stem weevil in perennial ryegrass (Popay et al., 1995). Subsequently, the release of perennial ryegrass cultivars infected with the endophyte AR1, producing peramine but neither lolitrem B nor ergovaline, was a major step forward in overcoming the ruminant toxicity associated with the wild-type endophytes. Ryegrass infected with AR1 has strong resistance to Argentine stem weevil (Popay et al., 1999) and pasture mealybug (Popay et al., 2000a; C. Pennell, A.J. Popay, O.J.-P. Ball, and D.E. Hume et al., 2004, unpublished data). Surprisingly, AR1 also has moderate activity against black beetle (Popay and Baltus, 2001) but, unlike the wild-type endophyte, AR1-infected ryegrass has little or no effect on porina (*Wiseana cervinata*), a New Zealand native foliar feeding caterpillar (Jensen and Popay, 2004) and shows increased susceptibility to the root aphid, *A. lentisci*, in pot trials (Popay et al., 2004).

In the search for alternative endophytes, a strain known as AR37 that produced none of the known common alkaloids was isolated from wild perennial ryegrass in France. This endophyte has no effect on Argentine stem weevil adults, unlike ryegrass with the peramine-producing endophytes, but has potent activity against the larvae, reducing damage to low levels (Popay and Wyatt, 1995). AR37 also reduces black beetle adult feeding (as Lp14 in Ball et al., 1994), pasture mealybug populations (C. Pennell, A.J. Popay, O.J.-P. Ball, and D.E. Hume et al., 2004, unpublished data), and survival of porina (Jensen and Popay, 2004). The most unusual feature of AR37, compared with the wild-type and AR1 endophytes in ryegrass, is the almost complete elimination of the root aphid, *A. lentisci* (Popay et al., 2004).

Other studies have similarly explored the biotic responses of a diverse range of endophytes. Resistance to fall armyworm is possible in the absence of undesirable alkaloids (Jones et al., 1997). Screening of several endophyte strains lacking the ability to produce lolitrem B identified one that had deleterious effects on the nematode *Meloidogyne marylandi* (Ball et al., 1997b). Clement et al. (2001) screened a range of endophyte accessions in tall fescue from the Mediterranean region and found that the majority were resistant to the bird cherry-oat aphid.

PLANT PATHOGEN RESPONSES IN TURF AND FORAGE GRASSES

Antifungal activity of endophyte isolates from both *Lolium* and *Festuca* has been demonstrated in vitro by several researchers (e.g., Christensen, 1996), but demonstration of in planta effects has proven to be more elusive. Vincelli and Powell (1991) reported that 'Manhattan II' turf perennial ryegrass had significantly less red thread disease (caused by *Laetisaria fuciformis*) when endophyte infection levels were high than when they were low. Endophyte infection also enhanced resistance to red thread disease in strong creeping and Chewings fescue, although the response varied with the host plant genotype and endophyte combination (Bonos et al., 2003). In tall fescue infected with two isolates of *Rhizoctonia zeae*, positive interactions between seedling survival and endophyte infection were reported (Gwinn and Gavin, 1992), but progress of *Rhizoctonia* blight or crop recovery was not altered by *N. coenophialum* in tall fescue in the field (Burpee and Bouton, 1993). Using a plant tissue assay, Christensen (1996) showed that meadow fescue infected with some *Neotyphodium* or *Epichloë* isolates inhibited growth of two plant pathogens, *Drechslera erythrospira* and *R. zeae*, but this activity was not necessarily the same as that shown in in vitro assays.

The mechanism of endophyte-enhanced resistance to fungal pathogens has not been elucidated. Epiphyllous mycelium was observed on leaf surfaces of *B. setifolius*, *F. ovina*, *F. rubra*, and *Poa ampla* (Moy et al., 2000) and, in the case of *P. ampla*, was identified as the endophyte *N. typhinum*. Mycelial nets such as these may act as competitors for space and infection pathways of plant pathogens. In addition, an endophytic β-1,6 glucanase, identified by Moy et al. (2002), could potentially function in cell wall degradation of pathogenic fungi.

Virus transmission in tall fescue can be reduced where endophyte infection interferes with survival of the virus vector (Mahmood, 1993). In roadside tall fescue in Tasmania, neither the incidence of *Barley yellow dwarf virus* (BYDV) nor the occurrence of the virus vector, *R. padi*, differed between E+ and E- plants (Guy and Davis, 2002). No effects of endophyte infection on incidence of

BYDV in ryegrass (Hesse and Latch, 1999) or on growth response of ryegrass plants infected with virus (either BYDV or *Ryegrass mosaic virus*) (Lewis, 2004) have been found.

MULTITROPHIC INTERACTIONS

Complex interactions not only occur between the fungal endophyte, its host grass, and insect herbivores, but also at higher trophic levels. In an interesting example of such interactions, fourth-instar black cutworm larvae feeding on perennial ryegrass with a prevalence of endophyte had a lower susceptibility to the entomopathogenic nematode, *Steinernema carpocapsae*, than those feeding on predominantly E– plants (Kunkel and Grewal, 2003). Infection status of perennial ryegrass did not appear to affect the ability of *S. carpocapsae* to attach or penetrate and subsequently develop into adults, but the alkaloid ergocristine reduced the infectivity of *S. carpocapsae* and the growth and pathogenicity of *Xenorhabdus nematophila*, the symbiotic bacterium carried by the nematodes (Kunkel et al., 2004).

In contrast to the above example, Grewal et al. (1995) observed increased susceptibility of *P. japonica* larvae to the entomopathogenic nematode *Heterorhabditis bacteriophora* from E+ plants of tall fescue and Chewings fescue (*F. rubra* subsp. *falax*) under greenhouse conditions. Koppenhoffer and Fuzy (2003) also found that tall fescue endophytes under greenhouse conditions had a weak and variable enhancing effect on nematode susceptibility of second- and third-instar *E. orientalis*, but had no effect on nematode susceptibility of third-instar *Cyclocephala borealis* and *P. japonica*. Under field conditions with natural white grub populations, however, E+ tall fescue had no significant effects on nematode efficacy against *E. orientalis*, *P. japonica*, and *C. borealis*. *Popillia japonica* larvae also do not have increased susceptibility to *Paenibacillus popilliae* when confined to E+ compared with E– tall fescue (Walston et al., 2001).

Endophyte effects on other plant competitors have also been evaluated with varying results. Clay and Holah (1999) reported that E+ tall fescue reduced species diversity in a long-term field experiment. They speculated that biotic or abiotic protection provided by the endophyte or suppression of mychorrizal associations may be responsible for the dominance of endophytic plants. Differential herbivory affected the outcome of plant competition between Kentucky bluegrass and infected tall fescue and perennial ryegrass with the latter producing nearly twice the biomass of uninfected plants when herbivores were present but a similar biomass in the absence of herbivory (Clay et al., 1993). Allelo-pathic effects of extracts from endophyte-infected ryegrass (Sutherland et al., 1999) and tall fescue (Springer, 1996) on clover have been reported and cannot be excluded as mechanisms behind the dominance of E+ plants in some plant communities.

Not all plant competitive interactions favor endophytic plants. In competition with dandelion (*Taraxacum officinale* aggr.), endophyte infection reduced perennial ryegrass biomass but increased that of tall fescue, independently of any effects of belowground herbivory (Richmond et al., 2004). Endophyte also reduced the ability of perennial ryegrass to compete with a fast growing warm-season grass, *Digitaria sanguinalis*, in newly established stands, suggesting that harboring the fungus may impose a physiological cost to the plant which can reduce its competitiveness under some circumstances (Richmond et al., 2003).

FACTORS AFFECTING BIOTIC RESPONSES

The distribution and concentration of alkaloids within plants are critical factors in determining endophyte-mediated resistance to herbivores. The particular alkaloids produced by *Neotyphodium* fungi are a characteristic of each different strain, although several factors modify the quantities that are produced (Clay and Schardl, 2002; Lane et al., 2000; Siegel and Bush, 1996). Plant genotype, tissue type, season, and various abiotic environmental conditions can all influ-ence alkaloid concentration in planta and associated biotic responses. Location of alkaloids within plants appears to be mainly an attribute of the compounds themselves, with the lolines, ergot alkaloids, and lolitrems occurring in the highest concentrations, in conjunction with that of the fungal mycelium, in the leaf sheaths, crown, developing inflorescences, and seed (Ball et al., 1997c; Keogh et al., 1996; Lane et al., 2000; Siegel and Bush, 1996). Unlike ergot alkaloids and the lolitrems, the loline alkaloids are also freely distributed into leaf lamina and, to a lesser extent, to the roots (Siegel and Bush, 1996). Peramine occurs in higher concentrations in the leaf lamina than the leaf sheaths in perennial ryegrass but only trace amounts have been recorded in the roots (Ball et al., 1997d).

Mechanisms

All four major classes of compounds isolated from E+ tall fescue and ryegrass have activity against insect herbivores. Ergovaline, lolitrem B, and the lolines, which are concentrated in the leaf sheath and crown where the strongest anti-insect activity of endophytic plants is observed, exhibit both antibiosis and antixenosis. The lolines in particular have broad spectrum toxic and deterrent activity against insects (Siegel and Bush, 1996) including aphids (Wilkinson et al., 2000), and white grubs (Patterson et al., 1991; Popay and Lane, 2000). Ergovaline can be toxic and/or deterrent to fall armyworm larvae, Japanese beetle larvae, and black beetle adults, while lolitrem B reduces growth and development of Argentine stem weevil larvae but has been tested against few other insects (Lane et al., 2000; Siegel and Bush, 1996). Peramine, on the other hand, is only known as a strong deterrent to Argentine stem weevil, although it has been associated with in planta inhibition of the greenbug. This alkaloid has not been shown to have major effects on other insects, suggesting it functions primarily as a marker for the presence of endophyte. Thus, detection of endophyte by Argentine stem weevil adults via the presence of peramine allows this insect to avoid exposure of the larvae to the toxic effects of lolitrem B.

Mechanisms behind suppression of nematode populations on E+ tall fescue are not understood, but probably relate to the mode of feeding of each species (Bernard et al., 1997). Changes in root morphology in the form of thickening of cell walls (Gwinn and Bernard, 1993) and increased chitinase activity (Roberts et al., 1992) are two possible mechanisms. Endophyte-produced alkaloids may also be toxic or deterrent to nematodes.

Alkaloids are always present in infected plants, albeit in variable concentrations, suggesting they are constitutive. Nevertheless, there is also evidence that endophytes may mediate an induced host response since fall armyworm pupae weighed substantially less after eating previously clipped, compared to un-clipped, E+ perennial ryegrass (Bultman and Ganey, 1995). Furthermore, it is becoming apparent as the biotic responses to more diverse endophytes are investigated that other secondary metabolites may also have a bioactive role in infected grasses. The AR37 endophyte in ryegrass, which produces none of the common alkaloids but has broad spectrum activity against insects, is probably the best example of this. Janthitrems have been isolated from AR37-infected ryegrass (Tapper and Lane, 2004), but their role in the bioactivity of this endophyte has yet to be elucidated. In the absence of peramine, however, adult Argentine stem weevil feed and oviposit freely on AR37-infected plants, and it is only the larvae that are severely affected (Popay and Wyatt, 1995). The AR1 endophyte is another case in point, since the mild deterrent activity against black beetle (*H. arator*) adults occurs in the absence of ergovaline, the only alkaloid known to affect this insect (Popay and Baltus, 2001).

Plant Genotype

Plant genotype plays an integral role in determining the outcome of the host plant–endophyte–insect interactions, primarily by modifying the expression of alkaloids produced in E+ ryegrass (Ball et al., 1995; Easton et al., 2002) and tall fescue (Hiatt and Hill, 1997). Prestidge and Ball (1995) suggested exploiting this inherent variation in alkaloid production to minimize the effects of endophyte infection on grazing animals by selecting for plant genotype–endophyte associations with low concentrations of toxic alkaloids. The success of such an option depends on identifying and maintaining alkaloid concentrations at the critical levels required for insect resistance. For instance, resistance to chinch bug was not compromised in different host plant genotypes expressing different levels of alkaloids in red and Chewings fescue (Yue et al., 2000b), which would allow selection of genotypes expressing low alkaloid concentrations to reduce toxicity to livestock. Variability in response to insects seems particularly apparent in plants hosting endophytes that have no toxicity to grazing animals. AR1 in ryegrass has highly variable effects on black beetle (Easton et al., 2000) and root aphid (*A. lentisci*) (Popay et al., 2004) that are related to plant genotype. The nontoxic endophyte, AR584, in tall fescue exhibits similar variability, conferring resistance to lesion nematodes in the tall fescue cultivar Georgia 5 but not in Jesup (Timper and Bouton, 2004). In these cases, understanding the reasons for the variation may allow selection for plant–endophyte associations with more robust resistance to pests.

Endophyte infection may enhance rather than inhibit insect performance in some plant genotypes (Bultman and Bell, 2003; Lopez et al., 1995; Popay et al., 2004; Saikkonen et al., 1999; Tibbets and Faeth, 1999). In these instances, plant chemistry factors may be altered by the presence of the endophyte in a way that improves individual insect fitness, where there is either no effect of the alkaloids on the insects concerned or the effects have been negated by increased host plant quality. Similarly, a tendency for E+ tetraploid and hybrid ryegrasses (*L. boucheanum* syn. *L. hybridum*) to sustain more damage by Argentine stem weevil than their diploid perennial counterparts was attributed not only to host genotype influencing the expression of peramine, but also to factors that increased the susceptibility of the tetraploid and hybrid plants to this insect, irrespective of the presence of endophyte (Popay et al., 2003b).

Environment

Although alkaloid levels fluctuate in response to environmental factors, few studies have investigated how this interacts with biotic effects of endophyte infection. Seasonal changes in endophyte concentration and alkaloid levels in perennial ryegrass (Ball et al., 1995) may reduce the effectiveness of the endophyte-mediated resistance to Argentine stem weevil (Popay and Mainland, 1991). Temperature and seasonal factors also interact with endophyte to alter the level of resistance in E+ ryegrass to the greenbug (Breen, 1992). In interactions between drought and infection status of tall fescue, reproduction of bird cherry-oat aphid was reduced by endophyte infection per se, but only by drought on E– plants and not on E+ plants (Bultman and Bell, 2003). In the same study, endophyte infection increased growth of fall armyworm in the absence of drought stress, whereas the reverse occurred in droughted plants. Plant nutrient status may also alter relative differences in insect performance between E– and E+ plants. Negative effects of endophyte in tall fescue on fall armyworm pupal mass were more pronounced with plants grown under low than under high fertility (Bultman and Conard, 1998). Elevated CO_2 reduced food conversion efficiency of fall armyworm fed on E+ tall fescue more than it did for those larvae fed E– tall fescue (Marks and Lincoln, 1996). In a further example of complex interactions, the feeding preferences of the locust *Schistocerca gregaria* for E– meadow fescue were altered by UV-B radiation with the direction of those changes dependent on whether comparisons were made with the ambient or the UV-A control (McLeod et al., 2001).

SUMMARY

The defensive mutualism hypothesis advanced by Clay (1988) suggests that benefits accrue to both plant and fungus as a result of protection from herbivory arising from the production of secondary metabolites by the fungus. Faeth and Sullivan (2003) and Saikkonen et al. (1998) have challenged this theory, arguing quite rightly that much of the evidence for it has come from research on cultivated tall fescue and ryegrass which have been artificially selected for maximum plant benefit. In this review we have endeavored to show that nature has produced a diversity of endophytes in grasses and, in concert with that, diversity in the dynamics of the symbiosis. Thus, although our knowledge of native grasses is still limited, there is evidence that species of *Epichloë* and *Neotyphodium* may provide their hosts, including tall fescue and ryegrass, with biotic protection in natural habitats. In some cases, such as for the wild-type endophytes in tall fescue and ryegrass, and for AR37 in the latter, selection pressures have resulted in the production of diverse alkaloids and a broad spectrum of biotic stress tolerance. In other situations, such as with *F. arizonica*, biotic responses appear to be largely absent.

Protection of plant and fungus is paramount where biotic stress tolerance has an integral role in the symbiosis. The nature of the secondary metabolites, their distribution, and that of the fungus within the plant are crucial to evaluating the importance of biotic responses in endophytic grasses. To be cost effective, the fungus is less likely to provide protection against herbivory that does not threaten its survival. Thus, it is no coincidence that the fungus is concentrated at the base of the plant where it is not only less accessible to herbivores but where secondary metabolites can be concentrated in order to protect both the fungus and the plant meristematic tissue from herbivores that feed in this region. Likewise, protection of developing inflorescences and seed is essential for survival of both fungus and plant.

The mechanisms underlying defense responses of endophytic grasses are not yet well understood, but are dependent on endophyte strain, secondary metabolite production, and specific interactions between fungus and plant genotype. Among the diversity of endophytes, there are strains that harbor new and highly bioactive compounds. As we gain more knowledge of these strains and their effects on insects and pathogens, there is more scope for combining the right fungal isolate with a complementary host genotype to target bioactivity and optimize plant performance of turf and forage grasses.

REFERENCES

Bacon, C.W., M.D. Richardson, and J.F. White, Jr. 1997. Modification and uses of endophyte-enhanced turfgrasses: A role for molecular technology. Crop Sci. 37:1415–1425.

Ball, O.J.-P., G.M. Barker, R.A. Prestidge, and D.R. Lauren. 1997d. Distribution and accumulation of the alkaloid peramine in *Neotyphodium lolii*-infected perennial ryegrass. J. Chem. Ecol. 23:1419–1434.

Ball, O.J.-P., G.M. Barker, R.A. Prestidge, and J.M. Sprosen. 1997c. Distribution and accumulation of the mycotoxin lolitrem B in *Neotyphodium lolii*-infected perennial ryegrass. J. Chem. Ecol. 23:1435–1449.

Ball, O.J.-P., E.C. Bernard and K.D. Gwinn. 1997b. Effect of selected *Neotyphodium lolii* isolates on root-knot nematode (*Meloidogyne marylandi*) numbers in perennial ryegrass. Proc. N. Z. Plant Protection Conf. 50:65–68.

Ball, O.J.-P., M.J. Christensen, and R.A. Prestidge. 1994. Effect of selected isolates of *Acremonium* endophyte on adult black beetle (*Heteronychus arator*) feeding. Proc. N. Z. Plant Protection Conf. 47:227–231.

Ball, O.J.-P., C.O. Miles, and R.A. Prestidge. 1997a. Ergopeptine alkaloids and *Neotyphodium lolii*-mediated resistance in perennial ryegrass against adult *Heteronychus arator* (Coleoptera: Scarabaeidae). J. Econ. Entomol. 90:1382–1391.

Ball, O.J.-P., R.A. Prestidge, and J.M. Sprosen. 1995. Interrelationships between *Acremonium lolii*, peramine, and lolitrem B in perennial ryegrass. Appl. Environ. Microbiol. 61:1527–1533.

Baltus, J.G. 2001. Earthworm feeding on perennial ryegrass seed and seedlings and the effect of endophyte (abstr.). N. Z. Plant Prot. 54:250.

Barger, J.L., and M.G. Tannenbaum. 1998. Consumption of endophyte-infected fescue seeds and osmoregulation in white-footed mice. J. Mammol. 79:464–474.

Bazely, D.R., M. Vicari, S. Emmerich, L. Filip, D. Lin, and A. Inman. 1997. Interactions between herbivores and endophyte-infected *Festuca rubra* from the Scottish islands of St Kilda, Benbecula and Rum. J. Appl. Ecol. 34:847–860.

Bell, N.L., and R.A. Prestidge. 1991. The effects of the endophytic fungus *Acremonium starii* on feeding and oviposition of the Argentine stem weevil. Proc. N. Z. Weed Pest Control Conf. 44:181–184.

Bell, N.L., and R.A. Prestidge. 1999. The effects of the endophytic fungus *Neotyphodium starii* on larval development of *Listronotus bonariensis* (Coleoptera: Curculionidae). Proc. Australasian Grassl. Invert. Ecol. Conf. 7:279–285.

Bernard, E.C., D. Gwinn Kimberly, C.D. Pless, and C.D. Williver. 1997. Soil invertebrate species diversity and abundance in endophyte-infected tall fescue pastures. p. 125–135. *In* C.W. Bacon and N.S. Hill (ed.) *Neotyphodium*/grass interactions. Plenum Press, New York.

Bonos, S.A. M.M. Mohr, W.A. Meyer, and C.R. Funk. 2003. The influence of fungal endophyte on red thread resistance in fine fescue. 2003 Annual meeting abstracts. ASA, CSSA, and SSSA, Madison, WI.

Braman, S.K., R.R. Duncan, M.C. Engelke, W.W. Hanna, K. Hignight, and D. Rush. 2002. Grass species and endophyte effects on survival and development of fall armyworm (Lepidoptera: Noctuidae). J. Econ. Entomol. 95:487–492.

Breen, J.P. 1992. Temperature and seasonal effects on expression of *Acremonium* endophyte-enhanced resistance to *Schizaphis graminum* (Homoptera: Aphididae). Environ. Entomol. 21:68–74.

Breen, J.P. 1993a. Enhanced resistance to fall armyworm (Lepidoptera: Noctuidae) in *Acremonium* endophyte-infected turfgrasses. J. Econ. Entomol. 86:621–629.

Breen, J.P. 1993b. Enhanced resistance to three species of aphids (Homoptera: Aphididae) in *Acremonium* endophyte-infected turfgrasses. J. Econ. Entomol. 86:1279–1286.

Breen, J.P. 1994. *Acremonium* endophyte interactions with enhanced plant resistance to insects. Annu. Rev. Entomol. 39:401–423.

Brem, D., and A. Leuchtmann. 2001. *Epichloë* grass endophytes increase herbivore resistance in the woodland grass *Brachypodium sylvaticum*. Oecologia 126:522–530.

Bultman, T.L., and G.D. Bell. 2003. Interaction between fungal endophytes and environmental stressors influences plant resistance to insects. Oikos 103:182–190.

Bultman, T.L., and N.J. Conard. 1998. Effects of endophtic fungus, nutrient level, and plant damage on performance of fall armyworm (Lepidoptera: Noctuidae). Environ. Entomol. 27:631–635.

Bultman, T.L., and D.T. Ganey. 1995. Induced resistance to fall armyworm (Lepidoptera: Noctuidae) mediated by a fungal endophyte. Environ. Entomol. 24:1196–1200.

Burpee, L.L., and J.H. Bouton. 1993. Effect of eradication of the endophyte *Acremonium coenophialum* on epidemics of *Rhizoctonia* blight in tall fescue. Plant Dis. 77:157–159.

Carriere, Y., A. Bouchard, S. Bourassa, and J. Brodeur. 1998. Effect of endophyte incidence in perennial ryegrass on distribution, host-choice and performance of the hairy chinch bug (Hemiptera: Lygaeidae). J. Econ. Entomol. 91:324–328.

Christensen, M.J. 1996. Antifungal activity in grasses infected with *Acremonium* and *Epichloë* endophytes. Australasian Plant Pathol. 25:186–191.

Clay, K. 1988. Clavicipitaceous fungal endophytes of grasses: Coevolution and the change from parasitism to mutualism. p. 79–105. *In* K.A. Pirozynski and D.L. Hawksworth (ed.) Coevolution of fungi with plants and animals. Academic Press, London.

Clay, K., G.P. Cheplick, and S. Marks. 1989. Impact of the fungus *Balansia henningsiana* on *Panicum agrostoides*: Frequency of infection, plant growth and reproduction, and resistance to pests. Oecologia 80:374–380.

Clay, K., and J. Holah. 1999. Fungal endophyte symbiosis and plant diversity in successional fields. Science (Washington, DC) 285:1742–1744.

Clay, K., S. Marks, and G.P. Cheplick. 1993. Effects of insect herbivory and fungal endophyte infection on competitive interaction among grasses. Ecology 74:1767–1777.

Clay, K., and C. Schardl. 2002. Evolutionary origins and ecological consequences of endophyte symbiosis with grasses. Am. Nat. 160:S99–S127.

Clement, S.L., L.R. Elberson, N.N. Youssef, C.M. Davitt, and R.P. Doss. 2001. Incidence and diversity of *Neotyphodium* fungal endophytes in tall fescue from Morocco, Tunisia, and Sardinia. Crop Sci. 41:570–576.

Clement, S.L., W.J. Kaiser, and H. Eichenseer. 1994. *Acremonium* endophytes in germplasms of major grasses and their utilisation for insect resistance. p. 185–200. *In* C.W. Bacon and J.F. White (ed.) Biotechnology of endophytic fungi of grasses. CRC Press, Boca Raton, FL.

Clement, S.L., D.G. Lester, A.D. Wilson, R.C. Johnson, and J.H. Bouton. 1996. Expression of Russian wheat aphid (Homoptera: Aphididae) resistance in genotypes of tall fescue harboring different isolates of *Acremonium* endophyte. J. Econ. Entomol. 89:766–770.

Clement, S.L., A.D. Wilson, D.G. Lester, and C.M. Davitt. 1997. Fungal endophytes of wild barley and their effects on *Diuraphis noxia* population development. Entomol. Exp. Appl. 82:275–281.

Crutchfield, B.A., and D.A. Potter. 1995. Damage relationships of Japanese beetle and southern masked chafer (Coleoptera: Scarabaeidae) grubs in cool-season turfgrasses. J. Econ. Entomol. 88:1049–1056.

Davidson, A.W., and D.A. Potter. 1995. Response of plant-feeding, predatory, and soil-inhabiting invertebrates to *Acremonium* endophyte and nitrogen fertilisation in tall fescue turf. J. Econ. Entomol. 88:367–379.

Easton, H.S., B.M. Cooper, T.B. Lyons, C.G.L. Pennell, A.J. Popay, B.A. Tapper, and W.R. Simpson. 2000. Selected endophyte and plant variation. p. 351–356. *In* V. H. Paul and P. D. Dapprich (ed.) 4th Int. *Neotyphodium*/Grass Interactions Symp., Soest, Germany.

Easton, H.S., G.C.M. Latch, B.A. Tapper, and O.J.-P. Ball. 2002. Ryegrass host genetic control of concentrations of endophyte-derived alkaloids. Crop Sci. 42:51–57.

Eerens, J.P.J., M.H.P.W. Visker, R.J. Lucas, H.S. Easton, and J.G.H. White. 1998. Influence of the ryegrass endophyte (*Neotyphodium lolii*) in a cool moist environment: IV. Plant parasitic nematodes. N. Z. J. Agric. Res. 41:209–217.

Eichenseer, H., and D.L. Dahlman. 1993. Survival and development of the true armyworm *Pseudaletia unipunctata* (Haworth) (Lepidoptera: Noctuidae), on endophyte-infected and endophyte-free tall fescue. J. Entomol. Sci. 28:462–467.

Elmi, A.A., C.P. West, R.T. Robbins, and T.L. Kirkpatrick. 2000. Endophyte effects on reproduction of a root-knot nematode (*Meloidogyne marylandi*) and osmotic adjustment in tall fescue. Grass Forage Sci. 55:166–172.

Faeth, S.H., and T.J. Sullivan. 2003. Mutualistic asexual endophytes in a native grass are usually parasitic. Am. Nat. 161:310–325.

Grewal, S.K., P.S. Grewal, and R. Gaugler. 1995. Endophytes of fescue grasses enhance susceptibility of *Popillia japonica* larvae to an entomopathogenic nematode. Entomol. Exp. Appl. 74:219–224.

Guy, P.L., and L.T. Davis. 2002. Variation in the incidence of *Barley yellow dwarf virus* and in the ability of *Neotyphodium* endophytes to deter feeding by aphids (*Rhopalosiphum padi*) on Australasian tall fescue. Australasian Plant Pathol. 31:307–308.

Gwinn, K.D., and E.C. Bernard. 1993. Interactions of endophyte-infected grasses with the nematodes *Meloidogyne marylandi* and *Pratylenchus scribneri*. p. 156–160. *In* D.E. Hume et al. (ed.) Proc. 2nd Int. Symp. *Acremonium*/Grass Interactions, Palmerston North, New Zealand.

Gwinn, K.D., and A.M. Gavin. 1992. Relationship between endophyte infestation level of tall fescue seed lots and *Rhizoctonia zeae* seedling disease. Plant Dis. 76:911–914.

Hesse, U., and G.C.M. Latch. 1999. Influence of *Neotyphodium lolii* and *Barley yellow dwarf virus*, individually and combined, on the growth of *Lolium perenne*. Australasian Plant Pathol. 28:240–247.

Hiatt, E.E.I., and N.S. Hill. 1997. *Neotyphodium coenophialum* mycelial protein and herbage mass effects on ergot alkaloid concentration in tall fescue. J. Chem. Ecol. 23:2721–2736.

Hunt, M.G., and J.A. Newman. 2004. Reduced herbivore resistance from a novel grass-endophyte association between AR542 and tall fescue. J. Appl. Ecol., in press.

Jensen, J.G., and A.J. Popay. 2004. Perennial ryegrass infected with AR37 endophyte reduces survival of porina larvae. N. Z. Plant Prot. 57:323–328.

Jones, R.S., O.J.-P. Ball, K.D. Gwinn, and T.A. Coudron. 1997. Feeding preferences of larval fall armyworm on *Neotyphodium*-infected grasses: Influence of host species and endophyte strain. p. 175–177. *In* C.W. Bacon and N.S. Hill (ed.) *Neotyphodium*/grass interactions. Plenum Press, New York.

Kanda, K., H. Hirai, H. Koga, and K. Hasegawa. 1994. Endophyte-enhanced resistance in perennial ryegrass and tall fescue to bluegrass webworm, *Parapediasia teterrella*. Jpn. J. Appl. Entomol. and Zool. 38:141–145.

Keogh, R.G., B.A. Tapper, and R.H. Fletcher. 1996. Distributions of the fungal endophyte *Acremonium lolii*, and of the alkaloids lolitrem B and peramine, within perennial ryegrass. N. Z. J. Agric. Res. 39:121–127.

Koga, H., Y. Hirai, K.I. Kanda, T. Tsukiboshi, and T. Uematsu. 1997. Successive transmission of resistance to bluegrass webworm to perennial ryegrass and tall fescue plants by artificial inoculation with *Acremonium* endophytes. Jpn. Agric. Res. Q. 31:109–115.

Koppenhoffer, A.M., R.S. Cowles and E.M. Fuzy. 2003. Effects of turfgrass endophytes (Clavicipitaceae: Ascomcyetes) on white grub (Coleoptera: Scarabaeidae) larval development and field populations. Environ. Entomol. 32:895–906.

Koppenhoffer, A.M., and E.M. Fuzy. 2003. Effects of turfgrass endophytes (Clavicipitaceae: Ascomcyetes) on white grub (Coleoptera: Scarabaeidae) control by the entomopathogenic nematode *Heterorhabditis bacteriophora* (Rhabditida: Heterorhabditidae). Environ. Entomol. 32: 392–396.

Koshino, H., S. Togiya, T. Yoshihara, S. Sakamura, T. Shimanuki, T. Sato, and A. Tajimi. 1987. Four fungitoxic C-18 hydroxy unsaturated fatty acids from stromata of *Epichloë typhina*. Tetra. Lett. 28:73–76.

Kunkel, B.A., and P.S. Grewal. 2003. Endophyte infection in perennial ryegrass reduces the susceptibility of black cutworm to an entomopathogenic nematode. Entomol. Exp. Appl. 107:95–104.

Kunkel, B.A., P.S. Grewal, and M.F. Quigley. 2004. A mechanism of acquired resistance against an entomopathogenic nematode by *Agrotis ipsilon* feeding on perennial ryegrass harboring a fungal endophyte. Biocontrol 29:100–108.

Lane, G.A., M.J. Christensen, and C.O. Miles. 2000. Coevolution of fungal endophytes with grasses: The significance of secondary metabolites. p. 341–388. *In* C.W. Bacon and J.F. White, Jr. (ed.) Microbial endophytes. Marcel Dekker, New York.

Leuchtmann, A., D. Schmidt, and L.P. Bush. 2000. Different levels of protective alkaloids in grasses with stroma- forming and seed-transmitted *Epichloë/Neotyphodium* endophytes. J. Chem. Ecol. 26:1025–1036.

Lewis, G.C. 2004. Effects of biotic and abiotic stress on the growth of three genotypes of *Lolium perenne* with and without infection by the fungal endophyte *Neotyphodium lolii*. Ann. Appl. Biol. 144:53–63.

Lewis, G.C., and R.O. Clements. 1986. A survey of ryegrass endophyte (*Acremonium loliae*) in the U.K. and its apparent ineffectuality on a seedling pest. J. Agric. Sci. (Cambridge) 107:633–638.

Lewis, G.C., and B. Vaughan. 1997. Evaluation of a fungal endophyte (*Neotyphodium lolii*) for control of leatherjackets (*Tipula* spp.) in perennial ryegrass. Tests Agrochem. Cult. 18:34–35.

Lopez, J.E., S.H. Faeth, and M. Miller. 1995. Effect of endophytic fungi on herbivory by redlegged grasshoppers (Orthoptera: Acrididae) on Arizona fescue. Environ. Entomol. 24:1576–1580.

Madej, C.W., and K. Clay. 1991. Avian seed preference and weight loss experiments: The effect of fungal endophyte-infected tall fescue seeds. Oecologia 88:296–302.

Mahmood, T., R.C. Gergerich, E.A. Milus, C.P. West, and C.J. D'Arcy. 1993. Barley yellow dwarf viruses in wheat, endophyte-infected and endophyte-free tall fescue, and other hosts in Arkansas. Plant Dis. 77:225–228.

Marks, S., and D.E. Lincoln. 1996. Antiherbivore defense mutualism under elevated carbon dioxide levels: A fungal endophyte and grass. Environ. Entomol. 25:618–623.

McDonald, G., A. Noske, R. van Heeswijk, and W.E. Frost. 1993. The role of perennial ryegrass endophyte in the management of pasture pests in south eastern Australia. Proc. Australasian Grassl. Invert. Ecol. Conf. 6:122–128.

McLeod, A.R., A. Rey, K.K. Newsham, G.C. Lewis, and P. Wolferstan. 2001. Effects of elevated ultraviolet radiation and endophytic fungi on plant growth and insect feeding in *Lolium perenne, Festuca rubra, F. arundinacea* and *F. pratensis*. J. Photochem. Photobiol., B 62:97–107.

Miles, C.O., M.E. di Menna, S.W.L. Jacobs, I. Garthwaite, G.A. Lane, R.A. Prestidge, S.L. Marshall, H.H. Wilkinson, C.L. Schardl, O.J.-P. Ball, and G.C.M. Latch. 1998. Endophytic fungi in indigenous Australasian grasses associated with toxicity to livestock. Appl. Environ. Microbiol. 64:601–606.

Moy, M., F. Belanger, R. Duncan, A. Freehoff, C. Leary, W. Meyer, R. Sullivan, and J.F. White. 2000. Identification of epiphyllous mycelial nets on leaves of grasses infected by Clavicipitaceous endophytes. Symbiosis 28:291–302.

Moy, M., H.M. Li, R. Sullivan, J.F. White, Jr., and F.C. Belanger. 2002. Endophytic fungal β-1,6-glucanase expression in the infected host grass. Plant Physiol. 130:1298–1308.

Murphy, J.A., S. Sun, and L.L. Betts. 1993. Endophyte-enhanced resistance to billbug (Coleoptera: Curculionidae), sod webworm (Lepidoptera: Pyralidae) and white grub (Coleoptera: Scarabaeidae) in tall fescue. Environ. Entomol. 22:699–703.

Nan, Z.B., and C.J. Li. 2000. *Neotyphodium* in native grasses in China and observations on endophyte/host interactions. p. 41–50. *In* V.H. Paul and P.D. Dapprich (ed.) 4th Int. *Neotyphodium*/Grass Interactions Symp., Soest, Germany.

Patterson, C.G., D.A. Potter, and F.F. Fannin. 1991. Feeding deterrency of alkaloids from endophyte-infected grasses to Japanese beetle larvae. Entomol. Exp. Appl. 61:285–289.

Pennell, C., and O.J.-P. Ball. 1999. The effects of *Neotyphodium* endophytes in tall fescue on pasture mealy bug (*Balanococcus poae*). Proc. N.Z. Plant Prot. Conf. 52:259–263.

Pennell, C., and M.P. Rolston. 2003. The effect of grass-endophyte associations on feeding of Canada geese (*Branta canadensis*). Proc. N. Z. Grassl. Assoc. 65:239–243.

Popay, A.J. 2004. Response of black beetle (*Heteronychus arator*), porina (*Wiseana cervinata*) and grass grub (*Costelytra zealandica*) to different endophytes in tall fescue. Proc. Australasian Grassl. Invert. Ecol. Conf. 8:(in press).

Popay, A.J., and J.G. Baltus. 2001. Black beetle damage to perennial ryegrass infected with AR1 endophyte. Proc. N. Z. Grassl. Assoc. 63:267–271.

Popay, A.J., J.G. Baltus, and C.G. Pennell. 2000a. Insect resistance in perennial ryegrass infected with toxin-free *Neotyphodium* endophytes. p. 187–193. *In* V.H. Paul and P.D. Dapprich (ed.) 4th Int. *Neotyphodium*/Grass Interactions Symp., Soest, Germany.

Popay, A.J., D.E. Hume, J.G. Baltus, G.C.M. Latch, B.A. Tapper, T.B. Lyons, B.M. Cooper, C.G. Pennell, J.P.J. Eerens, and S.L. Marshall. 1999. Field performance of perennial ryegrass (*Lolium perenne*) infected with toxin-free fungal endophytes (*Neotyphodium* spp.). p. 113–122. *In* D.R. Woodfield and C. Matthew (ed.) Ryegrass endophyte: An essential New Zealand symbiosis. Grasslands Res. and Practice Series No. 7. New Zealand Grassland Assoc., Wellington, New Zealand.

Popay, A.J., D.E. Hume, K.L. Davis, and B.A. Tapper. 2003b. Interactions between endophyte (*Neotyphodium* spp.) and ploidy in hybrid and perennial ryegrass cultivars and their effects on Argentine stem weevil (*Listronotus bonariensis*). N. Z. J. Agric. Res. 46:311–319.

Popay, A.J., D.E. Hume, R.A. Mainland, and C.J. Saunders. 1995. Field resistance to Argentine stem weevil (*Listronotus bonariensis*) in different ryegrass cultivars infected with an endophyte deficient in lolitrem B. N. Z. J. Agric. Res. 38:519–528.

Popay, A.J., and G.A. Lane. 2000. The effect of crude extracts containing loline alkaloids on two New Zealand insect pests. p. 471–475. *In* V.H. Paul and P.D. Dapprich (ed.) 4th Int. *Neotyphodium*/Grass Interactions Symp., Soest, Germany.

Popay, A.J., and R.A. Mainland. 1991. Seasonal damage by Argentine stem weevil to perennial ryegrass pastures with different levels of *Acremonium lolii*. Proc. N. Z. Weed Pest Control Conf. 44:171–175.

Popay, A., S. Marshall, and J. Baltus. 2000b. Endophyte infection influences disappearance of perennial ryegrass seed. N. Z. Plant Prot. 53:8–10.

Popay, A.J., and D.D. Rowan. 1994. Endophytic fungi as mediators of plant–insect interactions. p. 84–103. *In* E.A. Bernays (ed.) Insect–plant interactions. Vol. V. CRC Press, Boca Raton, FL.

Popay, A.J., W.B. Silvester, and P.J. Gerard. 2004. New endophyte isolate suppresses root aphid, *Aploneura lentisci*, in perennial ryegrass. p. 317. *In* R. Kallenbach et al. (ed.) 5th Int. Symp. on *Neotyphodium*/Grass Interactions, Fayetteville, Arkansas.

Popay, A.J., R.J. Townsend, and L.R. Fletcher. 2003a. The effect of endophyte (*Neotyphodium uncinatum*) in meadow fescue on grass grub larvae. N. Z. Plant Prot. 56:123–128.

Popay, A.J., and R.T. Wyatt. 1995. Resistance to Argentine stem weevil in perennial ryegrass infected with endophytes producing different alkaloids. Proc. N. Z. Plant Prot. Conf. 48:229–236.

Potter, D.A., C.G. Patterson, and C.T. Redmond. 1992. Influence of turfgrass species and tall fescue endophyte on feeding ecology of Japanese beetle and southern masked chafer grubs (Coleoptera: Scarabaeidae). J. Econ. Entomol. 85:900–909.

Prestidge, R.A., and O.J.-P. Ball. 1993. The role of endophytes in alleviating plant biotic stress in New Zealand. p. 141–151. *In* D.E. Hume et al. (ed.) 2nd Int. Symp. *Acremonium*/Grass interactions, Plenary Papers. AgResearch, Palmerston North, New Zealand.

Prestidge, R.A., and O.J.-P. Ball. 1995. A catch 22: The utilization of endophytic fungi for pest management. p. 171–192. *In* A.C. Gange and V.K. Brown (ed.) Multitrophic interactions in terrestrial systems: 36th Symp. British Ecological Society. Royal Holloway College, Univ. of London.

Prestidge, R.A., R.P. Pottinger, and G.M. Barker. 1982. An association of *Lolium* endophyte with ryegrass resistance to Argentine stem weevil. Proc. N. Z. Weed Pest Control Conf. 35:119–122.

Richmond, D.S., P.S. Grewal, and J. Cardina. 2003. Competition between *Lolium perenne* and *Digitaria sanguinalis*: Ecological consequences for harboring an endosymbiotic fungus. J. Vegetation Sci. 14:835–840.

Richmond, D.S., P.S. Grewal, and J. Cardina. 2004. Influence of Japanese beetle *Popillia japonica* larvae and fungal endophyte on competition between turfgrass and dandelion. Crop Sci. 44:600–606.

Richmond, D.S., H.D. Niemczyk, and D.J. Shetlar. 2000. Overseeding endophytic perennial ryegrass into stands of Kentucky bluegrass to manage bluegrass billbug (Coleptera: Curculionidae). J. Econ. Entomol. 93:1662–1668.

Richmond, D.S., and D.J. Shetlar. 1999. Larval survival and movement of bluegrass webworm in mixed stands of endophytic perennial ryegrass and Kentucky bluegrass. J. Econ. Entomol. 92:1329–1334.

Richmond, D.S., and D.J. Shetlar. 2000. Hairy chinch bug (Hemiptera: Lygaeidae) damage, population density, and movement in relation to the incidence of perennial ryegrass infected by *Neotyphodium* endophytes. J. Econ. Entomol. 93:1167–1172.

Roberts, C.A., S.M. Marek, T.L. Niblack, and A.L. Karr. 1992. Parasitic *Meloidogyne* and mutualistic *Acremonium* increase chitinase in tall fescue. J. Chem. Ecol. 18:1107–1116.

Saikkonen, K., M. Helander, S.H. Faeth, F. Schulthess, and D. Wilson. 1999. Endophyte–grass–herbivore interactions: The case of *Neotyphodium* endophytes in Arizona fescue populations. Oecologia 121:411–420.

Saikkonen, K., D. Ion, and M. Gyllenberg. 1998. The persistence of vertically transmitted fungi in grass metapopulations. Proc. R. Soc. London, Ser. B 269:1397–1403.

Schmidt, D., and P.L. Guy. 1997. Effects of the presence of the endophyte *Acremonium uncinatum* and of an insecticide treatment on seed production of meadow fescue. Rev. Suisse d'Agric. 29:97–99.

Siegel, M.R., and L.P. Bush. 1996. Defensive chemicals in grass-fungal endophyte associations. p. 81–119. *In* J.T. Romeo et al. (ed.) Phytochemical diversity and redundancy in ecological interactions. Vol. 30. Plenum Press, New York.

Siegel, M.R., G.C.M. Latch, L.P. Bush, F.F. Fannin, D.D. Rowan, B.A. Tapper, C.W. Bacon, and M.C. Johnson. 1990. Fungal endophyte-infected grasses: Alkaloid accumulation and aphid response. J. Chem. Ecol. 16:3301–3315.

Springer, T.L. 1996. Allelopathic effects on germination and seedling growth of clovers by endophyte-free and -infected tall fescue. Crop Sci. 36:1639–1642.

Stewart, T.M., C.F. Mercer, and J.L. Grant. 1993. Development of *Meloidogyne naasi* on endophyte-infected and endophyte-free perennial ryegrass. Australasian Plant Pathol. 22:40–41.

Stovall, M.E., and K. Clay. 1991. Adverse effects on fall armyworm feeding on fungus-free leaves of fungus-infected plants. Ecol. Entomol. 16:519–523.

Sutherland, B.L., D.E. Hume, and B.A. Tapper. 1999. Allelopathic effects of endophyte-infected perennial ryegrass extracts on white clover seedlings. N. Z. J. Agric. Res. 42:19–26.

Tapper, B.A., and G.A. Lane. 2004. Janthitrems found in a *Neotyphodium* endophyte of perennial ryegrass. p. 301. *In* R. Kallenbach et al. (ed.) 5th Int. Symp. *Neotyphodium/Grass* Interactions, Fayetteville, AR.

Tapper, B.A., and G.C.M. Latch. 1999. Selection against toxin production in endophyte-infected perennial ryegrass. p. 107–111. *In* D.R. Woodfield and C. Matthew (ed.) Ryegrass endophyte: An essential New Zealand symbiosis. Grasslands Res. and Practice Series No. 7. New Zealand Grassland Assoc., Wellington, New Zealand.

TePaske, M.R., R.G. Powell, and S.L. Clement. 1993. Analyses of selected endophyte-infected grasses for the presence of loline-type and ergot-type alkaloids. J. Agric. Food Chem. 41:2299–2303.

Tibbets, T.M., and S.H. Faeth. 1999. *Neotyphodium* endophytes in grasses: Deterrents or promoters of herbivory by leaf-cutting ants. Oecologia 118:287–305.

Timper, P., and J.H. Bouton. 2004. Effect of endophyte status and tall fescue cultivar on reproduction of lesion and stubby-root nematodes. p. 406. *In* R. Kallenbach et al. (ed.) 5th Int. Symp. *Neotyphodium/Grass* Interactions, Fayetteville, AR.

Vincelli, P., and A.J. Powell. 1991. Reaction of perennial ryegrass varieties to red thread, 1990. Biol. Cultural Tests Control Plant Dis. 6:102.

Walston, A.T., D.W. Held, N.R. Mason, and D.A. Potter. 2001. Absence of interaction between endophytic perennial ryegrass and susceptibility of Japanese beetle (Coleoptera: Scarabaeidae) grubs to *Paenibacillus popilliae* Dutky. J. Entomol. Sci. 36:105–108.

Watson, R.N., D.E. Hume, N.L. Bell, and F.J. Neville. 1995. Plant-parasitic nematodes associated with perennial ryegrass and tall fescue with and without *Acremonium* endophyte. Proc. N. Z. Plant Prot. Conf. 48:199–203.

Wheatley, W.M., D.E. Hume, H.W. Kemp, M.S. Monk, K.F. Lowe, A.J. Popay, D.B. Baird, and B.A. Tapper. 2003. Effects of fungal endophyte on the persistence and productivity of tall fescue at 3 sites in eastern Australia. Proc. 11th Aust. Agron. Conf., Geelong, VIC, Australia. 2–6 Feb. 2003. Australian Society of Agronomy, Brisbane.

White, J.F.J., A.E. Glenn, and K.F. Chandler. 1993. Endophyte-host associations in grasses. XVII. Moisture relations and insect herbivory of the emergent stromal leaf of *Epichloë*. Mycologia 85:195–202.

White, J.F., Jr., R.F. Sullivan, G.A. Balady, T.J. Gianfagna, Q. Yue, W.A. Meyer, and D. Cabral. 2001. A fungal endosymbiont of the grass *Bromus setifolius*: Distribution in some Andean populations, identification, and examination of beneficial properties. Symbiosis 31:241–257.

Wilkinson, H.H., M.R. Siegel, J.D. Blankenship, A.C. Mallory, L.P. Bush, and C.L. Schardl. 2000. Contribution of fungal loline alkaloids to protection from aphids in a grass–endophyte mutualism. Mol. Plant–Microbe Interact. 13:1027–1033.

Williamson, R.C., and D.A. Potter. 1997. Turfgrass species and endophyte effects on survival, development and feeding preference of black cutworms (Lepidoptera: Noctuidae). J. Econ. Entomol. 90:1290–1299.

Yoshimatsu, S., K. Arimura, and T. Shimanuki. 1998. Comparison of population growth rates of confused flour beetle, *Tribolium confusum* Jaquelin (Coleoptera: Tenebrionidae), on endophyte-infected or endophyte-uninfected seeds of ground tall fescue and perennial ryegrass. J. Appl. Entomol. and Zool. 42:227–229.

Yoshimatsu, S., T. Shimanuki, and K. Arimura. 1999. Influence of endophyte-infected tall fescue, *Festuca arundinacea* Shreb., seeds on the adult survival and reproduction of the grain beetle, *Oryzaephilus surinamensis* (L.) (Coleoptera: Silvanidae). Jpn. J. Entomol. 2:51–56.

Yue, Q., J. Johnson-Cicalese, T.J. Gianfagna, and W.A. Meyer. 2000b. Alkaloid production and chinch bug resistance in endophyte-inoculated chewings and strong creeping red fescues. J. Chem. Ecol. 26:279–292.

Yue, Q., C.J. Miller, J.F. White, Jr., and M.D. Richardson. 2000a. Isolation and characterization of fungal inhibitors from *Epichloë festucae*. J. Agric. Food Chem. 48:4687–4692.

ABIOTIC STRESSES IN ENDOPHYTIC GRASSES

D. P. Malinowski,[1] D. P. Belesky,[2] and G. C. Lewis[3]

Neotyphodium spp. endophytes form nonpathogenic, systemic, and usually intercellular symbiotic associations with several genera of cool-season grasses (Marshall et al., 1999; Tsai et al., 1994; Vinton et al., 2001). A shoot-located, intercellular fungus, first classified as *Epichloë typhina* (Bacon et al., 1977), later described by Morgan-Jones and Gams (1982) as *Acremonium coenophialum*, and finally renamed *N. coenophialum* by Glenn et al. (1996), was found in tall fescue (*Festuca arundinacea*) and led to many years of scientific inquiry of the livestock malady known as fescue toxicosis.

Within the past two decades, seven genera of grass fungal endophytes of the Clavicipitaceae (Ascomycetes) family were identified (White, 1994; White and Reddy, 1998). The associations between *Neotyphodium* spp. endophytes and cool-season forage grasses were investigated intensively because of the economic impact related to detrimental effects of endophyte-infected (E+) grasses on livestock performance. The most widely known *Neotyphodium* endophytes are *N. coenophialum*, *N. lolii*, and *N. uncinatum*, which colonize tall fescue, perennial ryegrass (*Lolium perenne*), and meadow fescue (*F. pratensis*), respectively. These grasses are native to Europe and the Mediterranean Basin. The ecological significance of cool-season grass associations with *Neotyphodium* endophytes seemed to be protection against herbivory (Schardl and Phillips, 1997); however, recent literature suggests that Arizona fescue (*F. arizonica*), a grass native to the southwestern USA, incurs a cost when hosting *Neotyphodium* spp. endophytes, and that antiherbivory expression is minimal (Faeth et al., 2004). On the other hand, abiotic stresses, drought, and high temperatures in particular seem to be the primary factors shaping the range of existence of E+ cool-season grasses in their native environments in Europe (Lewis, 2000b). There was no apparent herbivory or pest problem to cause the sort of selection

[1] Texas A&M University System, Vernon, TX, USA
[2] USDA-ARS, Appalachian Farming Systems Research Center, Beaver, WV, USA
[3] Institute of Grassland and Environmental Research, North Wyke Research Station, Okehampton, Devon, UK

pressure for E+ native grasses in Europe as it occurred in pastures with introduced cool-season perennial grasses in the USA and New Zealand. Thus, the high levels of endophyte infection in some populations of cool-season grasses indigenous to Europe may have resulted from selection pressure associated with climatic variables.

We present current knowledge of mechanisms and adaptations of economically important grasses and their associated endophytes to abiotic stresses. We also consider the agronomic significance of endophyte manipulation to maintain abiotic stress tolerance in forage grasses while minimizing impact on livestock performance.

DROUGHT STRESS

The ability of plants to tolerate drought relies on a variety of morphological and physiological adaptations and adjustments leading to efficient water use, tissue protection from oxidative stress, and rapid tissue regeneration after drought events. One important adaptation of E+ perennial ryegrass (Latch et al., 1985), tall fescue (De Battista et al., 1990), and meadow fescue (Malinowski et al., 1997) is generation of an extensive root system which may allow water uptake from a greater volume of soil. *Neotyphodium* spp. endophytes have not been found in roots of the host grasses, yet they do affect root morphology and functions. Malinowski et al. (1999) found smaller root diameter and longer root hairs in hydroponically grown E+ vs. endophyte-free (E-) tall fescue genotypes. This phenomenon has not yet been observed in soil-grown plants to date; however, it may improve water uptake efficiency in water-deficient soils.

Water status dynamics of E+ tall fescue (Buck et al., 1997; Elbersen and West, 1996; Elmi and West, 1995) and meadow fescue (Malinowski et al., 1997) plants differ from those of E- plants. Stomatal conductance of water-stressed E+ plants declined earlier and more rapidly than that of E- plants, suggesting involvement of endophyte or fungal metabolites in stomatal function. In contrast, stomatal function does not seem to be affected by infection with *N. lolii* in perennial ryegrass, and there are either inconsistent or negligible effects for both physiological (Barker et al., 1997; Amalric et al., 1999) and morphological traits (Cheplick et al., 2000; Lewis, 2000a). Lewis et al. (1997) found a higher incidence of E+ individuals in ryegrass populations with increasing soil water deficit across environments in France. Their findings were confirmed by Hesse et al. (2003), who evaluated endophyte effects on ryegrass ecotypes from habitats with contrasting soil water availability in Germany. Endophyte infection positively affected a number of morphological parameters associated with drought tolerance in ecotypes from dry or flood-dry environments, but reduced these parameters in ecotypes from wet habitats. These findings strongly suggest that the endophyte promotes population fitness in water-limited environments.

Enhanced accumulation of osmotically active metabolites was suggested as a mechanism of osmotic adjustment in E+ plants (Elmi and West, 1995). None of the known fungal metabolites in symbiotic grasses appear to be directly involved in regulation of osmotic adjustment. However, accumulation and metabolism of the following metabolites differ in E+ vs. E– cool-season grasses exposed to drought stress: glucose and fructose (Richardson et al., 1992), the fungal products mannitol and arabitol (Richardson et al., 1992), proline (Elbersen and West, 1996; Malinowski, 1995), amino acids (Lyons et al., 1990), and loline alkaloids (Bush et al., 1993; Kennedy and Bush, 1983). Interest in loline alkaloids centered on their insecticidal properties. The potential role of lolines in drought resistance has received recent attention (reviewed in Malinowski and Belesky, 2000; Spiering et al., 2002).

High concentrations of resveratrol, a phenolic compound, were associated in cool-season grasses with endophyte infection (Powell et al., 1994). Increased phenolic concentrations occurred in E+ vs. E– tall fescue (Malinowski et al., 1998) and perennial ryegrass (Zhou et al., 2003) grown in phosphorus-deficient media. This discovery opened an avenue for investigation of endophyte involvement in mineral nutrient acquisition and drought stress tolerance in cool-season grasses. Phenolic compounds are potent antioxidants and contribute to cell protection from oxidative stress during drought (Blokhina et al., 2003). Treatment of tall fescue, regardless of endophyte infection status, with Tasco-Forage[1] (Acadian Seaplants Ltd., Dartmouth, NS, Canada; a proprietary seaweed-based product rich in phenolic compounds) increased activity of the antioxidant superoxide dismutase (SOD) in herbage (Fike et al., 2001). Interestingly, E+ plants had greater SOD activity than E– plants even without treatment with Tasco-Forage when grown in Mississippi, but not in Virginia. Ayad (1998) reported similar response of SOD activity in E+ KY-31 tall fescue grown in the Southern High Plains of Texas. Ascorbate peroxidase activity in perennial ryegrass was greater in E+ compared with E– plants exposed to zinc stress, suggesting that *N. lolii* modifies plant metabolism by favoring H_2O_2

---

[1] Mention of trade names or commercial products is solely for the purpose of providing specific information and does not imply recommendation or endorsement by the Texas A&M University System and the U.S. Department of Agriculture.

scavenging throughout the catalase process (Bonnet et al., 2000). Tolerance to abiotic stresses in E+ cool-season grasses may be associated with tolerance of oxidative stress at the cellular level.

Relatively new research on dehydrins, a group of proteins synthesized in grasses in response to various abiotic stresses, including drought (Volaire, 2002), may be another new fruitful area for research on drought stress tolerance in E+ grasses. Carson et al. (2004) reported earlier detection of dehydrins in E+ than in E– tall fescue plants during drought in association with greater tiller survival in E+ plants.

LIGHT STRESS

Very limited information exists on responses of E+ grasses to light stress. Belesky et al. (1987) reported lower net photosynthesis (entire canopy) in E+ compared with E– tall fescue with increasing light intensity. Recently, Lewis (2004) found that shading did not suppress herbage growth of an E+ perennial ryegrass clone as much as it did the growth of its E– counterpart, suggesting that some perennial ryegrass/endophyte associations may have greater tolerance to shading and competition for light in certain situations.

Light quality (e.g., red/far red ratio, UV-A, or UV-B) determines many developmental processes in grasses (Skinner and Nelson, 1994). Elevated levels of UV radiation often accompany summer drought conditions, imposing an additional stress on cool-season grasses. Newsham et al. (1998) found that endophyte infection reduced seed production in a ryegrass clone in response to elevated UV-B radiation, but increased generation of vegetative propagules. In a subsequent experiment, no interactions occurred between endophyte status and elevated UV-B on selected morphological parameters in tall fescue, meadow fescue, red fescue (*F. rubra*), and perennial ryegrass (McLeod et al., 2001).

Feeding preferences of desert locust (*Schistocerca gregaria*) for E– or E+ leaves of meadow fescue were altered by UV-A ($\approx$ 320–400 nm) and UV-B ($\approx$ 280–320 nm) radiation. Under ambient UV radiation, feeding was less on E+ compared with E– leaves. The difference was attributed to loline alkaloids present in E+ plants. Feeding on E– plants was reduced when plants exposed to elevated UV-B were compared with plants grown under ambient UV. Because E– meadow fescue does not produce aminopyrrolizidine alkaloids (specifically loline), other metabolites synthesized in response to elevated UV-B radiation must have deterred locust feeding. If UV-B treatment was compared with UV-A treatment, feeding preferences for E+ plants were reduced while concentrations of loline alkaloids were not affected, suggesting an involvement of other metabolites. Plants often produce phenolic compounds in response to stress, including UV radiation (Blokhina et al., 2003) and endophyte infection in tall fescue (Malinowski et al., 1998) and perennial ryegrass (Zhou et al., 2003). Further investigation of the interaction of light and metabolism of E+ cool-season grasses is warranted since response to light environment could alter herbage composition and ultimately utilization by the grazing animal.

MINERAL STRESS

Understanding of mineral stress tolerance in E+ cool-season grasses increased rapidly during the past decade. At least three factors were explored, including relationships between alkaloid production and mineral nutrition (Arechavaleta et al., 1992; Belesky et al., 1988; Lyons and Bacon, 1984), the ability to thrive in marginal environments (Clay, 1990), and uses of E+ grasses for turf (Funk et al., 1994). We summarized progress in understanding mechanisms and adaptations of E+ grasses to soil mineral imbalances and their effects on plant communities in previous reviews (Belesky and Malinowski, 2000; Malinowski and Belesky, 2000). Additional findings since that review focus on mechanisms of mineral stress tolerance in E+ cool-season grasses.

Nitrogen fertilization was related to concentration of ergopeptine alkaloids in tall fescue (Belesky et al., 1988; Lyons and Bacon, 1984). Altered N metabolism in E+ tall fescue was discovered by Lyons et al. (1990), wherein E+ plants had greater glutamine synthetase activity than did E− plants. The mechanism controlled by the enzyme may have contributed to more efficient N use by E+ tall fescue (Arachevaleta et al., 1989). At present, it is not known if higher N use efficiency in E+ compared with E− cool-season grasses is directly related to fungal N metabolism or altered host plant enzyme activity (Ferguson et al., 1993; Kulkarni and Nielsen, 1986; Naffaa et al., 1998). Perennial ryegrass and Arizona fescue showed higher N use efficiency in response to endophyte infection when grown at low N availability (Lewis, 2004; Saikkonen et al., 1999; Schulthess and Faeth, 1998). Other studies with perennial ryegrass, however, showed minimal or inconsistent effects of endophyte at low N concentrations (Cheplick, 1998; Cheplick et al., 1989; Durand et al., 2002; Lewis et al., 1996; Marks and Clay, 1990; Ravel et al., 1997). Arachevalata et al. (1989) and Marks and Clay (1990) found beneficial effects of endophyte infection in tall fescue and perennial ryegrass at high N concentrations only. This suggests specificity of endophyte-host grass associations in terms of N use efficiency. Cheplick et al.

(1989) hypothesized that endophyte could compete with host grass for N and other nutrients when supply was limited, resulting in reduced plant growth. The authors presented evidence supporting their hypothesis for tall fescue but not perennial ryegrass. The interactions between endophytes and host grass nutrient status, therefore, are not yet well understood.

Similar to N nutrition, phosphorus availability influences ergot alkaloid production in E+ grasses (Porter, 1994). High P concentrations in the growth medium restricted activity of dimethylallyl tryptophan synthase, the first enzyme regulating biosynthesis of ergot alkaloids in *Claviceps* spp. (Flieger et al., 1991; Robbers, 1984). Pioneer research by Azevedo and Welty (1995) and Azevedo et al. (1993) confirmed reduced photosynthesis rates and growth of tall fescue in response to endophyte infection at high P availability and a relationship between ergovaline production and soil P availability. Malinowski et al. (1998) found that tall fescue genotypes with contrasting potential for ergot alkaloid production responded differently to P supply in terms of ergot alkaloid concentration. These E+ tall fescue genotypes also expressed reduced root and shoot growth when compared with E– counterparts where P availability increased in soil.

Endophyte presence in tall fescue shoots induces morphological and chemical modifications in root systems under P-deficiency stress. One morphological change results in increased root absorption area through reduced root diameter and increased root hair length (Malinowski et al., 1999). Other mechanisms involve exudation of phenolic-like compounds into the rhizosphere (Malinowski et al., 1998). A similar mechanism was reported in perennial ryegrass (Zhou et al., 2003). Plants with the ability to chemically modify the rhizosphere would have a competitive advantage in terms of nutrient acquisition and possible mechanisms of herbivory avoidance contributing to persistence.

Despite the involvement of phenolic compounds in P acquisition in E+ tall fescue (Malinowski and Belesky, 1999a), the chelating nature is responsible for tolerance to Al stress (Malinowski and Belesky, 1999b), a mechanism first suggested by Foy and Murray (1998). Sequestration of Al on root surfaces and in root free-spaces may explain tolerance to Al stress reported in other E+ cool-season grasses (Liu et al., 1996; Zaurov et al., 2001).

Infection of tall fescue with its endemic *N. coenophialum*-endophyte appears to reduce Cu concentrations in forage, and in serum of grazing animals, contributing to a range of immune-related disorders (Dennis et al., 1998; Saker et al., 1998). Malinowski et al. (2004) showed that phenolic compounds released from roots of tall fescue infected with a novel, non-ergot-alkaloid producing endophyte strain AR542 were able to chelate Cu in nutrient solution, although Cu concentrations in shoot tissues were not affected. This does not exclude, however, that such a mechanism may contribute to reduced Cu concentration of E+ tall fescue forage in field-grown plants. Interestingly, Cu chelating ability of root exudates of perennial ryegrass did not differ between plants infected with a novel strain of *N. lolii*-endophyte, AR1, and E– plants.

Phenolic compounds released into the rhizosphere of E+ tall fescue have the ability to reduce Fe outside of the root system (Malinowski et al., 1998). A mechanism of Fe reduction in the rhizosphere and subsequent uptake by roots operates in dicots, but is not known in monocots, where an alternative mechanism of Fe uptake occurs via phytosiderophore carriers (Marschner, 1986). At present, no data confirm an alternative mechanism of Fe uptake in E+ cool-season grasses, similar to that in dicots.

In perennial ryegrass, endophyte infection increased plant tolerance to excess Zn by preventing it from reaching shoot tissues, thereby protecting photosystem II activities (Monnet et al., 2001), and by improving SOD and ascorbate peroxidase activities, which minimize oxidative damage (Bonnet et al., 2000). Phenolic compounds produced by E+ perennial ryegrass (Zhou et al., 2003) could be involved in these phenomena, as in E+ tall fescue.

Several other studies indicated a close relationship between endophyte infection and mineral acquisition by tall fescue. Vazquez-de-Aldana et al. (1999) concluded that E+ plants accumulated more nutrients involved in protein synthesis (i.e., N and Mg) than did E– plants. Using a hydroponic system, Malinowski et al. (2000) reported mineral nutrient uptake rates of tall fescue clones that were strongly modified by endophyte infection when plants were grown in P-deficient nutrient solution. These results confirmed a regulatory effect of the leaf-located fungus on tall fescue root chemistry (Malinowski et al., 1998). Future research should identify metabolites directly involved in this process.

NOVEL ENDOPHYTES AND ABIOTIC STRESS TOLERANCE IN COOL-SEASON GRASSES

Within the past few years, a shift from quantitative to qualitative assessment of mechanisms regulating adaptation of *Neotyphodium*–grass associations to abiotic stresses has occurred. Selection of novel endophyte strains that produce little or no alkaloid(s) detrimental to grazing livestock, yet appear to retain host tolerance to abiotic and biotic stresses (Bouton et al., 2002a, 2002b; Tapper and

Latch, 1999; West and Gunter, 2004) is one example of a domestication process of the endophytes. Early results from grazing experiments on 'Jesup' tall fescue reinfected with MaxQ endophyte AR542 (Macoon et al., 2004), 'HiMag' tall fescue with endophyte UA4 (Nihsen et al., 2004), and perennial ryegrass rein-fected with the Plus AR1 (New Zealand Agriseeds, 2003) and Plus Nea2 (Heritage Seeds, 2003) novel endophytes confirm excellent animal performance. Research is currently under way to evaluate the persistence of grasses infected with novel endophytes in drought-prone environments. Bouton et al. (2002a) observed about 15% decline in stand survival of Jesup MaxQ when compared with Jesup infected with a wild-type endophyte, but both accessions survived much better than Jesup E– during 1997 through 2002 in Georgia. Tiller survival of an obligatory summer-dormant tall fescue 'Grasslands Flecha' with MaxQ endophyte AR542 was greater than that of E– plants during summer drought in the second and third growing seasons in North Texas (Malinowski et al., 2003), but endophyte infection had no effect on tiller survival during an extreme summer drought of 2003 (Malinowski et al., 2004, unpublished data). Under the same growth conditions, neither AR542 nor native endophyte strains contributed to the persistence of summer semidormant tall fescue cultivars Jesup and 'Geor-gia 5' that did not survive the first summer drought (Malinowski et al., 2003). West and Gunter (2004) reported persistence of HiMag tall fescue infected with novel endophytes comparable with E+ Kentucky 31 and enhanced relative to E– HiMag after 5 yr of field trials in southwest Arkansas.

Preliminary data (Malinowski et al., 2004, unpublished data) do not confirm any advantages of AR1 endophyte for drought tolerance of 'Aries' and 'Quartet' perennial ryegrass grown in North Texas. However, other data from a glass-house experiment (Lewis, 2003, unpublished data) show increased plant survival following drought stress in perennial ryegrass genotypes artificially inoculated with other strains of *N. lolii*-endophyte, compared with the uninocu-lated counterparts. In terms of mineral stress tolerance, the novel endophyte strain AR542 elicits similar responses in tall fescue to those shown for the wild-type endophyte strains (Malinowski et al., 2004).

Complete separation of biotic from abiotic responses is difficult, since some stresses may predispose the host–endophyte association to succumb to certain extrinsic stressors or express enhanced resistance to others. Because of the economic benefits of novel endophytes, future research should focus on stability of novel associations in managed pastures for delivering the benefits of abiotic stress tolerance. For example, selecting against ergot-alkaloid production (short-term economic return) may affect mechanisms of abiotic stress tolerance of endophyte–grass associations and ultimately persistence of the associations across time (long-term impact).

REFERENCES

Amalric, C., H. Sallanon, F. Monnet, A. Hitmi, and A. Coudret. 1999. Gas exchange and chlorophyll fluorescence in symbiotic and non-symbiotic ryegrass under water stress. Photosynthetica 37:107–112.

Arachevaleta, M., C.W. Bacon, C.S. Hoveland, and D.E. Radcliffe. 1989. Effect of the tall fescue endophyte on plant responses to environmental stress. Agron. J. 81:83–90.

Arechavaleta, M., C.W. Bacon, R.D. Plattner, C.S. Hoveland, and D.E. Radcliffe. 1992. Accumulation of ergopeptide alkaloids in symbiotic tall fescue grown under deficits of soil water and nitrogen fertilizer. Appl. Environm. Microbiol. 58:857–861.

Ayad, J.Y. 1998. The effect of seaweed extract (*Ascophyllum nodosum*) on antioxidant activities and drought tolerance of tall fescue (*Festuca arundinacea* Schreb.) Ph.D. thesis. Texas Tech Univ., Lubbock.

Azevedo, M.D., and R.E. Welty. 1995. A study of the fungal endophyte *Acremonium coenophialum* in the roots of tall fescue seedlings. Mycologia 87:289–297.

Azevedo, M.D., R.E. Welty, A.M. Creaig, and J. Bartlett. 1993. Ergovaline distribution, total nitrogen and phosphorus content of two endophyte-infected tall fescue clones. p. 59–62. *In* D.E. Hume et al. (ed.) Proc. 2nd Int. Symposium on *Acremonium*/ Grass Interactions, Palmerston North, New Zealand. 4–6 Feb. 1993. AgResearch, Grassland Research Centre, Palmerston North.

Bacon, C.W., J.K. Porter, J.D. Robbins, and E.S. Luttrell. 1977. *Epichloë typhina* from toxic tall fescue grasses. Appl. Environm. Microbiol. 35:576–581.

Barker, D.J., D.E. Hume, and P.E. Quigley. 1997. Negligible physiological responses to water deficit in endophyte-infected and uninfected perennial ryegrass. p. 137–139. *In* C.W. Bacon and N.S. Hill (ed.) *Neotyphodium*/grass interactions. Plenum Press, New York.

Belesky, D.P., O.J. Devine, and J.E. Pallas, Jr. 1987. Photosynthetic activity of tall fescue as influenced by a fungal endophyte. Photosynthetica 21:82–87.

Belesky, D.P., and D.P. Malinowski. 2000. Abiotic stresses and morphological plasticity and chemical adaptations of *Neotyphodium*-infected tall fescue plants. p. 455–484. *In* C.W. Bacon and J.F. White, Jr. (ed.) Microbial endophytes. Marcel Dekker, New York.

Belesky, D.P., J.A. Stuedemann, R.D. Plattner, and S.R. Wilkinson. 1988. Ergopeptine alkaloids in grazed tall fescue. Agron. J. 80:209–212.

Blokhina, O., E. Virolainen, and K.V. Fagerstedt. 2003. Antioxidants, oxidative damage, and oxygen deprivation stress: A review. Ann. Bot. (London) 91:179–194.

Bonnet, M., O. Camares, and P. Veisseire. 2000. Effects of zinc and influence of *Acremonium lolii* on growth parameters, chlorophyll *a* fluorescence and antioxidant enzyme activities of ryegrass (*Lolium perenne* L. cv Apollo). J. Exp. Bot. 51:945–953.

Bouton, J., R. Gates, N. Hill, and C. Hoveland. 2002a. Agronomic traits with MaxQ tall fescue. p. 40–41. Proc. Tall Fescue Toxicosis Workshop, SERAIEG-8, Wildersville, TN. 27–29 Oct. 2002. Missouri Forage and Grassland Council.

Bouton, J.H., G.C.M. Latch, N.S. Hill, C.S. Hoveland, M.A. McCann, R.H. Watson, J.A. Parish, L.L. Hawkins, and F.N. Thompson. 2002b. Reinfection of tall fescue cultivars with non-ergot alkaloid-producing endophytes. Agron. J. 94:567–574.

Buck, G.W., C.P. West, and H.W. Elbersen. 1997. Endophyte effect on drought tolerance in diverse *Festuca* species. p. 141–143. *In* C.W. Bacon and N.S. Hill (ed.) *Neotyphodium*/ grass interactions. Plenum Press, New York.

Bush, L.P., F.F. Fannin, M.R. Siegel, D.L. Dahlman, and H.R. Burton. 1993. Chemistry, occurrence and biological effects of saturated pyrrolizidine alkaloids associated with endophyte–grass interactions. Agric. Ecosyst. Environ. 44:81–102.

Carson, R.D., C.P. West, B. de los Reyes, S. Rajguru, and C.A. Guerber. 2004. Endophyte effects on dehydrin protein expression and membrane leakage in tall fescue. Poster No. 202. *In* C.F. Rosenkrans, Jr. (ed.) Proc. 5th Int. Symp. *Neotyphodium*/Grass Interactions, Fayetteville, AR. 23–26 May 2004. Plenum Press, New York.

Cheplick, G.P. 1998. Genotypic variation in the regrowth of *Lolium perenne* following clipping: Effects of nutrients and endophytic fungi. Functional Ecol. 12:176–184.

Cheplick, G.P., K. Clay, and S. Marks. 1989. Interactions between infections by endophytic fungi and nutrient limitation in the grasses *Lolium perenne* and *Festuca arundinacea*. New Phytol. 111:89–97.

Cheplick, G.P., A. Perera, and K. Koulouris. 2000. Effect of drought on the growth of *Lolium perenne* genotypes with and without fungal endophytes. Functional Ecol. 14:657–667.

Clay, K. 1990. Fungal endophytes of grasses. Annu. Rev. Ecol. Syst. 21:275–297.

De Battista, J.P., J.H. Bouton, C.W. Bacon, and M.R. Siegel. 1990. Rhizome and herbage production of endophyte removed tall fescue clones and populations. Agron. J. 82:651–654.

Dennis, S.B., V.G. Allen, K.E. Saker, J.P. Fontenot, J.Y.M. Ayad, and C.P. Brown. 1998. Influence of *Neotyphodium coenophialum* on copper concentration in tall fescue. J. Anim. Sci. 76:2687–2693.

Durand, J-L., M. Ghesquière, and C. Ravel. 2002. Effects of endophyte on long term productivity and tiller number in perennial ryegrass. Grassl. Sci. Eur. 7:528–529.

Elbersen, H.W., and C.P. West. 1996. Growth and water relations of field-grown tall fescue as influenced by drought and endophyte. Grass Forage Sci. 51:333–342.

Elmi, A.A., and C.P. West. 1995. Endophyte infection effects on stomatal conductance, osmotic adjustment and drought recovery of tall fescue. New Phytol. 131:61–67.

Faeth, S.H., M.L. Helander, and K.T. Saikkonen. 2004. Asexual *Neotyphodium* endophytes in a native grass reduce competitive abilities. Ecol. Lett. 7:304–313.

Ferguson, N.H., J.S. Rice, and N.G. Allgood. 1993. Variation in nitrogen utilization in *Acremonium coenophialum* isolates. Appl. Environm. Microbiol. 59:3602–3604.

Fike, J.H., V.G. Allen, R.E. Schmidt, X. Zhang, J.P. Fontenot, C.P. Bagley, R.L. Ivy, R.R. Evans, R.W. Coelho, and D.B. Wester. 2001. Tasco-Forage: I. Influence of a seaweed extract on antioxidant activity in tall fescue and in ruminants. J. Anim Sci. 79:1011–1021.

Flieger, M., P. Sedmera, J. Novak., L. Cvak, J. Zapletal, and J. Stuchlik. 1991. Degradation products of ergot alkaloids. J. Nat. Prod. 54:390–395.

Foy, C.D., and J.J. Murray. 1998. Developing aluminum-tolerant strains of tall fescue for acid soils. J. Plant Nutr. 21:1301–1325.

Funk, C.R., F.C. Belanger, and J.A. Murphy. 1994. Role of endophytes in grasses used for turf and soil conservation. p. 201–209. *In* C.W. Bacon and J.F. White, Jr. (ed.) Biotechnology of endophytic fungi of grasses. CRC Press, Boca Raton, FL.

Glenn, A.E., C.W. Bacon, R. Price, and R.T. Hanlin. 1996. Molecular phylogeny of *Acremonium* and its taxonomic implications. Mycologia 88:369–383.

Heritage Seeds. 2003. The Australian dairy pasture guide. 2003. Future prosperity... New Grass [Online]. Available at http://www.heritageseeds.com.au/ASSETS/pdfs/brochures/2003DairyBro.pdf [verified 15 Aug. 2004]. Heritage Seeds, Melbourne, Australia.

Hesse, U., W. Schöberlein, L. Wittenmayer, K. Förster, K. Warnstorff, W. Diepenbrock, and W. Merbach. 2003. Effects of *Neotyphodium* endophytes on growth, reproduction and drought-stress tolerance of three *Lolium perenne* L. genotypes. Grass Forage Sci. 58:407–415.

Kennedy, C.W., and L.P. Bush. 1983. Effect of environmental and management factors on the accumulation of *N*-acetyl and *N*-formyl loline alkaloids in tall fescue. Crop Sci. 23:547–52.

Kulkarni, R.K., and D. Nielsen. 1986. Nutritional requirements for growth of a fungus endophyte of tall fescue grass. Mycologia 78:781–786.

Latch, G.C.M., W.F. Hunt, and D.R. Musgrave. 1985. Endophytic fungi affect growth of perennial ryegrass. N. Z. J. Agric. Res. 28:165–168.

Lewis, G.C. 2000a. Effect of drought stress on genotypes of *Lolium perenne* and other grass species with and without *Neotyphodium/Epichloë* infection. Proc. Int. *Neotyphodium/* Grass Int. Symp. 4:201–205.

Lewis, G.C. 2000b. *Neotyphodium* endophytes: Incidence, diversity, and hosts in Europe. Proc. Int. *Neotyphodium/*Grass Int. Symp. 4:123–130.

Lewis, G.C. 2004. Effects of biotic and abiotic stress on the growth of three genotypes of *Lolium perenne* with and without infection by the fungal endophyte *Neotyphodium lolii*. Ann. Appl. Biol. 144:53–63.

Lewis, G.C., A.K. Bakken, J.H. Macduff, and N. Raistrick. 1996. Effect of infection by the endophytic fungus *Acremonium lolii* on growth and nitrogen uptake by perennial ryegrass (*Lolium perenne*) in flowing solution culture. Ann. Appl. Biol. 129:451–460.

Lewis, G.C., C. Ravel, W. Naffaa, C. Astier, and G. Charmet. 1997. Occurrence of *Acremonium* endophytes in wild populations of *Lolium* spp. in European countries and a relationship between level of infection and climate in France. Ann. Appl. Biol. 130:227–238.

Liu, H., J.R. Heckman, and J.A. Murphy. 1996. Screening fine fescues for aluminum tolerance. J. Plant Nutr. 19:677–688.

Lyons, P.C., and C.W. Bacon. 1984. Ergot alkaloids in tall fescue infected with *Sphacelia typhina*. Phytopathology 75:501.

Lyons, P.C., J.J. Evans, and C.W. Bacon. 1990. Effect of the fungal endophyte *Acremonium coenophialum* on nitrogen accumulation and metabolism in tall fescue. Plant Physiol. 92:726–732.

Macoon, B., R.C. Vann, J.D. Perkins III, and F.T. Withers, Jr. 2004. Steer performance and forage production on novel-endophyte fescue compared to ryegrass pastures. Poster No. 504. *In* C.F. Rosenkrans, Jr. (ed.) Proc. 5th Int. Symp. *Neotyphodium/*Grass Interactions, Fayetteville, AR. 23–26 May 2004. Plenum Press, New York.

Malinowski, D. 1995. Rhizomatous ecotypes and symbiosis with endophytes as new possibilities of improvement in competitive ability of meadow fescue (*Festuca pratensis* Huds.). Diss. ETH No. 11397. Swiss Federal Institute of Technology (ETH), Zurich.

Malinowski, D.P., G.A. Alloush, and D.P. Belesky. 1998. Evidence for chemical changes on the root surface of tall fescue in response to infection with the fungal endophyte *Neotyphodium coenophialum*. Plant Soil 205:1–12.

Malinowski, D.P., G.A. Alloush, and D.P. Belesky. 2000. Leaf endophyte *Neotyphodium coenophialum* modifies mineral uptake in tall fescue. Plant Soil 227:115–126.

Malinowski, D.P., and D.P. Belesky. 1999a. *Neotyphodium coenophialum*-endophyte infection affects the ability of tall fescue to use sparingly available phosphorus. J. Plant Nutr. 22:835–853.

Malinowski, D.P., and D.P. Belesky. 1999b. Tall fescue aluminum tolerance is affected by *Neotyphodium coenophialum* endophyte. J. Plant Nutr. 22:1335–1349.

Malinowski, D.P., and D.P. Belesky. 2000. Adaptations of endophyte-infected cool-season grasses to environmental stresses: Mechanisms of drought and mineral stress tolerance. Crop Sci. 40:923–940.

Malinowski, D.P., D.K. Brauer, and D.P. Belesky. 1999. *Neotyphodium coenophialum*-endophyte affects root morphology of tall fescue grown under phosphorus deficiency. J. Agron. Crop Sci. 183:53–60.

Malinowski, D., A. Leuchtmann, D. Schmidt, and J. Nösberger. 1997. Growth and water status in meadow fescue is affected by *Neotyphodium* and *Phialophora* species endophytes. Agron. J. 89:673–678.

Malinowski, D.P., H. Zuo, D.P. Belesky, and G.A. Alloush. 2004. Evidence for copper binding by extracellular root exudates of tall fescue but not perennial ryegrass infected with *Neotyphodium* spp. endophytes. Plant Soil (In press).

Malinowski, D.P., H. Zuo, W.E. Pinchak, J.P. Muir, and A.E. Stratton. 2003. Obligatory summer-dormant cool-season perennial grasses are drought resistant in semi-arid environments of the Texas Rolling Plains. Am. Forage Grassland Council Proc. 12:192–196.

Marks, S., and K. Clay. 1990. Effects of CO_2 enrichment, nutrient addition, and fungal endophyte infection on the growth of two grasses. Oecologia 84:207–217.

Marschner, H. 1986. Mineral nutrition in higher plants. Academic Press, Orlando, FL.

Marshall, D., B. Tunali, L.R. Nelson. 1999. Occurrence of fungal endophytes in species of wild *Triticum*. Crop Sci. 39:1507–1512.

McLeod, AR, A. Rey, K.K. Newsham, G.C. Lewis, and P. Wolferstan. 2001. Effects of elevated ultraviolet radiation and endophytic fungi on plant growth and insect feeding in *Lolium perenne*, *Festuca rubra*, *F. arundinacea* and *F. pratensis*. J. Photochem. Photobiol., B 62:97–107.

Monnet, F., N. Vaillant, P. Vernay, A. Coudret, H. Sallanon, and A. Hitmi. 2001. Relationship between PSII activity, CO_2 fixation, and Zn, Mn and Mg contents of *Lolium perenne* under zinc stress. J. Plant Physiol. 158:1137–1144.

Morgan-Jones, G., and W. Gams. 1982. Notes on Hyphomycetes. XLI. An endophyte of *Festuca arundinacea* and the anamorph of *Epichloë typhina*, new taxa in one of two new sections of *Acremonium*. Mycotaxon 15:311–318.

Naffaa, W., C. Ravel, and J.J. Guillaumin. 1998. Nutritional requirements for growth of fungal endophytes of grasses. Can. J. Microbiol. 44:231–237.

Newsham, K.K., G.C. Lewis, P.D. Greensdale, and A.R. McLeod. 1998. *Neotyphodium lolii*, a fungal leaf endophyte, reduces fertility of *Lolium perenne* exposed to elevated UV-B radiation. Ann. Bot. (London) 81:397–403.

New Zealand Agriseeds. 2003. 2004 endophyte options [Online]. Available at http://www. agriseeds.co.nz [updated 4 June 2004; verified 15 Aug. 2004]. NZA Limited, Christchurch, New Zealand.

Nihsen, M.E., E.L. Piper, C.P. West, R.J. Crawford, Jr., T.M. Denard, Z.B. Johnson, C.A. Roberts, D.A. Spiers, and C.F. Rosenkrans, Jr. 2004. Growth rate and physiology of steers grazing tall fescue inoculated with novel endophytes. J. Anim. Sci. 82:878–883.

Porter, J.K. 1994. Chemical constituents of grass endophytes. p. 103–123. *In* C.W. Bacon and J.F. White, Jr. (ed.) Biotechnology of endophytic fungi of grasses. CRC Press, Boca Raton, FL.

Powell, R.G., M.R. TePaske, R.D. Plattner, J.F. White, and S.L. Clement. 1994. Isolation of resveratrol from *Festuca versuta* and evidence for the widespread occurrence of this stilbene in the Poaceae. Phytochemistry 35:335–338.

Ravel, C., C. Courty, A. Coudret, and G. Charmet. 1997. Beneficial effects of *Neotyphodium lolii* on the growth and the water status in perennial ryegrass cultivated under nitrogen deficiency or drought stress. Agronomie 17:173–181.

Richardson, M.D, G.W. Chapman, Jr., C.S. Hoveland, and C.W. Bacon. 1992. Sugar alcohol in endophyte-infected tall fescue under drought. Crop Sci. 32:1060–1061.

Robbers, J.E. 1984. The fermentative production of ergot alkaloids. p. 197. *In* A. Mizrahi and A.L. van Wezel (ed.) Advances in biotechnological processes. Vol. 3. A.R. Liss, New York.

Saikkonen, K., M. Helander, S.H. Faeth, F. Schulthess, and D. Wilson. 1999. Endophyte-grass-herbivore interactions: The case of *Neotyphodium* endophytes in Arizona fescue populations. Oecologia 121:411–420.

Saker, K.E., V.G. Allen, J. Kalnitsky, C.D. Thatcher, W.S. Swecker, Jr., and J.P. Fontenot. 1998. Monocyte immune cell response and copper status in beef steers that grazed endophyte-infected tall fescue. J. Anim. Sci. 76:2694–2700.

Schardl, C.L., and T.D. Phillips. 1997. Protective grass endophytes: Where are they from and where are they going? Plant Dis. 81:430–437.

Schulthess, F., and S.H. Faeth. 1998. Distribution, abundances, and associations of the endophytic fungal community of Arizona fescue (*Festuca arizonica*). Mycologia 90:569–578.

Skinner, R.H., and C.J. Nelson. 1994. Effect of tiller trimming on phyllochron and tillering regulation during tall fescue development. Crop Sci. 34:1267–1273.

Spiering, M.J., H.H. Wilkinson, J.D. Blankenship, and C.L. Schardl. 2002. Expressed sequence tags and genes associated with loline alkaloid expression by the fungal endophyte *Neotyphodium uncinatum*. Fungal Genet. Biol. 36:242–254.

Tapper, B.A., and G.C.M. Latch. 1999. Selection against toxin production in endophyte-infected perennial ryegrass. p. 107–111. *In* D.R. Woodfield and C. Matthew (ed.) Ryegrass endophyte: An essential New Zealand symbiosis. Grassl. Res. and Practice Ser. 7. New Zealand Grassl. Assoc., Palmerston North, New Zealand.

Tsai, H.–F., J.-S. Liu, C. Staben, M.J. Christensen, G.C.M. Latch, M.R. Siegel, and C.L. Schardl. 1994. Evolutionary diversification of fungal endophytes of tall fescue grass by hybridization with *Epichloë* species. Proc. Natl. Acad. Sci. USA 91:2542–2546.

Vazquez-de-Aldana, B.R., B. Garcia-Criado, I. Zabalgogeazcoa, A. Garcia–Ciudad. 1999. Influence of fungal endophyte infection on nutrient element content of tall fescue. J. Plant Nutr. 22:163–176.

Vinton, M.A., E.S. Kathol, K.P. Vogel, and A.A Hopkins. 2001. Endophytic fungi in Canada wild rye in natural grasslands. J. Range Manage. 54:390–395.

Volaire, F. 2002. Drought survival, summer dormancy and dehydrin accumulation in contrasting cultivars of *Dactylis glomerata*. Physiol. Plant. 116:42–51.

West, C.P., and S.A. Gunter. 2004. Persistence of Hi-Mag tall fescue inoculated with nontoxic endophytes. Poster No. 518. *In* C.F. Rosenkrans, Jr. (ed.) Proc. 5th Int. Symp. *Neotyphodium*/Grass Interactions, Fayetteville, AR. 23–26 May 2004. Plenum Press, New York.

White, J.F., Jr. 1994. Taxonomic relationships among the members of the Balansiae (Clavicipitales). p. 3–20. *In* C.W. Bacon and J.F. White, Jr. (ed.) Biotechnology of endophytic fungi of grasses. CRC Press, Boca Raton, FL.

White, J.F., Jr., and P.V. Reddy. 1998. Examination of structure and molecular phylogenetic relationships of some graminicolous symbionts in genera *Epichloë* and *Parepichloë*. Mycologia 90:226–234.

Zaurov, D.E., S. Bonos, J.A. Murphy, M. Richardson, and F.C. Belanger. 2001. Endophyte infection can contribute to aluminum tolerance in fine fescues. Crop Sci. 41:1981–1984.

Zhou, F., Y.B. Gao, and W.J. Ma. 2003. Effects of phosphorus deficiency on growth of perennial ryegrass–fungal endophyte symbiont and phenolic content in root. Plant Physiol. Commun. 39:321–325.

GROWTH AND MANAGEMENT OF ENDOPHYTIC
GRASSES IN PASTORAL AGRICULTURE

David E. Hume[1] and David J. Barker[2]

S ince the discovery of the causal link between fungal endophyte (*Neotypho-dium* spp.) and animal disorders (Fletcher and Harvey, 1981; Hoveland et al., 1980) and plant tolerance of biotic and abiotic stresses (see Chapters 7 and 8, this volume), livestock farmers face a dilemma over whether to (a) incur the toxicity associated with feeding livestock tall fescue (*Festuca arundinacea*) or perennial ryegrass (*Lolium perenne*) infected with endophyte, or (b) incur reduced pasture production and persistence that frequently follows establishment of endophyte-free pastures (Bouton et al., 1993; Popay et al., 1999). In the former case, several options have been developed for alleviating or reducing toxicity through administering treatments to livestock (Stuedemann and Thompson, 1993), using grazing management to minimize exposure to endophyte (Keogh, 1983), and manipulating pasture composition (Ball, 1997; Cosgrove et al., 1996). A further option, now available to farmers in the USA, Australia, and New Zealand, is to establish tall fescue and perennial ryegrass that is infected with a selected endophyte that has low or no mammalian toxicity. These selected (or nontoxic) endophytes avoid the previous dilemma by essentially providing farmers with endophyte-infected grasses with good production and persistence combined with good animal health and productivity (Bouton et al., 2002; Fletcher, 1999; Popay et al., 1999). In this review we consider the options and consequences of using endophyte-free and selected endophytes as agronomic methods to avoid the use of toxic endophyte-infected pastures.

[1] AgResearch, Grasslands Research Centre, Private Bag 11008, Palmerston North, New Zealand
[2] Department of Horticulture and Crop Science, The Ohio State University, Columbus, OH 43210, USA

ENDOPHYTE CONTAMINATION IN PASTURES

The decision to use tall fescue or perennial ryegrass seed that is endophyte-free or infected with a selected endophyte (rather than a toxic endophyte in tall fescue or perennial ryegrass) must be preceded by consideration of the level of endophyte contamination that is acceptable in pasture, and the probability that the level of contamination will stay consistent between years and even between seasons within a year. We define *contamination* as the proportion of a tall fescue or perennial ryegrass tiller population that is infected with an endemic toxic endophyte. Because of the negative effects on animal health/productivity, such endophyte-infected tillers (or plants) are undesirable for a pastoral farmer. For tall fescue, this toxic endophyte (*N. coenophalium*) produces lysergyl alkaloids (primarily measured as ergovaline) in the host grass. For perennial ryegrass, endophyte (*N. lolii*) toxins produced include both lysergyl alkaloids (primarily measured as ergovaline) and the lolitrem subgroup of indole-diterpenes (primarily measured as lolitrem B). For lolitrem B, the threshold above which ryegrass staggers occurs in sheep and cattle is 1.8 to 2.5 ppm when measured in herbage to ground level (di Menna et al., 1992; Tor-Agbidye et al., 2001). A concentration of lolitrem B below this threshold is unlikely to be a problem in these animal species. In contrast, cattle, sheep and horse responses to lysergyls in tall fescue and perennial ryegrass are essentially linear and appear to have no safe threshold below which animal live weight gain (or abortion in horses) is unaffected (Layton et al., 2004; Schmidt and Osborn, 1993). In the USA, the rule of thumb is that for each 10% increase in endophyte infection in tall fescue there is a reduced potential weight gain in yearling cattle of 45 g d^{-1} (0.1 lb d^{-1}). Ultimately, the tolerable level of pasture contamination by lysergyl alkaloids is a matter of economics, where the benefit from financial gains in animal production will be balanced against the costs of the various methods of minimizing contamination of new pastures with toxic endophytes.

Increases in Endophyte Infection in New Zealand

Early New Zealand work on the agronomy of endophyte-infected perennial ryegrass pastures in the 1980s (Francis and Baird, 1989; Prestidge et al., 1984, 1985) soon recognized that pastures established from perennial ryegrass seed containing no endophyte, or a low proportion of endophyte-infected seed, could end up with a high proportion of endophyte contamination within 1 to 2 yr. Along with a potential increased risk of toxicity from such pastures, production of endophyte-free pastures may decline, particularly in summer/autumn (Pennell et al., 2001; Popay et al., 1999), and may result in pastures that have lower percentages of perennial ryegrass and higher percentages of weeds (Francis and Baird, 1989; Prestidge et al., 1985).

The initial level of contamination may be a significant factor affecting the subsequent rate of contamination. At two contrasting sites, Pennell et al. (2001) found that populations with an initial infection level of <4% endophyte remained relatively stable (with low infection), but populations where endophyte levels averaged 14% increased to 55% after 2 yr. This is not always true, and even where seed is sown that is free of endophyte, large increases in the proportion of endophyte-infected tillers have been recorded. For example, Popay et al. (1999) sowed the same endophyte-free seed line in a total of 10 trials over seven locations and 2 yr, and found that plots 2 to 3 yr later had perennial ryegrass infected with endophyte up to a rate of 54%. At other locations, contamination remained low (<10%) as it did during a 5-yr study reported by Eerens et al. (1998) at Gore. Although the sources of contamination or reasons for increases in percentage endophyte infection were not apparent in the study of Popay et al. (1999) (this will be discussed later), it was clear that there could be distinct differences between sites.

Increased rates of contamination by toxic endophyte can also be significant when initial infection levels are medium (e.g., 40–60%). In Taranaki, Popay et al. (2003) reported an increase in endophyte from 61 to 94% across 15 mo. Endophyte infection increased from 70 to 95% during 3 yr in the Manawatu, an environment considered to be climatically very favorable for perennial ryegrass growth but also sufficiently warm to allow several generations of insect pests per year (Hume and Brock, 1997a; Hume and Brock, 1998, unpublished data).

Endophyte-infected tall fescue has not been traded as commercially certified seed in New Zealand, but naturalized tall fescue growing on roadsides and near waterways is highly infected (Guy and Davis, 2002), and when it occurs in sufficient quantities in pastures it causes toxic effects in grazing animals (Kearns, 1986). Where a toxic endophyte has been a contaminant in agronomic plots, a large increase in endophyte infection has been recorded (from 30 to 80% during 3 yr) with a corresponding improvement in dry matter yields (Easton and Cooper, 1997). This, combined with the results of small-plot evaluations of the nontoxic endophyte AR542 in cultivars adapted to New Zealand (Cooper et al., 2002; Easton et al., 2004), indicates that a number of selective forces favor the growth and persistence of endophyte-infected tall fescue tillers.

Increases in Endophyte Infection in Australia

The case for increased endophyte infection in pastures in Australia is not as clear as that for New Zealand. However, a number of aspects point to similar trends.

Old perennial ryegrass pastures have high levels of infection in Tasmania (76%; Guy, 1992), Victoria/New South Wales (90%; Reed et al., 2000), and New South Wales's Northern (75%; Sen, 1995) and Central Tablelands (Wheatley, 1997). In addition, there are several insect pests of perennial ryegrass in Australia that can be affected by the presence of endophyte (McDonald et al., 1993). Given these two factors, it is not surprising that Australian cultivars developed from locally bred ecotypes that are persistent, such as 'Victorian', have high levels of seed infection (Cunningham et al., 1993; Heeswijck and McDonald, 1992; Valentine et al., 1993). Degree of infection of seed lots within a cultivar is likely to be driven by the age of seed when sold (see Chapter 17). Only one pasture study specifically monitored changes in percentage of endophyte-infected tillers across time (Cunningham et al., 1993). In this case, at a site with rainfall that was marginal for perennial ryegrass persistence, pastures sown with 79% endophyte-infected perennial ryegrass seed had 100% of the surviving plants infected 4 yr later.

Similar information is not available for tall fescue primarily because little or no endophyte-infected seed has been traded in the past (Easton et al., 1994; Valentine et al., 1993). However, where tall fescue is growing wild along roadsides and riverbanks in Tasmania, infection is high (78 to 100%; Guy and Davis, 2002). Similarly, in South Australia (Pulsford, 1950) and from collections in Victoria and New South Wales (Chapter 17, this volume; G.C.M. Latch, 1995, unpublished data), naturalized tall fescue in grazed pastures can be highly infected, which causes problems with animal health/productivity and grazing management. Recent trials evaluating the selected endophyte AR542 show that endophyte infection enhances production and persistence under conditions of soil water deficit and particularly when high populations of insect pests are present (Wheatley et al., 2003).

From the above information, there appears to be sufficient selective pressure in the high rainfall zone of southeastern Australia to favor endophyte-infected tillers. Both perennial ryegrass and tall fescue pastures with a low or moderate level of endophyte infection are likely to increase in infection level across time.

Increases in Endophyte Infection in Tall Fescue in the USA

With the discovery of the relationship between toxicosis and endophyte in tall fescue (Bacon et al., 1977; Hoveland et al., 1980, 1983) the option of excluding toxic tall fescue from pasture has been investigated. Many studies have reported no increases in the proportion of endophyte-infected tall fescue tillers across time in many states (Gwinn et al., 1998; Penrose et al., 2001; Read and Camp, 1986; Shelby and Dalrymple, 1993; Shelby et al., 1989; Siegel et al., 1984). However, numerous other reports have shown increases in infection percentage across time, and this may be dependent on various management factors (Gwinn et al., 1998; Penrose et al., 2001) or the length of the study. In Georgia, Belesky et al. (1987) recorded no increases for the first 2 yr, but then an increase of 16 percentage points in Year 3. Similarly, Pedersen et al. (1984) recorded a large increase in infection only between the fourth and fifth years. In a 12-yr study in Alabama, where the average increase was 4% per annum, there were periods of many years when no increases were observed (Shelby and Dalrymple, 1993).

As with the New Zealand studies reported for perennial ryegrass, the initial level of infection may affect the rate of increase. The results from 62 fields in Alabama and Tennessee (Shelby and Dalrymple, 1993; Shelby et al., 1989; Thompson et al., 1989) are summarized in two logistic responses in Fig. 9.1. At both low (<10%) and high (>60%) initial endophyte levels, the endophyte populations were relatively stable. Populations with moderate (10 to 60%) initial infection showed the greatest change in endophyte, with the highest rates (the point of inflection for respective curves) occurring at either 14 or 36% endophyte infection. The reason for the difference between these curves is not known, but might represent extremes in the potential invasion by toxic-endophyte-infected endemic tall fescue.

Population Stability of Nontoxic and Toxic Endophytes

With the commercial availability of specific endophyte strains within tall fescue and ryegrass in the USA, Australia, and New Zealand, the relative competitiveness of strains within a pasture has become important. From the perspective of animal production, the value of using low-toxin or nontoxic endophyte strains is that toxic strains are replaced. It is desirable that toxic strains will not rapidly reinvade the pasture, either reducing or eliminating the advantages of the newly established nontoxic endophytes. Selected nontoxic endophytes have a reduced number or range of alkaloids compared with the endemic toxic endophytes in these cultivars; therefore, some nontoxic endophyte–cultivar combinations may not offer the same degree of insect tolerance (see Chapter 7). For example, compared with the endemic toxic endophyte, ryegrasses infected with AR1 were more susceptible to African black beetle (*Heteronychus arator*) (Popay and Baltus, 2001) and root aphid (*Aploneura lentisci*) (Popay et al., 2004), and tall fescue infected with AR542 was more susceptible to nematodes (*Pratylenchus* spp.) (Timper et al., 2003) and shoot aphid (*Rhopalosiphum padi*) (Hunt, 2003). Therefore, tillers infected with nontoxic endophytes may not be as competitive as toxic endophyte-infected tillers in grazed pastures.

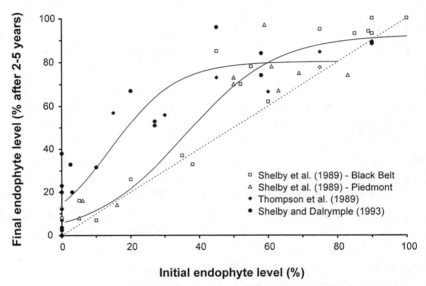

Fig. 9.1. *The relationship between final (after 2 to 5 yr) and initial endophyte levels for 62 fields in Alabama and Tennessee.*

Two New Zealand perennial ryegrass studies evaluating the nontoxic AR1 endophyte under self-contained dairy farmlet systems have reported minimal and/or stable levels of contamination with toxic endophytes. In the Waikato, Bluett et al. (2003) found that paddocks of AR1-infected perennial ryegrass had "almost undetectable" levels of lolitrem B and ergovaline (alkaloids produced by toxic endophytes in perennial ryegrass) during 3 yr, and at the end of the third year only four of 700 tillers were identified as producing these toxins. In a separate Northland study, pastures with the AR1 endophyte had measurable levels of contamination with toxic endophyte 1 yr after sowing (range = 0–18% contamination). These populations were stable and showed no significant changes 2 yr later (range = 2–17% contamination; Ussher, 2003; D.E. Hume et al., 2003, unpublished data).

While these trials are showing early signs that AR1 pastures might resist contamination from toxic endophytes, 'Grasslands Nui' perennial ryegrass plots purposely sown with a 50:50 blend of AR1 and the toxic endemic endophyte have not been stable (Fig. 9.2; D.E. Hume et al., 2003, unpublished data). On average, the proportion of the toxic endophyte has increased 10 percentage points per annum. As with the studies reported by Popay et al. (1999), there was variation between sites. More recent work with perennial ryegrass utilizing selected endophytes in mixtures also has exhibited changes in proportions of endophyte strains across time (D.E. Hume and A.J. Popay, 2003, unpublished data).

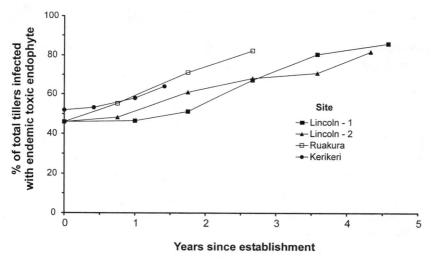

Fig. 9.2. *Changes in proportion of toxic endophyte-infected tillers since establishing mixtures of equal proportions of 'Grasslands Nui' perennial ryegrass infected with AR1 endophyte, and the same cultivar infected with toxic endemic endophyte, at four sites in New Zealand.*

The competitive ability of a selected nontoxic endophyte has been investigated under grazing and mowing at three sites in Ohio, USA (Barker et al., 2005). In this study, the endophyte strain AR542 in 'Jesup' tall fescue was sown in mixtures with varying proportions (0, 10, 20, 30, 40, and 50% of viable seed weight) of 'Kentucky 31' infected with its toxic endemic endophyte. At 18 mo after establishment, the proportion of tillers infected with the toxic endophyte remained similar to the sown proportion at one site, but had decreased to near-zero levels at the other two sites. The mechanisms by which AR542 might out-compete toxic endophyte are not known. The plots with 100% AR542 (actually 85% endophyte, 15% endophyte-free) remained 85% endophyte-infected after 18 mo with negligible toxic endophyte. In contrast, three plots sown with 100% endophyte-free tall fescue averaged 11% infection with toxic endophyte after 18 mo at the Jackson, OH, site and presumably this had resulted from endophyte-infected volunteer tall fescue.

Another U.S. study using mixtures of Jesup AR542 and its endemic toxic endophyte has been initiated in Georgia (J.H. Bouton et al., 2001, unpublished data). Mixtures were formulated to represent either low (10%) or medium (50%) contamination with toxic endophyte. Early results (17 mo after establishment) show no changes in the relative proportions of the endophyte strains. These researchers also tested paddocks of the selected endophytes AR501 and AR542 in three tall fescue cultivars used to evaluate animal health and productivity of cattle and lambs. After 3 yr, mean contamination was approximately 1% and no more than 8% in any one assessment.

Changes in Endophyte Infection within a Year

While there is strong evidence for changes in the percentage of endophyte-infected tillers between years, evidence for variation in contamination within a year is limited. When there was no increase in infection between years, Belesky et al. (1987) report no variation in the percentage of endophyte-infected tillers between spring and summer within the same year. Read and Camp (1986) did observe changes in one year only, being highest in late autumn, dropping again by spring/summer of the following year. Thompson et al. (1989) reported only slight variation from season to season within a year when they examined pastures that had >50% endophyte infection in tall fescue. Care should also be taken not to assess tiller infection in winter and early spring, when detection of fungal infection can be difficult (Barker et al., 2004).

Using the proportion of tillers infected with toxic endophyte as a gauge of contamination may not necessarily reflect the degree of contamination in terms of toxicity of the pasture through the growing season (Belesky et al., 1987, 1988). As ambient temperatures rise, increased growth of the fungus within an infected tiller may mean the same proportion of endophyte-infected tillers in the pasture produce more toxins, and have greater alkaloid concentrations per unit dry matter. These infected tillers also are likely to contribute a greater propor-tion to the grass herbage on-offer through greater tiller size/mass relative to the endophyte-free tillers in the pasture (Hill et al., 1991; Marks et al., 1991), especially as they are exposed to greater biotic and abiotic stresses. These stresses are often most pronounced during the summer and autumn.

SOURCES OF CONTAMINATION WITH TOXIC ENDOPHYTE-INFECTED PERENNIAL RYEGRASS AND TALL FESCUE

Contamination of pastures with toxic endophyte-infected tall fescue or perennial ryegrass can only occur through (i) seed and (ii) vegetative plants. The *Neotyphodium* endophytes infecting these grasses are not known to be transferred with pollen or by physical contact (Siegel et al., 1984). Contamination of pastures with toxic endophyte-infected plants may occur if (i) the seed lot used to establish new pastures contains seeds with viable toxic endophyte; (ii) natural reseeding occurs before or after pasture renovation; (iii) livestock that previously consumed grass seedheads containing toxic endophytes are moved to new pastures and deposit in dung any viable toxic endophyte in undigested seeds; or (iv) the farmer is negligent. Contamination from vegetative plants will occur primarily through the incomplete kill of existing pasture.

Sown Seed

Establishing new pastures with seed containing an unwanted viable toxic endophyte is among the most common mechanisms for reintroducing endophyte to renovated pastures. Seed should be clearly marked with its endophyte status and endophyte strain, and there must be an awareness of the issues regarding endophyte by the customer and the various sectors of the pastoral industry (see Chapter 17). Even during simple procedures such as blending cultivars and species for a seed mix, endophyte status and type should be considered.

In New Zealand, most cultivars of perennial and hybrid (*L. boucheanum* syn. *L. hybridum*) ryegrasses are available as *High* endophyte (>70% viable endophyte-infected seed) and *Nil* (≤ 5% viable endophyte-infected seed), and some are traded as *Low* endophyte. Given the above, Low implies between 5 and 70% infection, and this may be an option to reduce levels of endophyte infection. However, the dangers of using this category of seed have been highlighted by McCorkindale (1999). In this case, planting Low endophyte seed resulted in an endophyte infection level (70% tillers endophyte-infected) that was higher than observed in these pastures prior to renovation (55%). Nil seed ($\leq$ 5%) has a similar but considerably smaller risk of introducing toxic endophyte seed.

In the case of tall fescue, Ball (2001) discussed farmer practice in the USA of purchasing seed infected with viable toxic endophyte (often cheaper than endophyte-free seed), storing it under ambient conditions with an expectation that within 1 yr the endophyte will be nonviable, and then using this same seed lot to establish an endophyte-free pasture. Ball (2001) described this practice as a false economy; the seed not only may decline in germination and vigor, but also may contain viable toxic endophyte as not all the endophyte may be killed during a 1-yr period because of variable climatic conditions of ambient temperature and humidity.

Even when seed has been officially tested as 0% viable infection, low levels of infection (e.g., 1–2%) may escape detection (Shelby and Dalrymple, 1993; D.E. Hume, 2004, unpublished data). Most laboratory tests examine 20 to 100 seeds, and there is a statistical probability that low levels of infection may not be detected. However, given the evidence presented in the sections above, this may not contribute to significant increases in the percentage infection in a pasture for many years. The same issue of limits of detection and being within specified quality limits also applies to seed of selected endophytes with regard to contamination with toxic endophytes. Currently, commercial seed of AR1 shows little evidence of contamination (see Chapter 1.3); similar results have been observed for tall fescue infected with AR542 in the USA (A.E. Stratton, 2003, personal communication).

Natural Reseeding and Buried Seed

Grass seedheads produced in spring and summer that are not decapitated by grazing animals or mechanically defoliated by farmers will produce viable seed that will fall to the soil surface. In general, seeds of both tall fescue and perennial ryegrass have a short-term dormancy (postharvest dormancy) which allows them to form only a "transient seed bank" (Grime et al., 1988; Thompson and Grime, 1979). As temperatures decline and rainfall and soil moisture increase, most perennial ryegrass seeds germinate in the autumn–early winter period (Hume and Barker, 1991; L'Huillier and Aislabie, 1988), as does tall fescue although this may be delayed until the spring (Harris, 1961). Successful seedling establishment will depend on gaps that might have been created in pasture by summer–autumn drought, insect pests, and the death of summer-active annual grasses and weeds (Froud-Williams et al., 1984). Natural reseeding as a means of deliberately regenerating perennial ryegrass pastures through deferred grazing has been successfully demonstrated under dairying in New Zealand (L'Huillier and Aislabie, 1988; McCallum et al., 1991), under sheep grazing in summer-dry areas in Victoria, Australia (Waller et al., 1999), and in the hill country in New Zealand (Hume and Barker, 1991). It is likely that endophyte-infected plants in a pasture will produce more seed than endophyte-free plants (Hill et al., 1991), and once on the soil surface, endophyte-infected seeds have a higher chance of surviving seed predation (Madej and Clay, 1991; Popay et al., 2000) and early growth as a seedling through avenues such as improved insect tolerance (Stewart, 1985).

Through natural and cultural processes (Garwood, 1989), seed on the soil surface may become incorporated into the soil as buried seed. Subsequently, buried seed may be returned to the soil surface by the same processes, and may germinate and emerge to form new plants. At depth, seed may experience enforced dormancy (Froud-Williams et al., 1984) through lower temperatures, smaller temperature fluctuations, and less light and oxygen (Simpson, 1990). The greater long-term survival of buried seed compared with seed on the soil surface contributes to its relative importance as a vehicle for contamination of pastures.

The primary survival strategy of perennial ryegrass and tall fescue is by vege-tative means, that is, by tillering (Grime et al., 1988; Rice, 1989). Seeds of these grasses have few or none of the characteristics needed to create a significant reservoir of buried seed, and subsequently have poor survival when buried in the soil relative to other species (Rampton and Ching, 1970; Thompson and Grime, 1979). However, buried seed may be a concern in the relatively short time frames that are important in pastoral agriculture. For example, to rid paddocks of buried seed in the production of certified seed, New Zealand standards require a minimum of 2 yr between successive crops of ryegrass or tall fescue (AgriQuality New Zealand, 1999), and Oregon standards are for 2 to 5 yr depending on the class of seed produced (Rampton and Ching, 1970). In ex-perimental situations, small amounts of seed can be found under pasture 6 to 12 mo after the seed is deposited during summer (L'Huillier and Aislabie, 1988; Thompson and Grime, 1979; D.E. Hume, 1999, unpublished data), with some studies showing survival after burial for 1 to 2 yr (Harris, 1961; Hume et al., 1999, 2001b; Roberts, 1986), and up to 4 yr (Lewis, 1973; Rampton and Ching, 1970) depending on soil type and ryegrass cultivar. Survival for these time periods may only be 0.1%, but levels of 22% after 4-yr burial have been re-corded for perennial ryegrass (Lewis, 1973).

Survival of endophyte in buried seed has been examined in New Zealand for perennial ryegrass (Hume et al., 1999, 2001a) and tall fescue (D.E. Hume and M.P. Rolston, 2001, unpublished data). The most important finding has been that endophyte viability does not decline faster than seed viability, although this may have been expected from seed storage studies for endophyte-infected ryegrass and tall fescue (Hare et al., 1990; Rolston et al., 1986; Welty et al., 1987). After 1 yr, Hume et al. (1999) reported 4% seed survival, while the corresponding endophyte viability declined from 58 to 21%. While seed survival was higher with greater depth of burial, endophyte viability was unaffected by depth. Differences between years and sites indicate that soil moisture, at least during the first summer/autumn, affects survival of both the seed and endophyte, with drier soil conditions enhancing survival. Endophyte survival did not differ between endophyte strains (toxic or nontoxic). While Quigley et al. (1993) showed that it was possible for endophyte to provide some protective advantage to seed buried in the soil, the buried seed studies described here did not find that endophyte-infected seeds had a better survival than endophyte-free seeds because the percentage of viable endophyte-infected seeds did not increase with burial time.

Occurrence of viable endophyte in seed as determined from soil cores also indicates that endophyte does survive in seed under natural conditions. Studies in the Waikato, New Zealand, have recorded viable endophyte in 75% of buried perennial ryegrass seed (Burggraaf and Thom, 2000) and 28% when pastures were infected at a level of 85% (D.E. Hume et al., unpublished data from the study of Bluett et al., 2001). In the latter study, pastures also contained short-lived hybrid ryegrasses that were infected with *N. occultans* or *N. lolii* at a level of 25%. Buried seeds from these ryegrasses were infected with viable *N. occultans* or *N. lolii* at a rate of 9%. It is also worth noting that the New Zealand Seed Certification Scheme treats each endophyte–ryegrass cultivar association as if it was a new cultivar in order to maintain purity of seed lots for endophyte strain or endophyte absence (AgriQuality New Zealand, 1999; Rolston, 1993). This practice has been in use since 1991.

Transfer of Seed by Animals

Domestic livestock and wildlife, such as birds and rabbits, may transfer seed by attachment to their wool and hides, or by soil attached to their hooves. This may be more significant in tall fescue than perennial ryegrass due to the awn on the tall fescue seed (Grime et al., 1988). However, seed transfer by animals is primarily through consumed forage (or hay) and subsequent deposition in dung (Watkin and Clements, 1978). Animal manure has been used as a method to disseminate desirable forage plants such as legumes (Burton and Andrews, 1948; Suckling, 1952), but it also can spread undesirable weed seeds (Piggin, 1978; Watkin and Clements, 1978) or tall fescue and perennial ryegrass seed infected with toxic endophytes. Once viable seeds are deposited in the field in dung, emerging seedlings have a high potential for survival. The dung pat produced by cattle reduces competition from the surrounding pasture by creating a gap in the canopy, and developing seedlings benefit from the fertility of the dung (Shelby and Schmidt, 1991).

Four studies using ryegrass (*L. perenne, L. multiflorum,* or *L. rigidum*) suggest a very low survival rate of seed through the gut (<1%; Burggraaf and Thom, 2002; Gramshaw and Stern, 1977; Rolston et al., 2001; Yamada and Kawaguchi, 1971). In contrast, Lennartz (1957) reported 36% survival for perennial ryegrass. Two field studies in New Zealand that examined viable seeds in dung (see review by Hume, 1999) found seed deposition equivalent to 32 and 53 viable perennial ryegrass seeds m^{-2} with cattle and 8 viable perennial ryegrass seeds m^{-2} with sheep. In these cases, grazed pastures had high numbers of reproductive tillers because the intention was to use grazing animals to disseminate viable seed (Suckling, 1952). The situation under a high-producing dairy system with high-quality pastures in the Waikato region of New Zealand was somewhat different; 0.4 to 2.6 perennial ryegrass seedlings m^{-2} per annum emerged from dung (calculated from Burggraaf and Thom, 2000).

In controlled studies with tall fescue seed fed to cattle, Shelby and Schmidt (1991) recovered 12% of the seed fed as viable seedlings in dung, while Siegel et al. (1984) had a 50% recovery when seed was fed via a rumen cannula.

Variation between studies may be explained, in part, by known variation between animal species and the effects of diet quality. Comparative studies indicate that cattle digest fewer seeds than sheep (Harmon and Keim, 1934; Simao Neto et al., 1987). Increasing diet quality also increases amounts of viable seed passed, possibly through increasing the flow rate of digesta through the gut and reducing rumination time (Jones and Simao Neto, 1987). Level of feed intake, stage of lactation, and ambient temperature are examples of other factors that can affect the rate of passage by feed particles, and consequently affect the survival of viable seed.

Along with the reduction in the viability of seed passing through the animal's gut, survival of endophyte in tall fescue and ryegrass seeds is also reduced. Rolston et al. (2001) fed sheep perennial ryegrass seed infected with 80% viable endophyte, and the viable seed deposited in dung had only 18% viable endophyte. Similarly with cattle fed tall fescue seed, Shelby and Schmidt (1991) recorded a decline from 98 to 12% of seeds infected with viable endophyte. Siegel et al. (1984) reported a reduction from 63 to 25% when seed was fed via a rumen cannula. The sensitivity of endophyte to conditions in the gut was apparent in the studies by Rolston et al. (2001) and Shelby and Schmidt (1991). These studies showed that viable seed was passed in the dung up to 2 d after the last viable endophyte-infected seed was passed.

Attempts to field verify the results of controlled indoor seed feeding studies have produced varied results. In particular, Shelby and Schmidt (1991) found 85 and 89% (total range = 67 to 100%) endophyte infection in tall fescue seedlings germinating in dung deposited by steers and horses, respectively, under ad libitum grazing. Under intensive dairy grazing in New Zealand, the infection rate of perennial ryegrass seedlings in dung has been 38 to 40% consistently (Burggraaf and Thom, 2000, 2002); ingested seeds in these studies were infected at a 100% rate (Burggraaf and Thom, 2002). This was irrespective of whether the dung was sampled fresh or if seedlings were newly emerging from dung pats that had been deposited several months previous. However, under these same dairy grazing conditions, S.J. Bluett et al. (2000, unpublished data) have recorded 76% endophyte infection for perennial ryegrass seedlings in dung from ingested seed estimated to be >85% endophyte-infected.

Studies assessing the time taken for viable endophyte-infected seed to pass through the gut have been used to consider the length of time animals should be quarantined to ensure unwanted endophyte is not transferred in dung. During this quarantine period, animals should be fed forage that does not contain any viable endophyte-infected seed. While Rolston et al. (2001) reported that the last viable endophyte-infected perennial ryegrass seed appeared in dung more than 4 d after ingestion in sheep, they suggested a 1.5-d quarantine period as a practical guide. By 1.5 d after ingestion, 95% of all viable endophyte had passed through the sheep. Further protection could be gained by extending the quarantine period to 2 d; by this point, 98% of infected seed had been deposited in dung. While Shelby and Schmidt (1991) found no viable endophyte-infected tall fescue seeds after approximately 1.5 d, they recommended a quarantine period of 3 d to allow for reductions in the flow rate of digesta. Similar reasons were cited by Burggraaf and Thom (2002), who recommended a 3-d quarantine period for dairy cows grazing perennial ryegrass.

Transfer of Seed by Farmers

Seed containing unwanted toxic endophyte may be unwittingly transferred into paddocks by farmers on farm equipment and in straw and hay. This avenue for contamination has been well recognized in the production of certified seed (AgriQuality New Zealand, 1999). Observations by Siegel et al. (1984) on the location of endophyte contamination neighboring a seed-production paddock also supports this path of contamination for tall fescue. The greatest contamination results from feeding hay or straw in large bales that are left in a pasture and eaten by livestock during an extended period of time. In this situation, the pasture is damaged or completely killed where it is covered by the bale. The associated trampling by livestock as they feed from the bale creates a larger area where existing pasture is damaged or killed. Seed from the bale is also trampled into the ground, thereby improving chances of establishment. In a recently established pasture in Ohio, germinating tall fescue seedlings were observed at a density exceeding 1000 seedlings m^{-2} in the residue of hay that had been fed during winter (D.J. Barker, April 2001, unpublished data).

Vegetative Plants

The success of establishing a new pasture with no or minimal contamination with toxic endophyte will be negatively correlated with the proportion of any existing endophyte-infected tall fescue and perennial ryegrass that survives the renovation process. When pastures were established with AR1-infected perennial ryegrass by no-tillage seeding (also referred to as over-drilling or direct-drilling) and no herbicide application, Bluett et al. (2001) reported that 75% of the perennial ryegrass tillers in the new pasture were derived from plants that survived the renovation process. This was reduced to 9% when herbicides and/or cultivation were used. Even if a small number of plants survive these processes, they have an advantage over the new seedlings because they already have an established root system and regrowth occurs while new seedlings are still germinating and emerging (Bluett et al., 2001).

Many studies have examined the best methods of seedbed preparation using various cultivations and/or herbicide applications for tall fescue (Fribourg et al., 1988; Smith, 1989) and perennial ryegrass (Bluett et al., 2001; Hume and Lyons, 1992). Various desiccants, such as paraquat, or translocated herbicides, such as glyphosate, have been used alone, in combinations, and in repeated applications. In general, higher application rates and repeated applications are needed for high control (Burggraaf and Thom, 2000; Defelice and Henning, 1990). Tall fescue may be more difficult to kill than ryegrass. What appears to be successful herbicide treatment to the casual observer may be misleading because plants can survive and regrow from crowns (Smith, 1989) and underground rhizomes (Defelice and Henning, 1990). Rhizomes are a morphological feature that ryegrasses do not possess (Grime et al., 1988; Hume and Brock, 1997b). Burggraaf and Thom (2000) also have shown that effectiveness of herbicide applications may be compromised by dung pats that protect plants from contact with the herbicide. Even a double herbicide application did not completely kill all plants, suggesting that dung deposited by grazing dairy cattle should be broken up by harrowing prior to herbicide application.

While cultivation appears to guarantee a better success rate than application of herbicide (Hume and Lyons, 1992), farmers must ensure that cultivation techniques are thorough, burying and killing all turf, and this may also require application of herbicide to be completely effective (Hume and Lyons, 1992, 1993; McCorkindale, 1999). Cultivation has the additional advantage of burying any tall fescue or perennial ryegrass seed on the soil surface that may germinate and contaminate the new pasture. It also may improve the success rate of establishing the new pasture, making it more competitive, but cultivation also can increase erosion potential compared with no-till establishment.

Difficulty in achieving complete control of resident vegetative grass plants by cultivation and/or herbicides during a short period has prompted the option of planting a smother crop or successive smother crops prior to establishing the new pasture (Fribourg et al., 1988). For example, arable crops such as wheat (*Triticum aestivum*), maize (*Zea mays*), and pea (*Pisum sativum*), or green-feed forage crops such as *Brassica* spp., forage maize, and oat (*Avena sativa*) have been used for this purpose. Other options include perennial crops such as alfalfa (*Medicago sativa*). This approach not only allows greater time for control of existing grass plants through repeated cultivation and/or herbicide applications, but also prevents natural reseeding and reduces the viability of any buried seeds of tall fescue, perennial ryegrass, or weeds (Rice, 1989). A smother crop also can provide a high cash return or extra forage dry matter to offset pasture renewal costs.

Relative Importance of Sources of Contamination

It is difficult to define the relative importance of the various sources of contamination. No one study has established treatments that combine all these potential sources of contamination, or separate these sources from increases in the percentage of endophyte infection because of better tolerance of biotic and abiotic stresses by endophyte-infected tillers.

In the case of sown seed and destruction of existing vegetation, the importance of these sources for contaminating new pastures will be directly proportional to the percentage contamination within the seed lines and inversely proportional to the success rate of killing the existing vegetation. In contrast, practices such as hay and straw feeding and cleanliness of farm machinery are directly under farmer control.

The relative importance of depositing viable endophyte-infected perennial ryegrass seed via dung in a paddock, either during pasture renovation or after establishment, may be sizeable. While the New Zealand studies (Burggraaf and Thom, 2000; see also review by Hume, 1999) with perennial ryegrass indicate relatively high amounts of viable seed deposited per unit area, these values are only equivalent to sowing 0.009 to 0.600 kg perennial ryegrass seed ha^{-1}, which is small in comparison to establishing a new perennial ryegrass pasture with seeding rates of 15 to 20 kg seed ha^{-1}. However, this seed delivered within dung may establish earlier than the sown seed (Hume and Lyons, 1992), and therefore has a competitive advantage when emerging from large dung pats. It also can be a continual input across a number of years, and therefore may become a significant contributor to contamination of pastures. No or minimal endophyte contamination has occurred in studies where animals have been (a) confined to the same endophyte treatment (Bluett et al., 2003), (b) quarantined before entering paddocks (Gwinn et al., 1998), or (c) not allowed to graze (Penrose et al., 2001; Shelby and Dalrymple, 1993). These results, combined with field observations such as "inexplicable foci of endophyte infestation" attributed to deposition of dung pats (Shelby and Schmidt, 1991; Shelby et al., 1989), and survival of endophyte through the digestive tract, all indicate that dung, at least in cattle, has a significant impact across time and that quarantine periods are justified. If quarantine periods are not an option, then mature seedheads should be minimized through mechanical removal or by hard grazing to reduce the animal's intake of viable seed.

Natural reseeding and the occurrence of buried seed have been the subjects of a number of studies involving the establishment of new perennial ryegrass pastures (Bluett et al., 2001; Hume and Lyons, 1992, 1993; L'Huillier and Aislabie, 1988). Practices such as taking a hay crop have huge potential for contamination because up to 12 000 perennial ryegrass seedlings m^{-2} can arise from natural reseeding. This would represent a contamination of 84 to 94%, if seed was drilled no-till without application of herbicide during autumn. However, the establishment method, including cultivation and/or herbicide spraying, can overcome this worst-case scenario, particularly if attention is given to the timing of operations that potentially kill germinating seedlings in autumn (Bluett et al., 2001). In established pastures, natural reseeding also may be a significant source of contamination in tall fescue (Gwinn et al., 1998; Shelby and Dalrymple, 1993) and perennial ryegrass (Hume and Barker, 1991).

In a newly established pasture, Burggraaf and Thom (2000) identified contaminating sources after a single herbicide application prior to no-till seeding of endophyte-free perennial ryegrass. It can be calculated that the major contaminating source (59%) was plants arising from seeds germinating on the soil surface or seeds buried in the soil. Survival of existing plants contributed 12%, and seedlings germinating in dung contributed 29%. Endophyte infection of the pasture prior to spraying was 86%, with germinating seedlings, surviving plants, and seedlings in dung being 66, 50, and 39% endophyte-infected, respectively.

FACTORS TO CONSIDER IN SELECTING THE RIGHT ENDOPHYTE OPTION

The increase across time in the proportion of endophyte-infected tillers in pastures of tall fescue and perennial ryegrass is probably due to the advantages that endophytes afford these grasses through greater tolerance of biotic (see Chapter 7) and abiotic stresses (see Chapter 8) compared with endophyte-free plants or tillers. For the same reasons, the choice of whether to establish endophyte-infected (toxic or nontoxic) or endophyte-free perennial ryegrass or tall fescue will primarily be determined by the degree of biotic and abiotic stresses at each location. In general, as the proximity to the equator increases in Australia, New Zealand, or the USA, the mean temperature and associated evapotranspiration rate and likelihood of soil moisture deficits also increase. In addition, there is more frequent incidence and severity of damage by insect pests and plant diseases. Temperature and rainfall can differ in these countries with changing longitude; this clearly occurs when moving westward into the interior of the USA and Australia. In New Zealand, the mountain ranges impose a drier regime in the east of both islands with greater temperature ranges. While these are all generalizations that can be greatly modified by altitude, aspect, soil type, pest distribution, and summer-dominant rather than winter-dominant rainfall patterns, they give a broad perspective on the role of endophyte in pasture persistence and productivity and offer some insight into why increases in contaminating toxic endophytes are often a problem. Some illustrations of this are given below.

At Gore, in southern New Zealand, Eerens et al. (1998) reported no changes in endophyte contamination (<10%) during 5 yr, and no differences in perennial ryegrass tiller density, clover growth, total herbage production, and digestibility of perennial ryegrass between endophyte-infected and endophyte-free perennial ryegrass pastures. In this cool-temperate environment, temperature and rainfall are highly favorable for perennial ryegrass growth, with few summer droughts and a low population of the insect pest Argentine stem weevil (*Listronotus bonariensis*). In contrast, in northern New Zealand at Kerikeri, insects such as African black beetle attack endophyte-free tillers causing large differences in the production and persistence of both perennial ryegrass and tall fescue between endophyte-free and endophyte-infected plants (Cooper et al., 2002; Popay et al., 1999). In the USA, Bouton et al. (1993) clearly showed that on the stressful southern coastal plain of Georgia, endophyte-free tall fescue cultivars had poor productivity and persistence, and were rapidly invaded with toxic endophyte. In contrast, endophyte-free cultivars out-yielded endophyte-infected Kentucky 31 in Ohio (Penrose et al., 2000), and newer cultivars showed reduced reinvasion by toxic endophyte-infected fescue (Penrose et al., 2001). Clearly, there is considerable interaction between biotic and abiotic factors in these studies, and this may be better elucidated through modeling rather than site-specific studies or comparisons.

Another important factor in determining whether to establish endophyte-infected or endophyte-free cultivars is the willingness a farmer may have to manage endophyte-free pastures for stand persistence. While research studies have shown consistent outcomes and recommendations for tall fescue (Hoveland et al., 1997; Pedersen et al., 1990), implementing change in established farming practice can be a huge obstruction to establishing endophyte-free cultivars after years of experience with the more robust toxic endophyte-infected tall fescue (Ball, 2001). Again, this varies with the degree of environmental stress; poorly managed endophyte-free tall fescue pastures are likely to survive better under low than high biotic/abiotic stresses.

The availability of selected nontoxic endophytes now allows farmers the option of growing a pasture that is safe for grazing animals but considerably more productive and persistent than endophyte-free pastures. Although cultivars infected with these selected endophytes may not be quite as robust as endemic toxic endophytes in terms of pest tolerance (see Chapter 7), they still provide a very good option when farmers face the dilemma of choosing between animal production/health and agronomic performance. More recent work (Hume et al., 2004; Chapter 7, this volume) shows the potential for specific endophyte strains in the future that may enhance the agronomic performance beyond that of toxic endemic strains common in the perennial ryegrass and tall fescue cultivars currently used in the USA, Australia, and New Zealand.

Summary

Despite their appearance of stability, grasslands are in a constant state of change resulting from interactions with season, climate, insects, disease, livestock, and other plant species. Endophyte contributes to this complexity, and because most tall fescue and perennial ryegrass pastures contain some mixture of endophyte-free, nontoxic endophyte-infected, or toxic endophyte-infected plants, there are fluctuations in the plant populations that result from their associated endophyte. For the countries considered in this chapter (USA, Australia, New Zealand), evidence is overwhelming that livestock are responsive to the exclusion of toxic endophyte from pasture. Evidence is equally overwhelming that the endophyte-free status of tall fescue and perennial ryegrass pastures is usually short-lived in many environments. The use of nontoxic endophyte-infected tall fescue and perennial ryegrass seems to be an ideal compromise that optimizes livestock production and pasture persistence. Early evidence suggests successful application of nontoxic endophytes, but further work is required to fully evaluate this option. Exceptions to these generalities indicate that we do not fully understand all the processes occurring in endophyte-infected pastures, and this will be the focus of future studies. Most of our current research has been conducted in an environment with high background levels of endophyte contamination, and this ecology is certain to change as use of nontoxic endophytes and endophyte-free seed increases.

References

AgriQuality New Zealand. 1999. Seed Certification 1999–2000. Field and Laboratory Standards. AgriQuality, Christchurch, NZ.

Bacon, C.W., J.K. Porter, J.D. Robbins, and E.S. Luttrell. 1977. *Epichloe typhina* from toxic tall fescue grasses. Appl. Environ. Microbiol. 34:576–581.

Ball, D.M. 1997. Significance of endophyte toxicosis and current practices in dealing with the problem in the United States. p. 395–410. *In* C.W. Bacon and N.S. Hill (ed.) *Neotyphodium*/Grass Interactions. Proc. Int. Symp. *Acremonium*/Grass Interactions, 3rd, Athens, GA, USA. 28–31 May 1997. Plenum Press, New York.

Ball, D.M. 2001. Significance of endophyte toxicosis and current practices in dealing with the problem in the United States. p. 395–410. *In* V.H. Paul and P.D. Dapprich (ed.) The grassland conference 2000. Proc. Int. Symp. *Neotyphodium*/Grass Interactions, 4th, Soest, Germany. 27–29 Sept. 2000. Univ. Paderborn, Germany.

Barker, D.J., N. Hill, and J. Andrae. 2004. Measuring endophyte in tall fescue—Plants, fields and farms [Online]. Available at http://forages.oregonstate.edu/is/tfis/ (verified 18 Aug. 2004) Oregon State Univ., Forage Information System, Corvallis.

Barker, D.J., R.M. Sulc, T.L. Bultemeier, J. McCormick, R. Little, C.D. Penrose, and D. Samples. 2005. Contrasting toxic-endophyte contamination between endophyte-free and nontoxic-endophyte tall fescue pastures. Crop Sci. 45:(In press).

Belesky, D.P., J.D. Robbins, J.A. Stuedemann, S.R. Wilkinson, and O.J. Devine. 1987. Fungal endophyte infection-loline derivative alkaloid concentration of grazed tall fescue. Agron. J. 79:217–220.

Belesky, D.P., J.A. Stuedemann, R.D. Plattner, and S.R. Wilkinson. 1988. Ergopeptine alkaloids in grazed tall fescue. Agron. J. 80:209–212.

Bluett, S.J., V.T. Burggraaf, D.E. Hume, B.A. Tapper, and E.R. Thom. 2001. Establishment of ryegrass pastures containing a novel endophyte. Proc. N. Z. Grassl. Assoc. 63:259–265.

Bluett, S.J., E.R. Thom, D.A. Clark, K.A. MacDonald, and E.M.K. Minnee. 2003. Milksolids production from cows grazing perennial ryegrass containing AR1 or wild endophyte. Proc. N. Z. Grassl. Assoc. 65:83–90.

Bouton, J.H., R.N. Gates, D.P. Belesky, and M. Owsley. 1993. Yield and persistence of tall fescue in the southeastern coastal plain after removal of its endophyte. Agron. J. 85:52–55.

Bouton, J.H., G.C.M. Latch, N.S. Hill, C.S. Hoveland, M.A. McCann, R.H. Watson, J.A. Parish, L.L. Hawkins, and F.N. Thompson. 2002. Reinfection of tall fescue cultivars with non-ergot alkaloid-producing endophytes. Agron. J. 94:567–574.

Burggraaf, V.T., and E.R. Thom. 2000. Contamination and persistence of endophyte-free ryegrass pastures established by spray-drilling, and intensively grazed by dairy cows in the Waikato region of New Zealand. N. Z. J. Agric. Res. 43:163–173.

Burggraaf, V.T., and E.R. Thom. 2002. Minimising endophyte contamination from ryegrass seed in cow dung. Proc. N. Z. Grassl. Assoc. 64:97–101.

Burton, G.W., and J.S. Andrews. 1948. Recovery and viability of seeds of certain southern grasses and lespedeza passed through the bovine digestive tract. J. Agric. Res. 76:95–103.

Cooper, B.M., H.S. Easton, D.E. Hume, A.J. Popay, and D.B. Baird. 2002. Improved performance in Northland of tall fescue with a novel endophyte. Proc. Australasian Plant Breed. Conf. 12:379–381.

Cosgrove, G.P., C.B. Anderson, and T.R.N. Berquist. 1996. Fungal endophyte effects on intake, health and liveweight gain of grazing cattle. Proc. N. Z. Grassl. Assoc. 57:43–48.

Cunningham, P.J., J.Z. Foot, and K.F.M. Reed. 1993. Perennial ryegrass (*Lolium perenne*) endophyte (*Acremonium lolii*) relationships: The Australian experience. Agric. Ecosyst. Environ. 44:157–168.

Defelice, M.S., and J.C. Henning. 1990. Renovation of endophyte (*Acremonium coenophialum*)-infected tall fescue (*Festuca arundinacea*) pastures with herbicides. Weed Sci. 38:628–633.

di Menna, M.E., P.H. Mortimer, R.A. Prestidge, A.D. Hawkes, and J.M. Sprosen. 1992. Lolitrem B concentrations, counts of *Acremonium lolii* hyphae, and the incidence of ryegrass staggers in lambs on plots of *A. lolii*-infected perennial ryegrass. N. Z. J. Agric. Res. 35:211–217.

Easton, H.S., and B.M. Cooper. 1997. Field performance of tall fescue with low infection with *Neotyphodium* endophyte. p. 251–253. In C.W. Bacon and N.S. Hill (ed.) *Neotyphodium*/Grass Interactions. Proc. Int. Symp. *Acremonium*/Grass Interactions, 3rd, Athens, GA, USA. 28–31 May 1997. Plenum Press, New York.

Easton, H.S., C.K. Lee, and R.D. Fitzgerald. 1994. Tall fescue in Australia and New Zealand. N. Z. J. Agric. Res. 37:405–417.

Easton, H.S., C.G.L. Pennell, and D.E. Hume. 2004. Tall fescue cultivars and endophyte strains. In R. Kallenbach et al. (ed.) Proc. Int. Symp. *Neotyphodium*/Grass Interactions, 5th, Fayetteville, AR, USA. 23–26 May 2004. Univ. of Arkansas, Fayetteville.

Eerens, J.P.J., R.J. Lucas, H.S. Easton, and J.G.H. White. 1998. Influence of the ryegrass endophyte (*Neotyphodium lolii*) in a cool moist environment. I. Pasture production. N. Z. J. Agric. Res. 41:39–48.

Fletcher, L.R. 1999. "Non-toxic" endophytes in ryegrass and their effect on livestock health and production. p. 133–139. *In* D.R. Woodfield and C. Matthew (ed.) Ryegrass endophyte—An essential New Zealand symbiosis. Grassland Research and Practice Series No. 7. N. Z. Grassl. Assoc., Palmerston North, NZ.

Fletcher, L.R., and I.C. Harvey. 1981. An association of a *Lolium* endophyte with ryegrass staggers. N. Z. Vet. J. 29:185–186.

Francis, S.M., and D.B. Baird. 1989. Increase in the proportion of endophyte-infected perennial ryegrass plants in overdrilled pastures. N. Z. J. Agric. Res. 32:437–440.

Fribourg, H.A., S.R. Wilkinson, and G.N. Rhodes, Jr. 1988. Switching from fungus-infected to fungus-free tall fescue. J. Prod. Agric. 1:122–127.

Froud-Williams, R.J., R.J. Chancellor, and D.S.H. Drennan. 1984. The effects of seed burial and soil disturbance on emergence and survival of arable weeds in relation to minimal cultivation. J. Appl. Ecol. 21:629–641.

Garwood, N.C. 1989. Tropical soil seed banks: A review. p. 149–209. *In* M.A. Leek et al. (ed.) Ecology of soil seed banks. Academic Press, San Diego, CA.

Gramshaw, D., and W.R. Stern. 1977. Survival of annual ryegrass (*Lolium rigidum* Gaud.) in a Mediterranean type environment. I. Effect of summer grazing by sheep on seed numbers and seed germination in autumn. Aust. J. Agric. Res. 28:81–91.

Grime, J.P., J.G. Hodgson, and R. Hunt. 1988. Comparative plant ecology: A functional approach to common British species. Unwin Hyman, London.

Guy, P.L. 1992. Incidence of *Acremonium lolii* and lack of correlation with barley yellow dwarf viruses in Tasmanian perennial ryegrass pastures. Plant Pathol. 41:29–34.

Guy, P.L., and L.T. Davis. 2002. Disease notes or new records: Variation in the incidence of Barley yellow dwarf virus and in the ability of *Neotyphodium* endophytes to deter feeding by aphids (*Rhopalosiphum padi*) on Australasian tall fescue. Australasian Plant Pathol. 31:307–308.

Gwinn, K.D., H.A. Fribourg, J.C. Waller, A.M. Saxton, and M.C. Smith. 1998. Changes in *Neotyphodium coenophialum* infestation levels in tall fescue pastures due to different grazing pressures. Crop Sci. 38:201–204.

Hare, M.D., M.P. Rolston, M.J. Christensen, and K.K. Moore. 1990. Viability of *Lolium* endophyte fungus in seed and germination of *Lolium perenne* seed during five years of storage. p. 147–149. *In* S.S. Quisenberry and R.E. Joost (ed.) Proc. Int. Symp. *Acremonium*/Grass Interactions, 1st, New Orleans, LA, USA. 3 Nov. 1990. Louisiana Agric. Exp. Stn., Baton Rogue.

Harmon, G.W., and F.D. Keim. 1934. The percentage and viability of weed seeds recovered in the feces of farm animals and their longevity when buried in manure. J. Am. Soc. Agron. 26:762–767.

Harris, G.S. 1961. The periodicity of germination in some grass species. N. Z. J. Agric. Res. 4:253–260.

Heeswijck, R.V., and G. McDonald. 1992. *Acremonium* endophytes in perennial ryegrass and other pasture grasses in Australia and New Zealand. Aust. J. Agric. Res. 43:1683–1709.

Hill, N.S., D.P. Belesky, and W.C. Stringer. 1991. Competitiveness of tall fescue as influenced by *Acremonium coenophialum*. Crop Sci. 31:185–190.

Hoveland, C.S., R.L. Haaland, C.C. King, Jr., W.B. Anthony, E.M. Clark, J.A. Mc Guire, L.A. Smith, H.W. Grimes, and J.L. Holliman. 1980. Association of *Epichloe typhina* fungus and steer performance on tall fescue pasture. Agron. J. 72:1064–1065.

Hoveland, C.S., M.A. McCann, and N.S. Hill. 1997. Rotational vs. continuous stocking of beef cows and calves on mixed endophyte-free tall fescue-bermudagrass pasture. J. Prod. Agric. 10:245–250.

Hoveland, C.S., S.P. Schmidt, C.C. King, Jr., J.W. Odom, E.M. Clark, J.A. McGuire, L.A. Smith, H.W. Grimes, and J.L. Holliman. 1983. Steer performance and association of *Acremonium coenophialum* fungal endophyte on tall fescue pasture. Agron. J. 75:821–824.

Hume, D.E. 1999. Establishing and maintaining a toxin-free pasture: A review. p. 123–132. In D.R. Woodfield and C. Matthew (ed.) Ryegrass endophyte—An essential New Zealand symbiosis. Grassland Research and Practice Series No. 7. N. Z. Grassl. Assoc., Palmerston North.

Hume, D.E., and D.J. Barker. 1991. Natural reseeding of five grass species in summer dry hill country. Proc. N. Z. Grassl. Assoc. 53:97–104.

Hume, D.E., S.J. Bluett, V.T. Burggraaf, and M.J. Christensen. 2001a. Occurrence of *Neotyphodium* endophytes in Italian/hybrid ryegrasses in dairy pastures. p. 423–427. In V.H. Paul and P.D. Dapprich (ed.) The grassland conference 2000. Proc. Int. Symp. *Neotyphodium*/Grass Interactions, 4th, Soest, Germany. 27–29 Sept. 2000. Univ. Paderborn, Germany.

Hume, D.E., and J.L. Brock. 1997a. Increases in endophyte incidence in perennial ryegrass at Palmerston North, Manawatu, New Zealand. p. 61–63. In C.W. Bacon and N.S. Hill (ed.) *Neotyphodium*/Grass Interactions. Proc. Int. Symp. *Acremonium*/Grass Interactions, 3rd, Athens, GA, USA. 28–31 May 1997. Plenum Press, New York.

Hume, D.E., and J.L. Brock. 1997b. Morphology of tall fescue (*Festuca arundinacea*) and perennial ryegrass (*Lolium perenne*) plants in pastures under sheep and cattle grazing. J. Agric. Sci. (Cambridge) 129:19–31.

Hume, D.E., and T.B. Lyons. 1992. Establishment of new pastures free of ryegrass contamination. Proc. N. Z. Grassl. Assoc. 54:151–156.

Hume, D.E., and T.B. Lyons. 1993. Methods of establishing tall fescue and ryegrass in a dryland environment. Proc. N. Z. Grassl. Assoc. 55:105–111.

Hume, D.E., A.J. Popay, B.M. Cooper, J.P.J. Eerens, T.B. Lyons, C.G.L. Pennell, B.A. Tapper, G.C.M. Latch, and D.B. Baird. 2004. Effect of a novel endophyte on the productivity of perennial ryegrass (*Lolium perenne*) in New Zealand. In R. Kallenbach et al. (ed.) Proc. Int. Symp. *Neotyphodium*/Grass Interactions, 5th, Fayetteville, AR, USA. 23–26 May 2004. Univ. of Arkansas, Fayetteville.

Hume, D.E., M.P. Rolston, D.J. Baird, W.J. Archie, and M.R. Marsh. 1999. Survival of endophyte-infected ryegrass seed buried in soil. p. 151–156. In D.R. Woodfield and C. Matthew (ed.) Ryegrass endophyte—An essential New Zealand symbiosis. Grassland Research and Practice Series No. 7. N. Z. Grassl. Assoc., Palmerston North.

Hume, D.E., M.P. Rolston, D.J. Baird, W.J. Archie, and M.R. Marsh. 2001b. Endophyte-infected ryegrass seed in soil as a potential source of endophyte contamination of new pastures. p. 97–102. In V.H. Paul and P.D. Dapprich (ed.) The grassland conference 2000. Proc. Int. Symp. *Neotyphodium*/Grass Interactions, 4th, Soest, Germany. 27–29 Sept. 2000. Univ. Paderborn, Germany.

Hunt, M.G. 2003. Effects of environmental change on endophyte–plant–insect relationships. Ph.D. diss. Univ. of Oxford, UK.

Jones, R.M., and M. Simao Neto. 1987. Recovery of pasture seed ingested by ruminants. 3. The effects of the amount of seed in the diet and of diet quality on seed recovery from sheep. Aust. J. Agric. Res. 27:253–256.

Kearns, M.P. 1986. Tall fescue toxicity: An investigation of idiopathic bovine hyperthermia (IBH) in the North Auckland peninsula. Proc. N. Z. Grassl. Assoc. 47:183–186.

Keogh, R.G. 1983. Ryegrass staggers: Management and control. Proc. N. Z. Grassl. Assoc. 44:248–250.

Layton, D.L., L.R. Fletcher, A.J. Litherland, M.G. Scannell, J. Sprosen, C.J. Hoogendoorn, and M.G. Lambert. 2004. Effect of ergot alkaloids on liveweight gain and urine lysergol level in ewe hoggets and heifers. Proc. N. Z. Soc. Anim. Prod. 64:192–196.

Lennartz, H. 1957. The effect of the passage through the alimentary tract of cattle on the viability of seeds of grassland plants. (In German.) Z. Acker-Pflanzenbau 103:427–453.

Lewis, J. 1973. Longevity of crop and weed seeds: Survival after 20 years in soil. Weed Res. 13:179–191.

L'Huillier, P.J., and D.W. Aislabie. 1988. Natural reseeding in perennial ryegrass/white clover dairy pastures. Proc. N. Z. Grassl. Assoc. 49:111–115.

Madej, C.W., and K. Clay. 1991. Avian seed preference and weight loss experiments: The effect of fungal endophyte-infected tall fescue seeds. Oecologia 88:296–302.

Marks, S., K. Clay, and G.P. Cheplick. 1991. Effects of fungal endophytes on interspecific and intraspecific competition in the grasses *Festuca arundinacea* and *Lolium perenne*. J. Appl. Ecol. 28:194–204.

McCallum, D.A., N.A. Thomson, and T.G. Judd. 1991. Experiences with deferred grazing at the Taranaki Agricultural Research Station. Proc. N. Z. Grassl. Assoc. 53:79–83.

McCorkindale, B. 1999. Ryegrass endophyte—Can we successfully get rid of it? Report on Clutha Monitor Farm. Monitor Farm Programme. Meat N. Z./Woolpro, Balcutha.

McDonald, G., A. Noske, R. Van Heeswijck, and W.E. Frost. 1993. The role of the perennial ryegrass endophyte in the management of pasture pests in south eastern Australia. Proc. Australasian Conf. Grassl. Invertebrate Ecol. 6:295–302.

Pedersen, J.F., G.D. Lacefield, and D.M. Ball. 1990. A review of the agronomic characteristics of endophyte-free and endophyte-infected tall fescue. Appl. Agric. Res. 5:188–194.

Pedersen, J.F., M.J. Williams, E.M. Clark, and P.A. Backman. 1984. Indications of yearly variation of *Acremonium coenophialum* in seed from a permanent tall fescue sward. Crop Sci. 24:367–368.

Pennell, C.G.L., D.E. Hume, O.J.-P. Ball, H.S. Easton, and T.B. Lyons. 2001. Effects of *Neotyphodium lolii* infection in ryegrass on root aphid and pasture mealy bug. p. 465–469. *In* V.H. Paul and P.D. Dapprich (ed.) The grassland conference 2000. Proc. Int. Symp. *Neotyphodium*/Grass Interactions, 4th, Soest, Germany. 27–29 Sept. 2000. Univ. Paderborn, Germany.

Penrose, C.D., R.M. Sulc, and J.S. McCutcheon. 2001. Impact of management on endophyte free and endophyte infected tall fescue cultivars in Ohio. p. 610–612. *In* Proc. Int. Grassl. Congr., 19th, São Pedro, São Paula, Brazil. 11–21 Feb. 2001. Brazilian Society of Animal Husbandry, Brazil.

Penrose, C.D., R.M. Sulc, and E.M. Vollborn. 2000. A three year report of animal preference, stockpile, yield and quality of fescue and orchardgrass. Proc. Am. Forage Grassl. Counc. 9:31.

Piggin, C.M. 1978. Dispersal of *Echium plantagineum* L. by sheep. Weed Res. 18:155–160.

Popay, A.J., and J.G. Baltus. 2001. Black beetle damage to perennial ryegrass infected with AR1 endophyte. Proc. N. Z. Grassl. Assoc. 63:267–271.

Popay, A.J., D.E. Hume, J.G. Baltus, G.C.M. Latch, B.A. Tapper, T.B. Lyons, B.M. Cooper, C. Pennell, J.P.J. Eerens, and S.L. Marshall. 1999. Field performance of perennial ryegrass (*Lolium perenne*) infected with toxin-free fungal endophytes (*Neotyphodium* spp.). p. 113–122. *In* D.R. Woodfield and C. Matthew (ed.) Ryegrass endophyte—An essential New Zealand symbiosis. Grassland Research and Practice Series No. 7. N. Z. Grassl. Assoc., Palmerston North.

Popay, A.J., D.E. Hume, K.L. Davies, and B.A. Tapper. 2003. Interactions between endophyte (*Neotyphodium* spp.) and ploidy in hybrid and perennial ryegrass cultivars and their effects on Argentine stem weevil (*Listronotus bonariensis*). N. Z. J. Agric. Res. 46:311–319.

Popay, A., S. Marshall, and J. Baltus. 2000. Endophyte infection influences disappearance of perennial ryegrass seed. N. Z. Plant Prot. 53:398–405.

Popay, A.J., W.B. Silvester, and P.J. Gerard. 2004. Effect of different endophyte isolates in perennial ryegrass on a root aphid, *Aploneura lentisci*. *In* R. Kallenbach et al. (ed.) Proc. Int. Symp. *Neotyphodium*/Grass Interactions, 5th, Fayetteville, AR, USA. 23–26 May 2004. Univ. of Arkansas, Fayetteville.

Prestidge, R.A., M.E. di Menna, S. van der Zijpp, and D. Badan. 1985. Ryegrass content, *Acremonium* endophyte and Argentine stem weevil in pastures in the volcanic plateau. Proc. N. Z. Weed Pest Control Conf. 38: 41–44.

Prestidge, R.A., S. van der Zijpp, and D. Badan. 1984. Effects of Argentine stem weevil on pastures in the Central Volcanic Plateau. N. Z. J. Exp. Agric. 12:323–331.

Pulsford, M.F. 1950. A note on lameness in cattle grazing on tall meadow fescue (*Festuca arundinacea*) in South Australia. Aust. Vet. J. 26:87–88.

Quigley, P., X. Li, G. McDonald, and A. Noske. 1993. Effects of *Acremonium lolii* on mixed pastures and associated insect pests in south-eastern Australia. p. 177–180. In D.E. Hume et al. (ed.) Proc. Int. Symp. *Acremonium*/Grass Interactions, 2nd, Palmerston North, NZ. 4–6 Feb. 1993. AgResearch, Palmerston North.

Rampton, H.H., and T.M. Ching. 1970. Persistence of crop seeds in soil. Agron. J. 62:272–277.

Read, J.C., and B.J. Camp. 1986. The effect of the fungal endophyte *Acremonium coenophialum* in tall fescue on animal performance, toxicity, and stand maintenance. Agron. J. 78:848–850.

Reed, K.F.M., A. Leonforte, P.J. Cunningham, J.R. Walsh, D.I. Allen, G.R. Johnstone, and G. Kearney. 2000. Incidence of ryegrass endophyte (*Neotyphodium lolii*) and diversity of associated alkaloid concentrations among naturalised populations of perennial ryegrass (*Lolium perenne* L.). Aust. J. Agric. Res. 51:569–578.

Rice, K.J. 1989. Impact of seed banks on grassland community structure and population dynamics. p. 211–230. *In* M.A. Leek et al. (ed.) Ecology of soil seed banks. Academic Press, San Diego, CA.

Roberts, H.A. 1986. Persistence of seeds of some grass species in cultivated soil. Grass Forage Sci. 41:273–276.

Rolston, M.P. 1993. Use of endophyte in plant breeding and the commercial release of new endophyte–grass associations. p. 171–174. *In* D.E. Hume et al. (ed.) Proc. Int. Symp. *Acremonium*/Grass Interactions: Plenary Papers, 2nd, Palmerston North, NZ. 4–6 Feb. 1993. AgResearch, Palmerston North.

Rolston, M.P., M.D. Hare, K.K. Moore, and M.J. Christensen. 1986. Viability of *Lolium* endophyte fungus in seed stored at different moisture contents and temperature. N. Z. J. Exp. Agric. 14:297–300.

Rolston, M.P., L.R. Fletcher, C.G. Fletcher, and W.J. Archie. 2001. The viability of perennial ryegrass seed and its endophyte in sheep faeces. p. 405–408. *In* V.H. Paul and P.D. Dapprich (ed.) The grassland conference 2000. Proc. Int. Symp. *Neotyphodium*/Grass Interactions, 4th, Soest, Germany. 27–29 Sept. 2000. Univ. Paderborn, Germany.

Schmidt, S.P., and T.G. Osborn. 1993. Effects of endophyte-infected tall fescue on animal performance. Agric. Ecosyst. Environ. 44:233–262.

Sen, L.X. 1995. Incidence and significance of fungal endophytes (*Acremonium* spp.) in selected grass species on the Northern Tablelands of NSW. Post Graduate Diploma Sci. Agric., Univ. of New England, Armidale, Australia.

Shelby, R.A., and L.W. Dalrymple. 1993. Long-term changes of endophyte infection in tall fescue stands. Grass Forage Sci. 48:356–361.

Shelby, R.A., and S.P. Schmidt. 1991. Survival of the tall fescue endophyte in the digestive tract of cattle and horses. Plant Dis. 75:776–778.

Shelby, R.A., S.P. Schmidt, R.W. Russell, and W.H. Gregory. 1989. Spread of tall fescue endophyte by cattle. Leaflet 104. Alabama Agric. Exp. Stn., Auburn Univ., Auburn.

Siegel, M.R., M.C. Johnson, D.R. Varney, W.C. Nesmith, R.C. Buckner, L.P. Bush, P.B. Burrus, II, T.A. Jones, and J.A. Boling. 1984. A fungal endophyte in tall fescue: Incidence and dissemination. Phytopathology 74:932–937.

Simao Neto, M., R.M. Jones, and D. Ratcliff. 1987. Recovery of pasture seed ingested by ruminants. 1. Seed of six tropical pasture species fed to cattle, sheep and goats. Aust. J. Exp. Agric. 27:239–246.

Simpson, G.M. 1990. Seed dormancy in grasses. Cambridge Univ. Press, Cambridge, UK.

Smith, A.E. 1989. Herbicides for killing tall fescue (*Festuca arundinacea*) infected with fescue endophyte (*Acremonium coenophialum*). Weed Technol. 3:485–489.

Stewart, A.V. 1985. Perennial ryegrass seedling resistance to Argentine stem weevil. N. Z. J. Agric. Res. 28:403–407.

Stuedemann, J.A., and F.N. Thompson. 1993. Management strategies and potential opportunities to reduce the effects of endophyte-infested tall fescue on animal performance. p. 103–114. In D.E. Hume et al. (ed.) Proc. Int. Symp. *Acremonium*/Grass Interactions: Plenary Papers, 2nd, Palmerston North, NZ. 4–6 Feb. 1993. AgResearch, Palmerston North.

Suckling, F.E.T. 1952. Dissemination of white clover (*Trifolium repens*) by sheep. N. Z. J. Sci. Technol. 33:64–77.

Thompson, R.W., H.A. Fribourg, and B.B. Reddick. 1989. Sample intensity and timing for detecting *Acremonium coenophialum* incidence in tall fescue pastures. Agron. J. 81:966–971.

Thompson, K., and J.P. Grime. 1979. Seasonal variation in the seed banks of herbaceous species in 10 contrasting habitats. J. Ecol. 67:893–921.

Timper, P., J. Bouton, and R. Gates. 2003. Effect of tall fescue cultivar and endophyte status on reproduction of lesion (*Pratylenchus* spp.) and stubby root (*Paratrichodorus* spp.) nematodes. p. 21–22. *In* Proc. Tall Fescue Toxicosis Workshop. SERAIEG-8, Wildersville, TN.

Tor-Agbidye, J., L.L. Blythe, and A.M. Craig. 2001. Correlation of endophyte toxins (ergovaline and lolitrem B) with clinical disease: Fescue foot and perennial ryegrass staggers. Vet. Human Toxicol. 43:140–146.

Ussher, G. 2003. Northlands pasture toxin project. p. 62–64. *In* Annual Conf. N. Z. Large Herds Assoc., 34th, Paihia, NZ. N. Z. Large Herds Assoc.

Valentine, S.C., B.D. Bartsch, K.G. Boyce, M.J. Mathison, and T.R. Newbery. 1993. Incidence and significance of endophyte in perennial grass seed lines. Final report on project DAS33 prepared for the Dairy Research and Development Corporation. SARDI, Adelaide, Australia.

Waller, R.A., P.W.G. Sale, G.R. Saul, P.E. Quigley, and G.A. Kearney. 1999. Tactical versus continuous stocking for persistence of perennial ryegrass (*Lolium perenne* L.) in pastures grazed by sheep in south-western Victoria. Aust. J. Exp. Agric. 39:265–274.

Watkin, B.R., and R.J. Clements. 1978. The effects of grazing animals on pastures. p. 273–289. *In* J.R. Wilson (ed.) Plant relations in pastures. CSIRO, Melbourne, Australia.

Welty, R.E., M.D. Azevedo, and T.M. Cooper. 1987. Influence of moisture content, temperature, and length of storage on seed germination and survival of endophytic fungi in seeds of tall fescue and perennial ryegrass. Phytopathology 77:893–900.

Wheatley, W.M. 1997. Perennial ryegrass (*Lolium perenne*) staggers in the central tablelands, NSW, Australia. p. 447–449. In C.W. Bacon and N.S. Hill (ed.) *Neotyphodium*/Grass Interactions. Proc. Int. Symp. *Acremonium*/Grass Interactions, 3rd, Athens, GA, USA. 28–31 May 1997. Plenum Press, New York.

Wheatley, W.M., D.E. Hume, H.W. Kemp, M.S. Monk, K.F. Lowe, A.J. Popay, D.B. Baird, and B.A. Tapper. 2003. Effects of fungal endophyte on the persistence and productivity of tall fescue at 3 sites in eastern Australia [CD-ROM]. *In* M. Unkovich and G. O'Leary (ed.) Solutions for a better environment. Proc. Aust. Agron. Conf., 11th, Geelong, VIC, Australia. 2–6 Feb. 2003. Aust. Soc. Agron., VIDA, Horsham, VIC, Australia.

Yamada, T., and T. Kawaguchi. 1971. Dissemination of pasture plants by livestocks. I. Recovery and viability of some pasture plant seeds passed through digestive tract of goats. (In Japanese, with English abstract.) J. Jpn. Soc. Grassl. Sci. 17:36–47.

Section IV

Animal Toxicoses

MANAGING RYEGRASS–ENDOPHYTE TOXICOSES

Lester R. Fletcher[1]

In recent years, ryegrass (*Lolium perenne, L. multiflorum*, and hybrids) and endophyte (*Neotyphodium lolii*) research has focused on the discovery, development, and commercialization of nontoxic endophytes, such as AR1. While AR1 associations continue to produce peramine, they do not produce the toxins lolitrem B or ergovaline (Fletcher, 1999; Tapper and Latch, 1999).

Ryegrass associations with AR1 have now become the benchmark for New Zealand pastures (Easton and Tapper, 2004). Not all endemic endophyte–ryegrass associations can be replaced at once and some will never be replaced. This situation warrants a continuing commitment to exploring all avenues to minimize the impact of endemic endophytes on grazing livestock.

CONFIRMING THE SAFETY AND ANIMAL PRODUCTIVITY ON AR1 PASTURES

Industry standards demand that all ryegrass cultivars inoculated with AR1 meet selected animal health criteria before being passed for commercialization. Therefore, grazing trials with sheep have been conducted on all ryegrass cultivar–AR1 associations to confirm their nontoxicity before approval is granted for commercialization. Trials were conducted with three replicates with 10 lambs per unit (30 lambs per cultivar–endophyte association). Control treatments were 'Nui' ryegrass (*L. perenne*) free of endophyte and infected with its endemic endophyte. On a 0- to 5-point ascending scale, mean maximum ryegrass staggers scores must be no greater than 1 (Keogh, 1973). Rectal temperatures, respiration rates, and plasma prolactin concentrations, measured in ambient temperatures above 30°C, must be no different than those in similar animals grazing endophyte-free ryegrass. These criteria were determined from animal responses to grazing ryegrass with its endemic endophyte compared with those grazing the same ryegrass free of endophyte (Fletcher et al., 1999). Animal

---

[1] AgResearch Grasslands, Christchurch, New Zealand

safety (grazing) trials must therefore include those two treatments as standards. To date, 15 ryegrass cultivars with AR1 have undergone safety and efficacy testing under grazing and all have met the prescribed industry standards.

While the health and welfare of any animals grazing experimental ryegrass–endophyte associations are paramount, animal production advantages over those on ryegrass with endemic endophyte also need to be demonstrated if this technology is to be adopted by producers. Mean summer–autumn growth rates of weaned lambs were 170 g d^{-1}, 150 g d^{-1}, and 102 g d^{-1} for lambs grazing cultivars with AR1, nil, and endemic endophyte, respectively [LSD (0.05) = 48 g d^{-1}]. There were three reps of 10 lambs each for the controls (Nui nil and Nui with endemic endophyte) and AR1, but in eight different ryegrass cultivars. Thus, AR1 treatments are a mean of 240 lambs. In these trials, AR1 was in modern cultivars. These new cultivars have significant improvements, particularly in rust resistance, compared with the older control cultivar, Nui. These improvements possibly accounted for some of the enhanced performance in the lambs grazing these cultivars. Bluett et al. (2003) reported a 9% increase in milk production across 3 yr, from dairy cows grazing ryegrass pastures with AR1 compared with those grazing similar pastures with endemic endophyte. Other grazing trials on dairy farms have demonstrated milk production increases of up to 14% (Ussher, 2003). These increases in production, without endophyte-associated animal health problems, have led to an unprecedented uptake of this technology by New Zealand pastoral farmers. After only 2 yr in the market, 40% of proprietary ryegrass seed sold has AR1; this proportion is forecast to increase significantly. Reports indicate cultivars with AR1 are meeting farmers' expectations in terms of animal health and production.

RYEGRASS STAGGERS IN THE ABSENCE OF LOLITREM B

The selection criteria for nontoxic endophyte strains such as AR1 are the absence of ergovaline or lolitrem B production in conjunction with their host ryegrass, but with effective concentrations of peramine. Endophytes have been identified that do not produce any of the three metabolites, lolitrem B, ergovaline, and peramine, which have previously been the focus of attention. Despite the absence of peramine, or any other known endophyte-related invertebrate toxins or deterrents, the AR37 endophyte strain was found to confer resistance to a wide range of invertebrate pests (Hume et al., 2004; Popay and Bonos, 2004). The unexpected pest-resistance properties of AR37 prompted its study in a grazing trial. Lambs were grazed on AR37-infected pastures (two cultivars) in different experiments during summer and autumn for 6 yr. In some years, AR37 caused ryegrass staggers in grazing sheep with varying incidence and severity. Ryegrass staggers in the absence of lolitrem B had previously been reported in grazing sheep (Fletcher et al., 1993). In that case, the indole-diterpenoid tremorgen, paxilline, was associated with the outbreaks, especially those occurring late in the season. Since paxilline is considered to be a precursor to many other tremorgens, this correlation may only be indicative of the presence of other endophyte tremorgens rather than it being the causative tremorgen.

Severity of ryegrass staggers is scored for individual animals on a 0-to-5 ascending scale (Keogh, 1973). Because of individual animal variation in response, a mean flock score of 3.8 is very serious. The most affected animals (score of 5) required removal to nontoxic pastures for welfare reasons (Table 10.1).

Table 10.1. Mean maximum ryegrass staggers scores (0–5 ascending scale), 30 lambs/hoggets per treatment per grazing, minimum grazing period 30 d. In years 1999–2001, endophyte treatments in *Lolium perenne* cv. Grasslands Nui and in years 2002–2004 in an experimental line of *L. perenne*, GA66.

	Endemic endophyte	AR37	Endophyte-free
Summer			
1999	3.5	2.3	0
2000	2.7	2.1	0
2001†	2.6	1.0	0
2002†	2.8	0.0	0
2003†	2.8	0.0	0
2004	3.8	3.8	0
Autumn			
1999	1.0	2.0	0
2000	3.8	1.8	0
2004	3.8	0.4	

† Combined summer–autumn grazings.

Clinical signs of staggers in sheep grazing AR37 associations in the field are indistinguishable in nature from those of sheep grazing ryegrass with endemic endophyte ascribed to lolitrem B. Apart from autumn 1999 and summers of 2000 and 2004, clinical signs of ryegrass staggers in lambs grazing AR37 associations have not been as serious as clinical signs on comparable endemic associations. In autumn 1999 and summer 2000, ryegrass staggers on any treatment were only minor to moderate. There was also a time lag of 10 to 14 days for reaching the maximum ryegrass staggers score in lambs grazing AR37 associations compared with those on endemic associations. Severe ryegrass staggers has only been observed in sheep grazing AR37-infected herbage in one grazing sequence, in summer 2004. On this occasion, the affected sheep had to be removed from the trial along with affected sheep grazing endemic endophyte association. After all lambs were removed from both AR37 and endemic endophyte treatments, they were immediately replaced with groups of similar lambs which had been grazed on endophyte-free ryegrass for the previous 6 wk. After 14 d, the replacement lambs grazing the endemic endophyte ryegrass had again reached a mean ryegrass staggers score of 3.8, while those grazing AR37 associations had a mean score of only 0.4. Apart from the toxicity of AR37, the quantity or quality of herbage offered did not differ from that grazed by replacement lambs or the lambs that were removed. This rapid reduction in toxicity illustrates the specificity of conditions, albeit unknown, that are needed to produce and/or maintain toxicity in AR37 associations.

Epoxy-janthitrems have been isolated from ryegrass associations with AR37 (Tapper and Lane, 2004). These molecules are likely to be tremorgenic; however, a direct causative link with the staggers observed in sheep grazing AR37-infected ryegrass has not been established. Other, as yet unidentified, tremorgens may be active; these may be other mycotoxins from endophytes or saprophytic fungi acting as synergists or in concert with epoxy-janthitrems to cause or enhance tremorgenicity.

Despite the sporadic outbreaks of ryegrass staggers associated with AR37, only once have they been more serious than in endemic endophyte controls. Other parameters, such as body temperature, respiration rates, serum prolactin concentrations, and animal growth rates, have been significantly closer to normal level than when lambs are on ryegrass with endemic endophyte. Lambs grazing ryegrass with its endemic endophyte, with AR37, or without endophyte had mean (6 yr) summer–autumn growth rates of 44, 129, and 131 g d^{-1}, respectively ($P = 0.02$). In the absence of ryegrass staggers, growth rates of lambs on AR37 associations are at least similar to those on endophyte-free ryegrass. When other ryegrass–endophyte associations are not producing lolitrem B, ergovaline, or peramine, but yet have some bioprotective properties, and are grazed under the same conditions and management as AR37 when it is most toxic, they have not caused ryegrass staggers. As a model, AR37 has highlighted the potential to extend the range of bioprotection of pastures with selected endophytes. However, it also raises awareness to risks involving other toxins associated with selected endophytes which could impact animal health and production.

MANAGING RISK

The superior bioprotective characteristics of AR37 associations, relative to endemic endophyte associations, are attractive to the pastoral industry, especially in areas of high invertebrate pest and (or) drought challenge. Animal growth rates are significantly better on AR37 associations than on endemic endophyte associations in the absence of staggers. However, all these benefits must be weighed against the potential to cause sporadic but sometimes serious ryegrass staggers under occasional and, as yet undetermined, conditions. When conditions causing or exacerbating ryegrass staggers in animals grazing AR37 associations are determined, it is likely that options to control or manage them will be limited. The industry dilemma will be to decide whether the risk of sporadic outbreaks of staggers outweighs the bioprotective benefits of AR37 associations.

MINIMIZING THE IMPACT OF ENDEMIC ENDOPHYTE IN GRAZING SYSTEMS

Variation in Pasture Toxicity

Although nontoxic endophytes such as AR1 offer considerable advantages to the producer, a proportion of ryegrass based pastures with the endemic endophytes will never be replaced for various reasons, including topography (Fletcher, 1999). In these situations, minimizing the impact of endemic endophyte on animal health and production is the only option. Various prophylactics and *cures* for animal health problems associated with ryegrass–endophyte toxicosis are promoted, but there is little sound evidence for their efficacy. Management to minimize the intake of toxins appears to be the only viable alternative. To develop effective management strategies and (or) packages requires an understanding of seasonal variations in toxin concentration in fresh and conserved pastures, the factors affecting the toxin concentration and the acquisition and excretion of the toxin by the animals.

The use of the generic terms *wild-type*, *standard*, or *endemic* endophyte in ryegrass–endophyte associations can be misleading. Ryegrass with endemic endophyte (*Neotyphodium lolii*) normally produces peramine and the toxins ergovaline and lolitrem B. However, it should not be assumed that, apart from minor variations, these compounds are produced in similar concentrations by all endemic ryegrass–endophyte associations and throughout the season. Seasonal variation has been reported before (Fletcher et al., 2000; Reed et al., 2000). Recent research has also revealed significant differences in concentrations and seasonal patterns of lolitrem B and ergovaline among 10 different ryegrass cultivars with their endemic endophyte (Fig. 10.1).

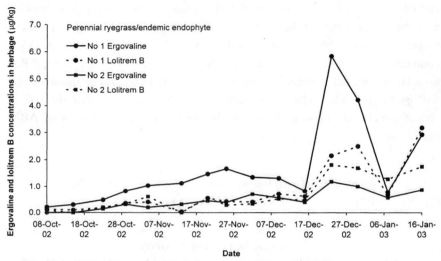

Fig. 10.1. *Ergovaline and lolitrem B concentration profiles of two perennial rye-grass cultivars with their endemic endophytes. Means of the same 10 plants were sampled every 7 d, and grown and managed under the same conditions.*

This variability increases the complexity in determining periods of maximum toxicity as well as the development of simple generic management strategies. Concentrations of lolitrem B and ergovaline also vary considerably from year to year in some pastures (Welty et al., 1994). While these differences are clearly climate driven, specific conditions which influence them are difficult to identify (Fletcher et al., 2000).

High levels of nitrogen fertilizer are known to increase some ryegrass–endophyte alkaloids, but little is known about other plant nutritional factors which influence toxin production or concentration. The reproductive state of the host ryegrass and associated morphological changes have a major influence on the spatial distribution of toxins and may influence overall toxicity (Keogh et al., 1996). However, ingestion of toxins is unlikely to be at the same concentration as that in the total plant; since there is usually the potential for selection by the grazing animal (Cosgrove et al., 2002). All these factors and potential interactions need to be taken into account in devising effective grazing strategies to minimize toxin intake.

Threshold Concentrations of Toxins

Published safe, or threshold level, concentrations of lolitrem B and ergovaline in forage for different classes of livestock have been based on the manifestation of clinical symptoms (Tor-Agbidye et al., 2001). This, however, does not take into consideration subclinical effects on production or factors such as diet selection, physiological state of the animal, or environmental conditions; all of which can significantly impact the grazing animal (Fisher et al., 2004; Pearson et al., 1996). On intensive, high-production grazing systems with liveweight targets and associated contract deadlines, anything less than optimum growth rates or milk production can significantly impact profit.

Residual Toxins in Conserved Forage

Significant concentrations of lolitrem B and ergovaline have been detected in hay and silage, but concentrations of these alkaloids in the parent herbage are not always reported (Fink-Gremmels and Bull, 2000; Roberts et al., 2002; Turner et al., 1991). Most ryegrass hay and silage are made in spring and early summer when concentrations of ergovaline and lolitrem B in the fresh herbage are low. As a consequence, the concentrations of these alkaloids in silage and hay tend to be relatively low. However, in some regions, surplus pasture is conserved as hay or silage in late summer and autumn when alkaloid concentrations are likely to be highest in ryegrass infected with endemic endophyte.

A paddock of perennial ryegrass with a high endophyte percentage (>90%) which had previously been harvested for seed was fertilized with 100 kg urea ha^{-1} and irrigated twice at 35 mm during a 6-wk regrowth period from 4 Feb. 2003 until 18 Mar. 2003. This resulted in a pure ryegrass pasture with 3000 kg ha^{-1} dry matter, including a significant proportion of aftermath seedheads. The pasture was cut with a mower on 18 Mar. 2003, and triplicate samples were taken directly behind the mower to determine the ergovaline and lolitrem B concentrations in the parent material. The following treatments were then applied:

1. Ensile herbage immediately after cutting (silage Day 1)
2. Ensile herbage 2 d after cutting (silage Day 2)

3. Ensile herbage 3 d after cutting (silage Day 3)

4. Bale herbage 5 d after cutting (hay Day 6)

5. Bale herbage 7 d after cutting (hay Day 8)

Silage was made by passing the cut herbage through a silage chopper, then compressing this material into 20-L plastic buckets with a plastic liner and airtight lids. Six buckets were made on each day, three with silage inoculum and three without. Hay was made in the conventional manner by baling the naturally dried herbage with a small baler. The bales of hay were stored in an open barn along with other hay while the buckets of silage were stored under cover out of direct sunlight.

Samples were taken from one bale under each treatment and from two buckets of silage (with and without inoculum) from each treatment at 111 and 207 d after conservation. Samples of hay and silage, along with original samples from the parent material, were freeze dried and analyzed for ergovaline, lolitrem B, and peramine (B.A. Tapper and E. Davies, personal communication). Ergovaline concentrations in finely chopped silage made 1 d after cutting and sampled on Days 111 and 207 did not differ significantly from freeze-dried parent material sampled fresh at harvest. Ergovaline concentrations in silage made on Days 2 and 3 tended to be slightly lower than the parent material. In contrast, ergovaline concentrations in hay made from the same material on Day 6 and Day 8 were much lower at Day 111. In terms of potential toxicity, ergovaline was almost negligible at Day 207 (Fig. 10.2).

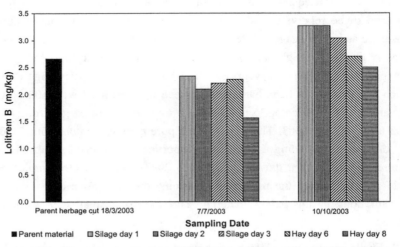

Fig. 10.2. *Effect of conservation method and time, after conservation, on lolitrem B concentration in silage and hay.*

Lolitrem B concentrations were not significantly different from those in the parent material at both sampling dates whether conserved as hay or silage (Fig. 10.3).

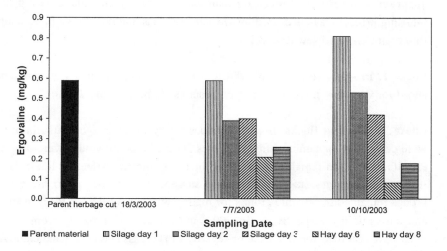

Fig. 10.3. *Effect of conservation method and time, after conservation, on ergovaline concentration in silage and hay.*

Apart from the ergovaline in the hay, concentrations of these endophyte toxins appear to be stable in conserved forage and remain at levels which are likely to result in toxicity if fed as a major part of an animal's diet (Clark et al., 1996).

Prediction of Toxicosis—Diagnostic Assays

Currently, the best predictive measure of impending toxicity in grazing animals is the concentration of toxins in the pasture being grazed. Selection, grazing height, accessibility, and variation in toxin concentrations between paddocks in the rotation can all influence toxin intake by grazing animals. Although more diagnostic than predictive, measuring toxins or their metabolites in body fluids such as urine is likely to be a better indicator of toxin challenge. Mean urine lysergols (ergot alkaloid derivatives), measured by ELISA and quoted as lysergol/creatinine ratio, show a positive relationship to increasing ergovaline intake (Layton et al., 2004).

A similar negative relationship has been developed for decreasing mean liveweight gains with increasing intake of ergovaline, allowing prediction of liveweight gain conceded per unit of ergovaline ingested.

Breeding Animals Resistant to Toxicosis

Individual animal variation in response to ryegrass staggers toxins provides an opportunity to select and breed for resistance in grazing animals. A two-phase breeding project was conducted by Dr. Chris Morris and colleagues at Ruakura Research Centre in New Zealand.

Phase 1. Progeny tests. From 1988 to 1992, 18 different rams were progeny tested for variation in ryegrass staggers response (Amyes et al., 2002).

Phase 2. Selection flocks. Breeding values for ryegrass staggers susceptibility were calculated for lambs, rams, and ewes. These breeding values were used to classify 18-mo-old females, to be mated in 1993, into two selection lines: those resistant to ryegrass staggers and those susceptible. Females from these two groups remained in separate groups for mating with rams selected for resistance and susceptibility respectively. In 2004, more than 200 progeny from seven rams were recorded for incidence or severity of ryegrass staggers. Challenge via controlled dosing of lolitrem B has not been possible because of the difficulty and cost of extracting it in a purified form. Selection pressure has, therefore, depended on the degree of challenge to the lambs in the field each year. This, in turn, is dependent on toxin concentrations in the pasture which, as discussed earlier, varies considerably through the season and from year to year. There is now considerable differential between the two flocks (Fig. 10.4).

Until 2001, the divergence was primarily due to increased susceptibility in the susceptible flock. However, under greater challenge from 2003 onward, resistance has increased significantly.

There has been interest, and some research, into determining the physiological mechanisms involved in individual animal variation in response to lolitrem B and for ergovaline challenge. There is some evidence that animals resistant to one toxin may also have enhanced resistance to other, apparently unrelated, toxins. This presents the hypothesis that the mechanism for resistance to toxins may be more general than specific (Hohenboken et al., 2000; Morris et al., 1995).

The ultimate goal of this research is to identify the gene(s) responsible for resistance to ryegrass staggers or tolerance of lolitrem B, the markers for them, and to screen for those during ram selection. This will facilitate the development of resistance in the national flock. To this end, DNA samples have been collected from individual animals throughout this breeding program.

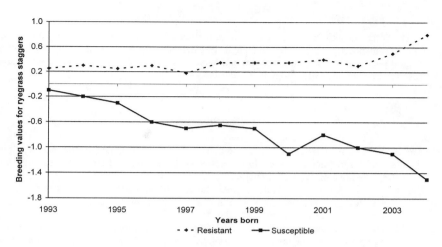

Fig. 10.4. *Ryegrass staggers breeding values for lambs born 1993–2004. Extended from Amyes et al. 2002 (C.A. Morris, 2004, personal communication).*

SUMMARY

The greatest progress toward eliminating or minimizing ryegrass endophyte toxicosis has been through the development and use of nontoxic endophytes. There is likely to be further potential using this approach in the short term, and genetic modification of the plant and (or) endophyte genome in the longer term. Knocking out genes responsible for some toxins has already been achieved, and we look forward to the application of this to grassland farming. Progress in breeding resistant animals has been slow, and is unlikely to make an impact on ryegrass–endophyte toxicosis in the near future. However, if the gene(s) responsible for resistance can be identified, then application of this technology may aid in ram selection.

Minimizing the impact of endemic endophyte in farm systems where it cannot be replaced involves grazing management strategies to reduce exposure to toxins. Improved strategies will come with the ability to predict peak concentrations. The impact on animals by out-of-season ryegrass–endophyte toxins in conserved ryegrass warrants further investigation.

ACKNOWLEDGEMENTS

The author thanks Dr. Chris Morris for current data for Fig. 10.4, Dr. Syd Easton for help with this manuscript, the wider endophyte research team in New Zealand who are all an integral part of this research, and Meat and Wool New Zealand Innovations and Foundation Research Science and Technology for funding.

REFERENCES

Amyes, N.C., C.A. Morris, and N.R. Towers. 2002. Ryegrass staggers: Genetics accounts for a six-fold difference in susceptibility between selection lines of lambs at Ruakura. Proc. N. Z. Soc. Anim. Prod. 62:191–194.

Bluett, S.J., E.S. Kolver, M.J. Auldist, E.R. Thom, S.R. Davis, V.C. Farr, and B.A. Tapper. 2003. Perennial ryegrass endophyte effects on plasma prolactin concentration in dairy cows. N. Z. J. Agric Res. 46:9–14.

Clark, D.A., E.R. Thom, and C.D. Waugh. 1996. Milk production from pastures and pasture silage with different levels of endophyte infection. Proc. N. Z. Soc. Anim. Prod. 56:292–296.

Cosgrove, G.P., C.B. Anderson, M. Phillot, D. Nyfeler, D.E. Hume, A.J. Parsons, and G.A. Lane. 2002. The effect of endophyte alkaloids on diet selection by sheep. Proc. N. Z. Soc. Anim. Prod. 62:167–170.

Easton, H.S., and B.A. Tapper. 2004. *Neotyphodium* down under—Research developments in New Zealand. *In* R. Kallenbach et al. (ed.) Proc. Int. Symp. *Neotyphodium*/Grass Interactions, 5th, Fayetteville, AR, USA. 23–26 May 2004. Univ. Arkansas, Fayetteville, AR.

Fink-Gremmels, J., and S. Bull. 2000. Prevalence of the ryegrass stagger syndrome in horses in the Netherlands. Book of abstracts. *In* V.H. Paul and P.D. Dapprich (ed.) The Grassland Conf. 2000. Proc. Int. Symp. *Neotyphodium*/Grass Interactions, 4th, Soest, Germany. 27–29 Sept. 2000. Univ. Paderborn, Germany.

Fisher, M.J., D.W. Bohnert, C.J. Ackerman, C.S. Schauer, T. DelCurto, A.M. Craig, E.S. Vanzant, D.L. Harmon, and F.N. Schrick. 2004. Evaluation of perennial ryegrass straw as a forage source for ruminants. J. Anim. Sci. 82:2175–2814.

Fletcher, L.R. 1999. "Non-toxic" endophytes in ryegrass and their effect on livestock health and production. p. 133–139. *In* D.R. Woodfield and C. Matthew (ed.) Ryegrass endophyte—An essential New Zealand symbiosis. Grassland Res. Pract. Ser. No. 7. New Zealand Grassl. Assoc., Palmerston North, NZ.

Fletcher, L.R., I. Garthwaite, and N.R. Towers. 1993. Ryegrass staggers in the absence of lolitrem B. p. 119–121. *In* D.E. Hume et al. (ed.) Proc. Int. Symp. *Acremonium*/Grass Interactions, 2nd, Palmerston North, New Zealand. Feb. 1993. AgResearch, Palmerston North, New Zealand.

Fletcher, L.R., G.A. Lane, D.B. Baird, and E. Davies. 2000. Seasonal variations of alkaloid concentrations in two perennial ryegrass–endophyte associations. p. 535–542. *In* V.H. Paul and P.D. Dapprich (ed.) The Grassland Conf. 2000. Proc. Int. Symp. *Neotyphodium*/Grass Interactions, 4th, Soest, Germany. 27–29 Sept. 2000. Univ. Paderborn, Germany.

Fletcher, L.R., B.L. Sutherland, and C.G. Fletcher. 1999. The impact of endophyte on the health and productivity of sheep grazing ryegrass-based pastures. p. 11–17. *In* D.R. Woodfield and C. Matthew (ed.) Ryegrass endophyte—An essential New Zealand symbiosis. Grassland Res. Pract. Ser. No. 7. New Zealand Grassl. Assoc., Palmerston North.

Hohenboken, W.D., J.L. Robertson, D.J. Blodgett, C.A. Morris, and N.R. Towers. 2000. Sporidesmin-induced mortality and histological lesions in mouse lines divergently selected for response to toxins in endophyte-infected fescue. J. Anim. Sci 78:2157–2163.

Hume, D.E., A.J. Popay, B.M. Cooper, J.P.J. Eerens, T.B. Lyons, C.G.L. Pennell, B.A. Tapper, G.C.M. Latch, and D.B. Baird. 2004. Effect of a novel endophyte on the productivity of perennial ryegrass (*Lolium perenne*) in New Zealand. *In* R. Kallenbach et al. (ed.) Proc. Int. Symp. *Neotyphodium*/Grass Interactions, 5th, Fayetteville, AR, USA. 23–26 May 2004. Univ. Arkansas, Fayetteville.

Keogh, R.G. 1973. Induction and prevention of ryegrass staggers in grazing sheep. N. Z. J. Exp. Agric. 1:55–57.

Keogh, R.G., B.A. Tapper, and R.H. Fletcher. 1996. Distributions of the fungal endophyte *Acremonium lolii*, and of the alkaloids lolitrem B and peramine, within perennial ryegrass. N. Z. J. Exp. Agric. 39:121–127.

Layton, D.L., L.R. Fletcher, A.J. Litherland, M.G. Scannell, J. Sprosen, C.J. Hoogendoorn, and M.G. Lambert. 2004. Effect of ergot alkaloids on liveweight gains and urine lysergol levels in ewe hoggets and heifers. Proc. N. Z. Soc. Anim. Prod. 64:192–196.

Morris, C.A., N.R. Towers, M. Wheeler, and N.C. Amyes. 1995. A note on the genetics of resistance or susceptibility to ryegrass staggers in sheep. N. Z. J. Agric. Res. 38:367–371.

Pearson, E.G., C.B. Andreasen, L.L. Blythe, and A.M. Craig. 1996. Atypical pneumonia associated with ryegrass staggers in 17 calves. J. Am. Vet. Med. Assoc. 209(6):1137–1142.

Popay, A.J., and S.A. Bonos. 2004. Biotic responses in endophytic grasses. *In* R. Kallenbach et al. (ed.) Proc. Int. Symp. *Neotyphodium*/Grass Interactions, 5th, Fayetteville, AR, USA. 23–26 May 2004. Univ. Arkansas, Fayetteville.

Reed, K.F.M., A. Leonforte, P.J. Cunningham, J.R. Walsh, D.I. Allen, G.R. Johnstone, and G. Kearney. 2000. Incidence of ryegrass endophyte (*Neotyphodium lolii*) and diversity of associated alkaloid concentrations among naturalised populations of perennial ryegrass (*Lolium perenne* L.). Aust. J. Agric. Res 51:569–578.

Roberts, C., R. Kallenbach, and N. Hill. 2002. Harvest and storage method affects ergot alkaloids concentration in tall fescue [Online]. Available at www.plant managementnetwork.org/cm/. Crop Manage. DOI 10.1094/CM-2002-0917-01-BR.

Tapper, B.A., and G.A. Lane. 2004. Janthitrems found in a *Neotyphodium* endophyte of perennial ryegrass. *In* R. Kallenbach et al. (ed.) Proc. Int. Symp. *Neotyphodium*/ Grass Interactions, 5th, Fayetteville, AR, USA. 23–26 May 2004. Univ. Arkansas, Fayetteville.

Tapper, B.A., and G.C.M. Latch. 1999. Selection against toxin production in endophyte-infected perennial ryegrass. p. 107–111. *In* D.R. Woodfield and C. Matthew (ed.) Ryegrass endophyte—An essential New Zealand symbiosis. Grassland Res. Pract. Ser. No. 7. New Zealand Grassl. Assoc., Palmerston North, NZ.

Tor-Agbidye, J., L.L. Blythe, and A.M. Craig. 2001. Correlation of endophyte toxins (ergovaline and lolitrem B) with clinical disease: Fescue foot and perennial ryegrass staggers. Vet. Human Toxicol. 43:140–146.

Turner, K.E., C.P. West, A.S. Moubarack, and E.L. Piper. 1991. Chemical changes in tall fescue silages harvested at two stages of maturity. p. 110–113. *In* Proc. Am. Forage Grassl. Conf., Columbia, MO. 1–4 Apr. 1991. American Forage Grasslands Council. Georgetown, TX.

Ussher, G. 2003. Northlands Pasture Toxin Project. N.Z. Large Herds Assoc. Pahia, Annu. Conf. 34:62–64.

Welty, R.E., A.M. Craig, and M.D. Azevedo. 1994. Variability of ergovaline in seeds and straw and endophyte infection in seeds among endophyte-infected genotypes of tall fescue. Plant Dis. 78:845–849.

INTERACTION BETWEEN THERMAL STRESS AND FESCUE TOXICOSIS: ANIMAL MODELS AND NEW PERSPECTIVES

Donald E. Spiers,[1] Tim J. Evans,[2] and George E. Rottinghaus[2]

Fescue toxicosis, associated with tall fescue (*Festuca arundinacea*), is a complex, multifaceted disease syndrome that still has many questions regarding its pathophysiology and management. In this review, we will describe the interactions between environmental disruption of the thermoregulatory system and fescue toxicosis. Thermal stress will be described in terms of its impact on the thermoregulatory profile, and the complex nature of fescue toxicosis will be explained using an animal model scenario and control feedback approach. Recent developments in the treatment and management of fescue toxicosis will be discussed based on the comparative results of short- and long-term experiments using rodent and bovine models.

RELATIONSHIP BETWEEN FESCUE TOXICOSIS AND ENVIRONMENTAL CONDITIONS

Adverse effects on animal health and productivity have long been associated with the intake of endophyte-infected (E+) tall fescue. Fescue toxicosis-related health problems, particularly those in cattle, are exacerbated by extreme ambient temperatures, and the level of environmental stress experienced by livestock exposed to E+ tall fescue is an important determinant of which clinical signs are observed in cases of fescue toxicosis (Hemken et al., 1981). In fact, environmental conditions have the potential to lower the minimum ergot alkaloid levels necessary to produce disease, thereby predisposing animal populations to fescue toxicosis.

Summer slump is the most common and economically significant syndrome associated with fescue toxicosis in livestock. This particular endophyte-related

[1] Animal Science Research Center, University of Missouri, Columbia, MO, USA
[2] Veterinary Medical Diagnostic Laboratory, University of Missouri, Columbia, MO, USA

syndrome results from alterations in the thermoregulatory ability (i.e., reduced peripheral heat transfer) and changes in the endocrine and neurotransmitter milieus of affected animals (Burrows and Tyrl, 2001; Kerr and Kelch, 1999; Paterson et al., 1995; Porter and Thompson, 1992). It has recently been reported that the exacerbated poor performance in beef cattle under the conditions of heat stress commonly seen with exposure to E+ tall fescue is associated with the disruption of the normal patterns of diurnal core temperature (Al-Haidary et al., 2001; Burke et al., 2001). In cattle and sheep, summer slump is more severe when environmental temperatures exceed 32°C, especially with high humidity, and is characterized by hyperthermia, rough hair coat, tachypnea, increased salivation, hypoprolactinemia, and reductions in reproductive performance, growth rate, feed intake, and milk production (Aldrich et al., 1993; Burrows and Tyrl, 2001; Kerr and Kelch, 1999; Paterson et al., 1995; Porter and Thompson, 1992; Thompson et al., 1987).

In contrast to summer slump, fescue foot is characterized by tissue necrosis and dry gangrene of the distal extremities, including the tips of the ears and tail, and is generally associated with ergot alkaloid intake during periods of cold environmental temperatures during the late fall and especially the winter. Fescue foot is reported to develop from one to several weeks after introduction to tall fescue hay or pasture, and most commonly affects the hind limbs. In its early stages, the condition is characterized by swelling and reddening at the coronary band, knuckling of the pastern joint, arching of the back, and shifting of weight from one hind limb to the other. In more severe cases of fescue foot, animals will appear unthrifty, with increasing lameness, and inflammation and ischemic necrosis of the hooves, associated phalanges, joints, and soft tissues (Burrows and Tyrl, 2001; Kerr and Kelch, 1999; Tor-Agbidye et al., 2001; T.J. Evans, 2001, unpublished data). Although the critical level of thermal stress required for summer slump has been identified, there has been no determination of the required lower level that produces fescue foot.

THERMOREGULATORY TERMINOLOGY

Any attempt to understand the complexity of the problems associated with fescue toxicosis should take into consideration that fescue toxicosis is essentially a series of disruptions in an animal's homeostatic ability. Cannon (1929) recognized that living organisms utilize a complex collection of coordinated physiological reactions to minimize systemic disruptions. In other words, the steady-state internal environment is maintained despite changes in the animal's external environment. He created the term *homeostasis* to characterize the physicochemical interactions within the organism that establish a consistent internal environment. Thermoregulation, or control of body temperature, is a key regulatory system that underlies many of the other homeostatic mechanisms within the body (e.g., endocrine, cardiovascular). Homeothermy, or the maintenance of a constant body temperature in different thermal environments, depends on the regulation of the heat content of the body and is achieved by delicately balancing heat production and heat loss (IUPS Thermal Commission, 2001).

The homeothermic ability or thermal status of an animal can be visualized schematically using its thermoregulatory profile, which contains plots representing core body temperature, metabolic heat production, and evaporative heat loss (Fig. 11.1). Key elements of the profile have been defined by the IUPS Thermal Commission (2001), and include the lower critical temperature (i.e., the ambient temperature below which the animal must increase its metabolic rate to maintain heat balance) and the upper critical temperature (i.e., the ambient temperature above which the animal must increase evaporative heat loss rate to maintain thermal balance). The region between these critical temperatures is denoted as the *thermoneutral zone*, where there is minimal expenditure of energy to maintain homeothermy and, conversely, more potential energy available for production. The thermoneutral zone is the central component of the thermoregulatory profile, and major determinants of this zone include body size, quantity of thermal insulation, and basal metabolic rate. For example, the lower critical temperature may decrease from 0°C in the one-month-old calf to −35°C in the finishing feedlot cow as it increases thermal insulation and thermogenic ability (Yousef, 1985). Likewise, both lower and upper critical temperatures may shift upward (Fig. 11.2) with a reduction in thermal insulation (e.g., loss of winter hair coat) and adaptation to heat stress (Bligh, 1985).

The concept of the thermoregulatory profile and the factors that affect an animal's thermal status can be applied to the problem of fescue toxicosis. Animal age, breed, and species are major determinants of the limits of the thermoneutral zone (i.e., upper and lower critical temperatures) and, in turn, the thermoregulatory response to E+ tall fescue toxins. Since an ambient condition constituting a heat stress for one animal may represent a thermoneutral condition for another animal (Fig. 11.2), potential variations in thermal status between animals becomes extremely important when comparing animal responses to E+ tall fescue. Comparisons of these responses should always be conducted between animals in the same relative region of their thermoregulatory profile.

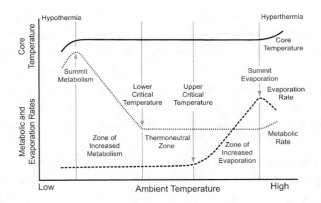

Fig. 11.1. *General thermoregulatory profile for a homeother-mic mammal which shows thermal status, metabolic heat production, and evaporative heat loss.*

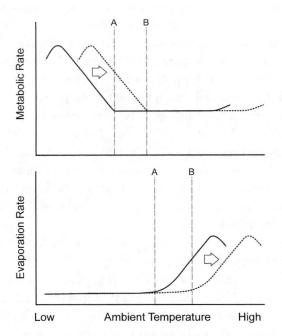

Fig. 11.2. *Shifts in lower and upper critical tempera-tures of the thermoregulatory profile are denoted by movement from A to B positions, with metabolic rate (top graph) representing the lower critical temperature and evaporation rate (bottom graph) identifying the upper critical temperature.*

ANIMAL MODELS AND CONTROL SYSTEM TERMINOLOGY

Horton and Bicak (1987) note that "models are abstract constructs and simplifications of real systems: the model approximates, but does not duplicate, reality. The value for decision making lies in models' capacity to contain the conceptually important information without all the details." The American National Research Council Committee on Animal Models for Research and Aging defines an animal model as "one in which normative biology or behavior can be studied, or in which a spontaneous or induced pathological process can be investigated, and in which the phenomenon in one or more respects resembles the same phenomenon in humans or other species of animals." (http://www.ccac.ca/English/educat/Module04E/module04-04.html). This committee has identified two categories of models that include: (i) spontaneous or natural models and (ii) experimental models. The first is appropriate for naturally occurring animal diseases or conditions, and the second, although similar to the first, is artificially created by the researcher. Likewise, there are short-term studies (i.e., static models of an instant in time) that last for only minutes or hours and long-term studies (i.e., dynamic models of the time element) that last for days or weeks.

Animal models are a central theme of animal physiology and combine the concepts of homeostasis with feedback control systems. Wiener (1961) first emphasized the use of animal/machine analogies and noted that the maintenance of homeostasis is achieved through negative feedback control. The concept was originally borrowed from engineers in an attempt to compartmentalize and, in turn, simplify complex physiological processes and systems. "A control system can be viewed as a set of communication channels interconnecting subsystems that process information" (Houk, 1988). The advantage of using this approach for the analysis of a problem is that the input and output at each level of a model can be controlled and measured to identify components of a physiological system. An animal's potential for maintenance of homeostasis is dependent on its ability to sense and respond to changes in its internal and external environments.

The terms *stress* and *strain* are extremely important in the context of a control system approach. They are used to clearly identify the specific effector sites that are responsible for disruptions in homeostasis. Although stress and strain are often used interchangeably in reference to interactions between animals and environmental conditions, the terms are not equivalent. Environmental conditions that shift a system from its resting or base level are referred to as the *stressors* (Fregly, 1996). *Stress* or *adaptagent* can be any factor that produces an adaptive response, whether that response is positive or negative (Curtis, 1983). *Eustress* refers to a system input that has neutral or even beneficial effects on an organism (Yousef, 1985). In contrast, *distress* would refer to any system input that produces a negative response. The actual displacement from resting or baseline level by a stressor is referred to as the *strain* (or *adaptate*) (Fregly, 1996; Yousef, 1985) and may consist of either adaptive or nonadaptive responses (Curtis, 1983).

Labeling of the stressor is dependent on the system being discussed. For example, internal body temperature might be the stressor or input into the system that affects central and peripheral blood flow. However, in a system designed to control body temperature, internal body temperature represents the effect or output. Ultimately, the goal of control system analysis is to derive a strain/stress index (i.e., the reciprocal of Young's modulus of elasticity; Fregly, 1996) to identify an animal's ability to adapt to a particular stressor and to characterize different components of the acclimation process (e.g., duration, frequency, and magnitude of response).

APPLICATION OF ANIMAL MODELS TO FESCUE TOXICOSIS

Fescue toxicosis is an extremely complex problem in that the magnitude of its impact varies with environmental conditions, dose of toxin(s), and animal status. It might be argued that the only approach to understanding and treating this problem is to conduct studies under controlled conditions. Otherwise, there may be alterations in multiple variables that impact the system in opposite directions negating each other and resulting in minimal net change. One goal is to develop animal models for fescue toxicosis that are based on a combination of controlled studies and field trials. An animal model of fescue toxicosis allows us to simplify a complex problem through an approximation of the actual scenario. Addressing the problem through the framework of a model, we are able to identify and evaluate key components that may be extremely important in the development of new treatment(s).

The use of terms like *stress* and *strain* represents a continuation of the application of control system terminology to clearly identify the specific effector sites that are responsible for fescue toxicosis. The animal model/feedback control approach is ideal for studies of thermal stress and strain. Thermal stress is unique among stressors in that it can be easily quantified for a precise determination of the strain/stress relationship. Ambient temperature can be manipulated to generate different levels of input, and thermal strain can be reliably measured with continued development of new technologies. In addition, the stressor input into a system can be cycled to simulate the outdoor environment.

Cost-effective animal models and technologies are important for the evaluation of potential treatments for reduction of thermal strain associated with fescue toxicosis. It is imperative that there be a clear understanding of thermoregulatory processes affected by fescue toxicosis prior to development of effective models. Ultimately, critical endpoints can be identified for the development of challenge tests that are needed for the generation of effective treatments and eventual identification of animal ability to perform well on pastures containing E+ tall fescue.

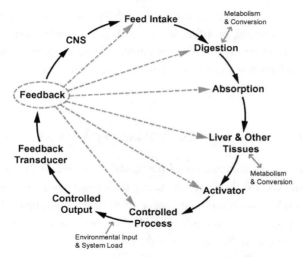

Fig. 11.3. *General model of fescue toxicosis that includes potentially important sites of toxin action with associated input and output signals.*

The complex problem of fescue toxicosis and its disruption of basic homeostatic processes within the body can be better visualized using a control system approach that includes both passive and active elements of the designated system, and allows one to identify sites that are potentially vulnerable to the effects of toxins found in E+ tall fescue. Figure 11.3 shows one possible control system that is associated with fescue toxicosis, with inputs and outputs from numerous sensitive sites where information is processed. Any of these sites could be a key effector responsible for many symptoms associated with fescue toxicosis, and, in every case, each site contains a number of unanswered questions. The model begins with feed intake and ends with feedback to an unknown *black box*, likely found in the central nervous system (CNS) and other sites already noted within the model. These key sites include digestion, absorption, and specific tissues.

Since it is still unclear which specific toxins are directly responsible for the various manifestations of the syndrome, there is limited information regarding utilization, modification, and processing of toxins that enter this system. However, the model used in Fig. 11.3 allows one to visualize the effect of toxins associated with fescue toxicosis on peripheral vasculature (i.e., activator or target) to cause vasoconstriction and reduction of blood flow to these sites (i.e., controlled processes) that, in effect, limit heat loss from the skin (i.e., controlled output). Such change might stimulate thermal receptors in the tissue and eventually feed back to the hypothalamus (i.e., CNS), which is a central site for thermoregulatory control, or affect key processing sites previously noted in the model. Our interest throughout the remainder of this chapter is the impact of thermal stressors on this system and potential treatments to reduce the overall load.

SHORT-TERM RESPONSES: BOVINE AND RODENT MODELS

The thermoregulatory components of the model presented in Fig. 11.3, relating to fescue toxicosis, can be analyzed using both short- and long-term studies. Each type of study reveals different aspects of the complex relationship between fescue toxicosis and thermoregulation. Short-term or static studies using the whole animal are useful for identifying the direct effect of E+ tall fescue toxins on different avenues of heat exchange. Hyperthermia is often used to identify the presence of summer slump. Any shift in core body temperature is a reflection of the body heat content, which in turn is a balance between heat production and heat loss components, as noted in Fig. 11.1. Identification of the cause for a shift in body temperature associated with fescue toxicosis must examine these components.

One important consideration in short-term studies is which toxin(s) should be tested. Ergovaline (EV) is considered by some investigators to be the most toxic of the ergopeptine alkaloids (Bacon et al., 1986; Yates et al., 1985), but is generally unavailable in quantities needed for experiments involving large animals. Many earlier animal studies utilized other less-expensive and more-readily available alkaloids. Ergotamine tartrate is known to decrease feed intake in domestic animals (Greatorex and Mantle, 1973, 1974; Osborn et al., 1992) and lower skin temperature of appendages (e.g., tail, ear), resulting in hyperthermia (Carr and Jacobson, 1969; McCollough et al., 1994; Osborn et al., 1992). Diethylamide of lysergic acid produces hyperthermia in mammals (Murakami et al., 1980).

Initial short-term studies centered on determining if injection of ergot alkaloids into cattle affects body temperature. Carr and Jacobson (1969) showed that compounds found in E+ tall fescue were capable of producing responses in Holsteins that were characteristic of fescue toxicosis (i.e., decreased skin temperature, increased rectal temperature). They injected ergotamine tartrate [(5'α)-12'-hydroxy-2'-methyl-5'-(phenylmethyl)-ergotaman-3',6',18-trione] [intramuscular; 35–92 µg kg^{-1} body weight (BW)] at an air temperature of 18.5°C and monitored skin temperature of the distal portion of the tail. Within several hours, skin temperature at this site decreased 8°C, suggesting a reduction in blood flow to this region. In addition, the animals received an 80% ethanol extract of E+ tall fescue under the same environmental condition, using both oral and intraperitoneal administrations. In both cases, there were reductions in skin temperature and an increase in rectal temperature to suggest that compounds found in E+ tall fescue were capable of producing symptoms associated with fescue toxicosis. In summary, they suggested that reduction in skin blood flow at higher air temperature might be responsible for decreased heat loss and hyperthermia, with similar shifts in blood flow at cold air temperature, leading to a reduction in skin temperature. It is likely that such blood-flow reductions to extremities in the cold might also lead to reduced nutrient flow to these regions and ultimately result in tissue necrosis. Browning et al. (1998) injected Angus heifers intravenously with either ergotamine tartrate (5–7 mg per animal) or ergonovine maleate (5–7 mg per animal) at an average air temperature of 35.2°C. Both produced an increase in respiration rate (60–90 breaths min^{-1}) postinjection and a reduction in skin temperature. However, neither treatment had a significant effect on rectal temperature.

Injection of EV into cattle may produce thermoregulatory symptoms of fescue toxicosis. Al-Haidary et al. (1995) injected EV intraperitoneally (5.2 µg kg^{-1} BW d^{-1}) into cattle exposed to heat stress (31°C) for 3 d to determine if this toxin is capable of producing short-term changes in thermoregulatory ability. This resulted in a significant increase in core body temperature and respiration rate during the 3-d treatment period (Fig. 11.4). Likewise, skin temperature in certain regions (i.e., hip and back) was lower than the control level, which suggests reduced blood flow and heat loss. These results show that EV is also capable of producing symptoms associated with fescue toxicosis and at a lower dose than reported for ergotamine. McCollough et al. (1994) measured skin temperature of calves, as an estimate of skin blood flow, for a 4-h period following intravenous injections of several ergopeptide alkaloids (ergotamine, ergine, and EV) to compare the relative potency of the three compounds. They found that EV had a greater activity than the others in rate and magnitude of reduction in tail skin temperature, and attributed it to a reduction in peripheral blood flow.

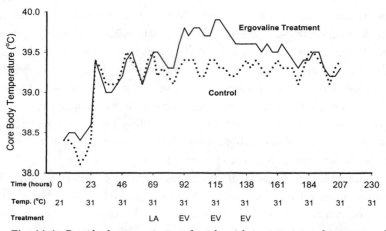

Fig. 11.4. *Core body temperature of cattle with intraperitoneal injection of either lactic acid vehicle (LA) or ergovaline (EV; 5.2 μg kg^{-1} BW d^{-1}).*

Rodents have been used successfully for many years as an initial, inexpensive animal model to evaluate both physiological and behavioral responses to a number of toxins. Several studies have been conducted using a rodent model to determine short-term effects of toxins associated with fescue toxicosis and to identify avenues of heat production and loss that might be responsible for shifts in thermal status. In an early study, Roberts et al. (1949) injected rats intraperitoneally with ergotoxine (4.5 mg kg^{-1} BW) at an air temperature of 28 to 30°C. Rats were tested at 12 and 37 d of age, where this air temperature represents slightly cold and hot environments, respectively (Spiers and Adair, 1986). At the younger age, there was a hypothermic response to ergotoxine and an accompanying reduction in metabolism. Interestingly enough, the older animals displayed a hyperthermic response.

It is entirely possible that the rats responded differently to the toxin at different ages because 28 to 30°C simply represented different stressors, and the specific response to ergot alkaloids is highly stress dependent. It was suggested (Roberts at al., 1949) that the hypothermia might be due to nonselective inhibition/stimulation of thermogenic and thermolytic centers. Loew et al. (1978) reported a hypothermic response in rabbits to intravenous ergopeptine alkaloid injection and indicated that the ergot alkaloid-induced hypothermia might be attributed to either central or peripheral α-adrenergic blockade. Neal and Schmidt (1985) fed a diet containing 50% E+ tall fescue for 15 d at an ambient temperature of 24 to 32°C and noted that the rats exhibited significant hypothermia. In fact, it was concluded that, because of the hypothermic response, the rat is probably not an appropriate model for fescue toxicosis.

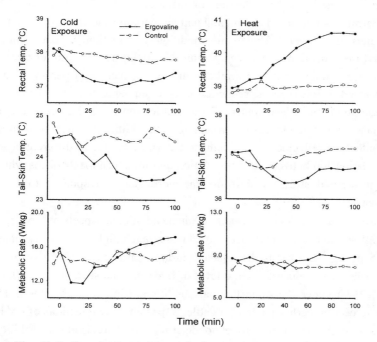

Fig. 11.5. *Rectal temperature, tail skin temperature, and metabolic rate of rats following intraperitoneal injection of ergovaline (15 µg kg⁻¹ BW) under hot (31–33°C; right graph) and cold (7–9°C; left graph) conditions. (Adapted from Spiers et al., 1995.)*

It is entirely possible that the response to ergot alkaloids is dependent on the magnitude and direction (e.g., hot vs. cold response) of thermal strain. The rat model was used to evaluate these possibilities (Fig. 11.5). Adult rats were tested at ambient temperatures that are known to represent cold (7–9°C), thermoneutral (22°C), and hot (31–33°C) conditions (Gordon, 1990) to determine heat production and loss factors that might contribute to the shift in thermal status following a single intraperitoneal injection of EV (15 µg kg^{-1} BW; Spiers et al., 1995; Zhang et al., 1994). As expected, hyperthermia occurred in the hot environment shortly after injection of EV (Spiers et al., 1995). The increase in core temperature was preceded by a reduction in tail skin temperature, which was thought to be due a decrease in peripheral blood flow and heat loss. There was no indication of a shift in heat production that might contribute to the hyperthermic response. Injection of EV under thermoneutral condition resulted in a decrease in core temperature that was due to both an increase in peripheral heat loss, as suggested by an increase in tail skin temperature, and a reduction in heat production, as noted by the reduction in metabolic rate (Zhang et al., 1994). Hypothermia also occurred at cold ambient temperature shortly after injection of EV (Spiers et al., 1995). Under this condition, a reduction in heat production preceded the reduction in core temperature. Tail skin temperature decreased at the cold temperature, but was unlikely due to a reduction in peripheral blood flow. Instead, it followed the reduction in core temperature and was likely due to a reduction in body heat content. These studies show that the direct responses to EV change with the level of thermal stress. In addition, similar shifts in body heat content, as reflected by core temperature, may be due to different changes in the relationship between heat production and loss. The mechanisms whereby different thermal stressors shift the thermoregulatory responses to EV are unknown.

The rodent model has been used to evaluate additional aspects of the response to E+ tall fescue toxins. It has been noted that a much higher dose of ergotamine (i.e., 75 µg kg^{-1} BW) must be injected during heat stress (31°C) (Nesbitt and Spiers, 1995) to produce the same level of hyperthermia seen with EV (i.e., 15 µg kg^{-1} BW) under the same thermal conditions (Spiers et al., 1995). These results agree with earlier reports for cattle. Repeated daily injections of EV have been used with the rodent model to determine if there is the possibility of adaptation to the thermoregulatory effects of EV under heat stress (31°C) conditions (Nesbitt and Spiers, 1995). Rats were injected daily for 6 d with either 35 µg EV kg^{-1} BW or lactate vehicle. On the last day of treatment, both groups were injected with EV. Both rectal and tail skin temperatures were monitored for 60 min following each injection. Comparisons were made between the first- and sixth-day responses for the EV group and between the sixth-day response of treatment and control groups to EV injection. There was neither a change in response from day to day, nor a shift in thermoregulatory response to a single injection on Day 6. This suggests that there is little adaptation of the thermal response to EV during a short period.

Cattle and rats appear to exhibit similar short-term thermoregulatory responses to E+ tall fescue toxins when tested under comparable thermal conditions representative of similar levels of strain. Short-term or static studies can be useful in identifying the direct effect of E+ tall fescue toxins on thermoregulatory processes. It remains to be seen if these changes in performance shift with continued exposure through dietary treatment, or if there is a time factor that must be considered when predicting the response to these compounds. An unexplored use of the short-term procedure is as a challenge test to evaluate potential treatments for fescue toxicosis or identify animal sensitivity to the toxins and possible adaptation across time.

LONG-TERM RESPONSES: BOVINE AND RODENT MODELS

Long-term or dynamic studies are more complicated than static studies in that they must consider the time element and the possibility of adaptation to the situation in question. Likewise, irrelevant issues in static studies, such as circadian rhythms and the indirect effect of reduced feed intake during treatment, become extremely important for long-term studies. The most successful natural models for understanding fescue toxicosis and identifying treatments must be based on studies that have monitored thermal and nutritional status during periods of days to weeks under laboratory and field conditions. Several studies have attempted to detail the complex relationship between productivity and thermoregulatory ability that occurs in heat-stressed cattle fed a diet containing E+ tall fescue seed. Reduced intake of E+ tall fescue is more prevalent at air temperature above 31°C (Hemken et al., 1984). As a result of reduced intake, there is decreased average daily gain under these conditions (Hoveland et al., 1983; Schmidt et al., 1982). In terms of the change in thermal status, the general conclusion is that the E+ tall fescue-induced hyperthermia results from a reduction in heat dissipation due to increased peripheral vasoconstriction (Osborn et al., 1992; Rhodes et al., 1991).

Different studies have identified both early and late shifts in thermal status of cattle when fed a diet containing E+ tall fescue. Al-Haidary et al. (2001) administered diets containing E+ tall fescue seed (5 μg EV kg^{-1} BW d^{-1}) for several days to cattle maintained at thermoneutrality (21°C). This treatment continued for 2 d at 31°C. At the end of this period, measurements were made of heat production and loss to identify avenues that were responsible for the E+ tall fescue-induced hyperthermia. These measurements included metabolic rate, core and skin temperatures, and respiratory and skin vaporizations. In general, the hyperthermic response to E+ tall fescue treatment and constant heat stress was most noted during the night period when there had been a constant accumulation of metabolic heat and no increase in the thermal gradient for heat dissipation at night. The increase in internal body temperature was associated with an increase in respiration rate that likely aided heat dissipation. In contrast, E+ tall fescue treatment produced no significant change in metabolic heat production or skin and respiratory vaporization rates for water. There was no increase in skin temperature with the rise in internal body temperature as anticipated. The lack of a parallel change in internal and external values suggests a lack of an increase in peripheral blood to augment heat loss under these conditions (Osborn et al., 1992; Rhodes et al., 1991; Solomons et al., 1989). Aldrich et al. (1993) fed Holstein steers an E+ tall fescue diet (285 μg kg^{-1} EV) for approximately 20 d at an air temperature of 32°C. Once again, the E+ tall fescue-treated animals exhibited a core body temperature that was above that of control animals, with no effect on metabolic heat production or respiratory vaporization. In contrast, skin vaporization was reduced in the E+ tall fescue-treated animals by 50%, and this reduction was used to explain the noted differences in core temperature. The difference in skin vaporization between the two studies of E+ tall fescue effects on heat exchange could be attributed to differences in sample time between the studies or shifts in response to E+ tall fescue treatment across time. It is possible the long-term E+ tall fescue treatment (i.e., >2 wk) could cause a change in skin vaporization that is not seen with shorter treatment durations. Unfortunately, there are few controlled studies of this nature that extend beyond several weeks to verify these possible changes in heat exchange due to E+ tall fescue treatment.

One important consideration is the relationship between the circadian change in internal body temperature and environmental thermal conditions that affect the response to E+ tall fescue toxins. Recent studies using implanted sensors have allowed for a more complete determination of complex daily rhythms. In one study of fescue toxicosis, Angus heifers were exposed to daily cycling heat stress (i.e., 5 h at 33°C and 5 h at 25°C) for 10 d (Fig. 11.6; Snyder et al., 1998). Both the mean and amplitude of the daily core temperature cycles of the heat-exposed cattle increased from the thermoneutral level. Such change has been reported for cattle during heat stress (Hahn, 1995) and is often accompanied by reduced feed intake, which is different for the acute, chronic, and adaptive phases of response (Hahn et al., 1992). After a recovery period of 21 d at thermoneutrality, the same animals were fed an E+ tall fescue diet (5 µg EV kg^{-1} BW d^{-1}) and exposed to the same heat stress for 10 d (Fig. 11.6). Mean core temperature was shifted to a higher level than at thermoneutrality, but the amplitude of the daily cycle was reduced as a result of E+ tall fescue treatment. It is important to note that major differences occurred during the high and low periods of the daily temperature cycle. In fact, a comparison of core temperature during the transition period in the daily cycle shows little difference between groups.

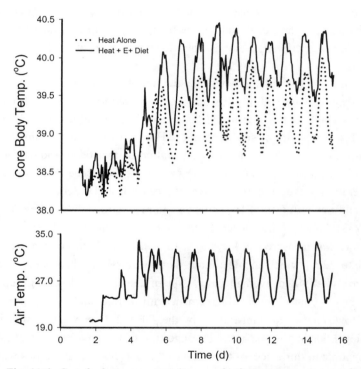

Fig. 11.6. *Core body temperature (top graph) during exposure to cycling heat stress (bottom graph) for cattle consuming control and E+ tall fescue diets.*

Circadian differences in thermoregulatory performance are even seen at a constant level of heat stress. In one study, beef calves were exposed to 31°C for 3 d and then fed a diet containing E+ tall fescue seed (5 µg EV kg^{-1} BW d^{-1}) for 5 d (Al-Haidary, 1995; Al-Haidary et al., 1995) to determine shifts in thermo-regulatory ability due to fescue toxicosis at a constant level of thermal stress. All measurements reported previously (Al-Haidary et al., 2001) were performed at the end of each phase of this study, with each sample date including measurements at both noon and midnight. This approach was to determine if heat production and loss differences could occur at extreme points of the daily cycle to explain differences in core temperature status. Once again, the effects of E+ tall fescue treatment on core temperature were increased at midnight. Likewise, there was an increase in respiration rate at this time. None of the other variables (i.e., skin temperature, skin or respiratory vaporization, metabolic rate) were affected by E+ tall fescue treatment. This would suggest that differences in heat production and water vaporization, at least during peak periods of the daily thermal cycle, are not responsible for noted shifts in body heat content. These shifts, during early days of E+ tall fescue treatment, are likely due to the inability to increase peripheral blood flow to augment heat loss under these conditions.

Use of the rodent model to determine the long-term impact of fescue toxicosis on thermoregulatory ability has not been explored to any extent. It remains, however, a viable option due to the faster growth rate and shorter life span of rodents, lower quantity of seed, and the reduced cost in terms of animal value and maintenance cost. Any future consideration of the rodent model for long-term studies of mechanisms of action or treatments for fescue toxicosis rests on a consideration of the similarity of response for cattle and rodents.

Neal and Schmidt (1985) fed adult female rats various diets, one containing 50% KY-31 infected seed, to evaluate response during a 15-d period. Rectal temperature was measured on Days 3, 6, 13, and 15, and air temperature ranged from 24 to 32°C, which includes both thermoneutral and heat stress regions. Daily feed intake and average daily gain of the E+ tall fescue group were reduced below control level in agreement with bovine studies (Jackson et al., 1984). However, rectal temperature of the E+ tall fescue group was below control level, leading to the authors' suggestion of the possibility "that rats are not susceptible to this effect of toxic fescue." It was noted that the rats were not heat stressed at the time of measurement. More recent results of the short-term response to EV at different ambient temperatures (Spiers et al., 1995; Zhang et al., 1994) actually confirm this finding of a hypothermic response to E+ tall fescue toxins at thermoneutrality. This does not eliminate the fact that there may be a hyperthermic response, similar to that in cattle, when the toxins are administrated during heat stress.

A rodent model for fescue toxicosis has been generated to evaluate the response of rats to a diet containing E+ tall fescue toxins (D.E. Spiers, P.A. Eichen, and G.E. Rottinghaus, 2004, unpublished data) and used as preliminary

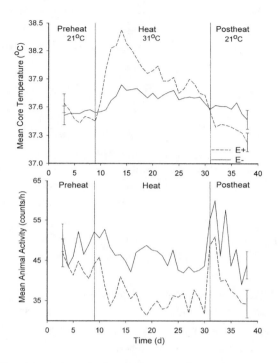

Fig. 11.7. *Core body temperature (top graph) and activity (bottom graph) of rats fed endophyte-infected (E+) or endophyte-free (E−) seed diets before, during, and after exposure to heat stress.*

studies to identify potential treatments for this condition (Roberts et al., 2002). Adult male rats, implanted with transmitters to monitor core temperature and general activity, were fed diets containing either E+ tall fescue seed (41% seed for 165 μg EV kg^{-1} BW d^{-1}) or endophyte-free (E−) fescue seed. Animals were administered these diets for 1 wk at thermoneutrality (21°C), 23 d of heat stress (31°C), followed by 1 wk of recovery at 21°C. Rats receiving the E+ tall fescue diet decreased feed intake and weight gains throughout the study. Hypothermia occurred under thermoneutral condition, and hyperthermia was found during heat stress (Fig. 11.7). The thermoregulatory results agree with studies of acute response to E+ tall fescue toxins at different air temperatures (Spiers et al., 1995; Zhang et al., 1994), and likewise support long-term studies using cattle and rats. A new finding is that activity level was lower in rats fed the E+ tall fescue diet compared with controls during all periods. This suggests the possibility that rats may be exhibiting sickness behavior (Avitsur et al., 1997; Luker et al., 2000) on the E+ tall fescue diet. Additional studies are needed that include other animal species to verify this possibility. These results show that the rodent model can be used with the bovine model to evaluate mechanisms of action for fescue toxicosis and development of potential treatments.

USE OF BOVINE AND RODENT MODELS TO EVALUATE NEW APPROACHES TO REDUCE THE IMPACT OF FESCUE TOXICOSIS ON THERMOREGULATORY ABILITY

The rationale for the development of procedures to reduce problems associated with fescue toxicosis begins within the framework of the model presented in Fig. 11.3. The approaches summarized below utilize the collective potential of the bovine and rodent models presented earlier at different sites within this fescue toxicosis model. The initial approach addresses the question and challenge of identifying animal sensitivity to fescue toxicosis and (or) heat stress. This work is only in the early stages of development, but has the potential to integrate into several sites within the model (e.g., intake, absorption, and metabolism of toxins, activator responsiveness, and system load). Studies that have used both animal models to evaluate fescue seed containing novel endophytes are summarized, relative to thermoregulatory benefit. The information from these studies would enter the model at the level of toxin input into the system. Finally, two treatments are presented that appear to alter an animal's ability to respond to heat stress and, in turn, improve its performance on E+ tall fescue.

Breed/Strain Differences within Models and Selection for Sensitivity

There are reportedly breed differences in the susceptibility of cattle to fescue foot and summer slump. For example, Brahman-type cattle are more likely than European breeds to develop gangrene of the extremities associated with fescue foot, whereas cattle breeds of European origin may be more predisposed to ergot alkaloid-related heat stress (Burrows and Tyrl, 2001; Kerr and Kelch, 1999). One strategy, which has received little study, is to select animals that are more tolerant of environmental stressors (i.e., heat, E+ tall fescue toxins, etc.), based on production or physiological endpoints (Siegel, 1993). Brahman breeds, which are more resistant to heat stress, exhibit less reduction in daily BW gain (Goetsch et al., 1988; McMurphy et al., 1990) and daily milk yield (Brown et al., 1993) when fed E+ tall fescue during heat stress. Interpretation of these studies remains a challenge since the impact level for fescue toxicosis is correlated with severity of heat stress. Because Brahman breeds are more heat tolerant and at a different point in their thermoregulatory profile (Fig. 11.2), they would often be experiencing a lower level of heat strain than European breeds when tested at air temperatures toward the upper end of the thermoregulatory profile (i.e., Sites B vs. A in Fig. 11.2). In other words, it is difficult to separate heat sensitivity from E+ tall fescue toxin sensitivity due to their interdependence.

Another approach is to use a challenge test to evaluate an animal's resistance to E+ tall fescue toxins. This is similar to the classical halothane test for porcine stress syndrome in pigs that have the autosomal recessive gene (Mabry et al., 1981; Reik et al., 1983). These animals exhibit a rapid rise in core temperature (i.e., malignant hyperthermia), increased metabolic rate, and cardiovascular distress when anesthetized with halogenated anesthetics (Gronert, 1986). Likewise, they display more extensive physiological changes when exposed to thermal stress (D'Allaire and DeRoth, 1986). Similar tests might be designed for detecting animals that are susceptible to fescue toxicosis to yield an immediate economic benefit. Change in plasma prolactin level might be used as a detector of exposure to heat stress and E+ tall fescue toxin and serve as an endpoint in selecting for sensitive/insensitive animals. A correlation between low plasma prolactin level and elevated rectal temperature has been reported for cattle fed E+ tall fescue (Kerr et al., 1990).

The short-term tests described earlier can be used with bovine and rodent models to identify animal sensitivity to E+ tall fescue toxins. In one study, Angus heifers were selected from a summer field trial on E+ tall fescue pasture for sensitivity to fescue toxicosis based on ADG (Snyder et al., 1998). Insensitive and sensitive groups were first tested in an environmental chamber study to determine response to 10 d of cycling heat stress (i.e., 33°C day and 25°C night temperatures) followed after 21 d of recovery by a second chamber study where the same level of heat stress was applied together with an E+ tall fescue seed diet (5 μg kg^{-1} BW d^{-1}). Core temperatures were different for the two groups. The sensitivity, however, was to heat stress alone and not to the heat stress/E+ tall fescue combination. Within-breed variability in tolerance to E+ tall fescue has also been reported for both Angus (Lipsey et al., 1992) and Hereford (Gould and Hohenboken, 1993) breed types. Likewise, there are field reports of individual differences in cattle grazing on E+ tall fescue pasture. The rodent model also has been used as a tool to examine the possibility of breeding for tolerance to fescue toxicosis. Hohenboken and Blodgett (1997) divergently selected mice for resistance and susceptibility to fescue toxicosis. The primary variable used in this selection was growth rate. Adult mice from both E+ tall fescue sensitive and resistant lines have been fed an E+ tall fescue diet (165 µg EV kg^{-1} BW d^{-1}) during exposure to thermoneutral (24°C) and heat stress (34°C) conditions to determine differences in feed intake, growth rate, and core temperature control (Eichen et al., 2001). There were no differences between lines for any measured variable in either environment. More importantly, intake of the E+ tall fescue diet by mice did not affect rate of feed consumption, as previously reported for the rat. For this reason, the murine model might not be appropriate as a model for fescue toxicosis in cattle.

Development of Novel Endophytes

One approach to reduce the occurrence of fescue toxicosis is the selection of an endophyte strain that provides the tall fescue with heat, drought, and insect resistance, but produces few ergot alkaloids. Two HiMag tall fescue lines (HiMag4 and HiMag9) were recently produced to contain novel (EV-deficient) endophytes. Both rodent and bovine models were used to determine if intake of diets containing these lines would reduce production (i.e., lower feed intake and growth) and hyperthermia problems associated with fescue toxicosis. The benefit of the rodent model is that it allows for rapid testing of a small quantity of new seed prior to the large-scale production and time needed to generate pastures containing these lines. In the initial study, rats were fed diets containing either standard E+ tall fescue seed (KY31) or novel E+ tall fescue seed (HM4 or HM9) diets for 13 d (Roberts et al., 2002). Ergovaline content of KY31, HM4, and HM9 diets were 3700, 50, and 50 µg kg^{-1}, respectively. On Day 5, air temperature was increased from thermoneutrality (21°C) to heat stress (31°C) condition. Feed intake (Fig. 11.8) and BW gain for rats fed the KY31 diet were immediately depressed below the level for the other diets and remained at this level through the remainder of the study. These results quickly demonstrated that there were no toxins in either HiMag diet that would compromise feed intake and consequently BW gain. As a result of this study, pastures were planted with HM4, HM9, KY31, and endophyte-free HiMag (HiMag-) seed for eventual cattle testing (Nihsen et al., 2004) through summer conditions. On KY31 pasture, BW gains were lower and both respiration rate and rectal temperature were higher than on any of the other pastures. These results, that novel

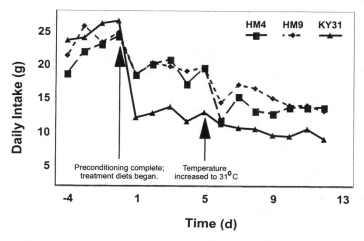

Fig. 11.8. *Change in feed intake of rats consuming diets containing HiMag (HM4, HM9) fescue seed or Kentucky 31 (KY31) seed during a 13-d period.*

endophytes may reduce problems associated with fescue toxicosis, can be used to benefit both tall fescue and grazing animals during summer months.

Improved Responsiveness to Heat Load

Ivermectin. The anthelmintic ivermectin (Ivomec, Merck & Co., Inc., Rahway, NJ), when given to cattle on E+ tall fescue pastures, results in more weight gains than expected for parasite control alone (Bransby, 1997; Oakley, 1990), suggesting that ivermectin partially reduces weight loss caused by fescue toxicosis. Rats pretreated with ivermectin exhibit diminished core temperature response to acute EV challenge (Spiers et al., 1997). Ivermectin increases γ-aminobutyric acid (GABA) release and binding to postsynaptic receptors (Oakley, 1990). Likewise, stimulation of GABA receptors reduces core temperature (Sancibrian et al., 1991; Serrano et al., 1986; Zarrindast and Oveissi, 1988) and prevents increased core temperature during heat stress (Biswas and Poddar, 1990). A study was conducted to determine if ivermectin can reduce some symptoms in cattle resulting from the intake of E+ tall fescue. Ivermectin (12 mg d^{-1}) or blank osmotic pumps were placed in the reticulum of Simmental heifers, followed by dietary administration of E+ tall fescue diet (5 µg EV kg^{-1} BW d^{-1}) and exposure to a progressive increase in heat stress (29°C for 12 d; 30°C for 6 d; 31°C for 4 d) (Spiers, 1998). Feed intake of ivermectin-treated animals was greater than controls from Days 5 through 7 at 29°C, with core temperature being lower than controls from Days 6 to 12 at 29°C and for the first 3 d at 30°C (Fig. 11.9). Both rodent and bovine studies suggest that ivermectin may reduce some impact of fescue toxicosis on feed intake and body temperature control.

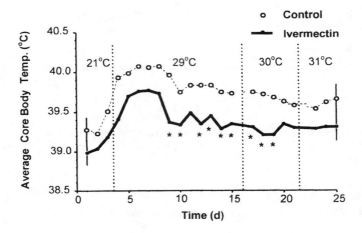

Fig. 11.9. *Core body temperature of heifers with intraperitoneal blank (negative control) or ivermectin pumps and exposed to thermoneutral and heat stress environments.*

Nitric oxide. A primary cause of the hyperthermia associated with fescue toxicosis is excessive vasoconstriction of peripheral vasculature that reduces heat loss from the skin and, in turn, produces heat accumulation in the summer environment (Dyer, 1993). Primary E+ tall fescue toxins such as EV may be responsible for this effect on specific blood vessels. Endothelial-derived nitric oxide (NO) plays a major role in maintaining vasomotor tone in major vascular beds, such as the arteriovenous anastomoses. Likewise, NO is a component of heat defense mechanisms. It is likely that ergopeptine alkaloids act at the level of NO production and activity, since these toxins are known to affect endothelial cells lining blood vessels (Strickland et al., 1996). Studies have been performed using rodent and cattle models to determine if NO production is altered by fescue toxicosis and to evaluate the benefit of supplemental NO donors to animals experiencing fescue toxicosis.

Rats were transdermally administered a NO (N+) donor [nitroglycerin (1,2,3-propanetriol trinitrate); 50 mg kg^{-1} BW d^{-1}] in combination with E+ tall fescue

(160 μg EV kg^{-1} BW d^{-1}) or E– diets during a 15-d exposure to constant heat stress (31°C). The N+ treatment alleviated hyperthermia associated with E+ tall fescue intake during the night, suggesting that depression of NO plays a role in hyperthermia associated with fescue toxicosis. It appears that NO supplementation to rats can override the extreme vasoconstrictive effects of E+ tall fescue toxins.

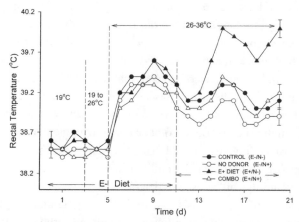

Fig. 11.10. *Rectal temperature of cattle under thermoneutral and heat stress conditions when administered a NO donor (i.e., nitroglycerin; N+) and fed E+ and E– diets.*

Angus/Simmental crossbred steers were used to determine if NO plays a role in some of the thermal strain symptoms observed in cattle experiencing fescue toxicosis during heat stress. As in the rodent study, some animals received transdermal N+ administration (2.5 mg nitroglycerin kg^{-1} BW d^{-1}) while on E– or E+ tall fescue (44 μg EV kg^{-1} BW d^{-1}) during exposure to cycling daily heat stress (26–36°C) (Fig. 11.10). The E+ tall fescue treatment, together with heat stress, produced an increase in core temperature above control levels. Treatment with N+ failed to alleviate hyperthermia in control E– tall fescue animals, but reduced the more excessive hyperthermia associated with fescue toxicosis (i.e., E+ tall fescue treatment) and returned animals to control levels. Moreover, N+ did not appear to enhance other performance characteristics, including average daily feed intake and gain, which was similar to what was observed using the rodent model.

Summary

A functional thermoregulatory system is required for maintenance of homeo-thermy, which is essential for normal growth, development, and productivity of many animals. A comparative approach is presented that utilizes this system in both small nonruminant and large ruminant species to determine the action of toxins associated with fescue toxicosis, as well as for identification of species-specific activities related to these compounds. The logistical and economic advantages of using a rodent model are discussed, together with the challenges of this comparative approach. Several symptoms characteristic of fescue toxico-sis are seen in both rats and cattle (e.g., hyperthermia, lowered feed intake, and reduced growth), and evaluations of several treatments, using these responses as endpoints, are presented. The use of fescue containing a novel endophyte and treatments with a nitric oxide donor (i.e., nitroglycerin) and a commercially-available anthelmintic (i.e., ivermectin) have all shown potential for limiting the adverse effects of endophyte-infected fescue in an experimental setting and under a variety of environmental conditions. New and existing models of thermoregulatory ability can be used to determine the impact of fescue toxicosis on the maintenance of homeostasis and to evaluate potential prophylactic or therapeutic approaches to this significant agricultural problem. In addition, animal models can be used to identify susceptible and resistant individuals within a population. Our research efforts with these experimental models will continue to identify a broader range of physiologic effects and alterations in gene expression associated with fescue toxicosis in order to develop effective prophylactic and/or therapeutic strategies for this economically significant disease syndrome.

References

Aldrich, C.G., M.T. Rhodes, J.L. Miner, M.S. Kerley, and J.A. Paterson. 1993. The effects of endophyte-infected tall fescue consumption and use of a dopamine antagonist on intake, digestability, body temperature, and blood constituents in sheep. J. Anim. Sci. 71:158–163.

Al-Haidary, A. 1995. Effect of metabolites of endophyte-infected tall fescue on the heat stress responses of beef calves. Ph.D. diss. Univ. of Missouri, Columbia.

Al-Haidary, A., D.E. Spiers, G.E. Rottinghaus, and G.B. Garner. 1995. Effect of administration of ergovaline on the thermoregulatory functions of beef calves under heat stress. J. Anim. Sci. 73(Suppl. 1):132.

Al-Haidary, A., D.E. Spiers, G.E. Rottinghaus, G.B. Garner, and M.R. Ellersieck. 2001. Thermoregulatory ability of beef heifers following intake of endophyte-infected tall fescue during controlled heat challenge. J. Anim. Sci. 79:1780–1788.

Avitsur, R., Y. Pollak, and R. Yirmiya. 1997. Administration of interleukin-1 into the hypothalamic paraventricular nucleus induces febrile and behavioral effects. Neuroimmunomodulation 4:258–265.

Bacon, C.W., P.C. Lyons, J.K. Porter, and J.D. Robbins. 1986. Ergot toxicity from endophyte-infected grasses: A review. Agron. J. 78:106–116.

Biswas, S., and M.K. Poddar. 1990. Does GABA act through dopaminergic-cholinergic interaction in the regulation of higher environmental temperature-induced change in body temperature? Meth. Find. Exp. Clin. Pharm. 12:303–308.

Bligh, J. 1985. Temperature Regulation. p. 75–96. *In* M. Yousef (ed.) Stress physiology in livestock. Vol. I: Basic principles. CRC Press, Boca Raton, FL.

Bransby, D. 1997. Steer weight gain responses to ivermectin when grazing fescue. Large Anim. Pract. 18(3):16–19.

Brown, M.A., L.M. Tharel, A.H. Brown, Jr., W.G. Jackson, and J.R. Miesner. 1993. Milk production in Brahman and Angus cows on endophyte-infected fescue and common bermudagrass. J. Anim. Sci. 71:1117–1122.

Browning, Jr., R., M.L. Leite-Browning, H.M. Smith, and T. Wakefield, Jr. 1998. Effect of ergotamine and ergonovine on plasma concentrations of thyroid hormones and cortisol in cattle. J. Anim. Sci. 76:1644–1650.

Burke, J.M., D.E. Spiers, F.N. Kojima, G.A. Perry, B.E. Salfen, S.L. Wood, D.J. Patterson, M.F. Smith, M.C. Lucy, W.G. Jackson, and E.L. Piper. 2001. Interaction of endophyte-infected fescue and heat stress on ovarian function in the beef heifer. Biol. Reprod. 65:260–268.

Burrows, G.E., and R.J. Tyrl. 2001. Toxic plants of North America. Iowa State Univ. Press, Ames.

Cannon, W.B. 1929. Organization for physiological homeostasis. Physiol. Rev. 9:399–431.

Carr, S.B., and D.R. Jacobson. 1969. Bovine physiological responses to toxic fescue and related conditions for application in a bioassay. J. Dairy Sci. 52:1792–1999.

Curtis, S. 1983. Environmental management in animal agriculture. Iowa State Univ. Press, Ames.

D'Allaire, S., and L. DeRoth. 1986. Physiological responses to treadmill exercise and ambient temperature in normal and malignant hyperthermia susceptible pigs. Can. J. Vet. Res. 50:78–83.

Dyer, D.C. 1993. Evidence that ergovaline acts on serotonin receptors. Life Sci. 53(14):PL223–228.

Eichen, P.A., S.A. Clark, M.J. Leonard, and D.E. Spiers. 2001. Identification of animal sensitivity to fescue toxicosis using mouse and cattle studies. *In* Proc. Tall Fescue Toxicosis Workshop, SERAIEG-8, Chapel Hill, TN.

Fregly, M.J. 1996. Adaptations: Some general characteristics. p. 3–15. *In* M.J. Fregly and C.M. Blatteis (ed.) Handbook of physiology, Section 4, Environmental physiology, Vol. 1. Oxford Univ. Press, New York.

Goetsch, A.L., K.L. Landis, G.E. Murphy, B.L. Morrison, Z.B. Johnson, E.L. Piper, A.C. Hardin, and K.L. Hall. 1988. Supplements, parasite treatments and growth implants for Brahman or English crossbred steers grazing endophyte-infected or noninfected fescue in the spring and fall. Prof. Anim. Sci. 4:32–38.

Gordon, C.J. 1990. Thermal biology of the laboratory rat. Physiol. Behav. 47:963–991.

Gould, L.S., and W.D. Hohenboken. 1993. Differences between progeny of beef sires in susceptibility to fescue toxicosis. J. Anim. Sci. 71:3025–3032.

Greatorex, J.C., and P.G. Mantle. 1973. Experimental ergotism in sheep. Res. Vet. Sci. 15:337–346.

Greatorex, J.C., and P.G. Mantle. 1974. Effects of rye ergot on the pregnant sheep. J. Reprod. Fertil. 37:33–41.

Gronert, G.A. 1986. Malignant hyperthermia. p. 1763. *In* A.G. Engel and B.Q. Banker (ed.) Myology. McGraw-Hill, Minneapolis, MN.

Hahn, G.L. 1995. Environmental influences on feed intake and performance of feedlot cattle. P. 207–225. *In* F.N. Owens (ed.) Proc. Symp.: Intake by feedlot cattle. Oklahoma State Univ., Stillwater.

Hahn, G.L., Y.R. Chen, J.A. Nienaber, R.A. Eigenberg, and A.M. Parkhurst. 1992. Characterizing animal stress through fractal analysis of thermoregulatory responses. J. Therm. Biol. 17(2):1–6.

Hemken, R.W., J.A. Boling, L.S. Bull, R.H. Hatton, R.C. Buckner, and L.P. Bush. 1981. Interaction of environmental temperature and anti-quality factors on the severity of summer fescue toxicosis. J. Anim. Sci. 52:710–714.

Hemken, R.W., J.A. Jackson, Jr., and J.A. Boling. 1984. Toxic factors in tall fescue. J. Anim. Sci. 58:1011–1016.

Hohenboken, W.D., and D.J. Blodgett. 1997. Growth and physiological responses to toxicosis in lines of mice selected for resistance or susceptibility to endophyte-infected tall fescue in the diet. J. Anim. Sci. 75:2165–2173.

Horton, J.C., and C.J. Bicak. 1987. Modeling for biologists. Bioscience 37:808–809.

Houk, J.C. 1988. Control strategies in physiological systems. Fed. Am. Soc. Exp. Biol. J. 2:97–107.

Hoveland, C.S., S.P. Schmidt, C.C. King, Jr., J.W. Odom, E.M. Clark, J.A. McGuire, L.A. Smith, H.W. Grimes, and J.L. Holliman. 1983. Steer performance and association of *Acremonium coenophialum* fungal endophyte on tall fescue pasture. Agron. J. 75:821–824.

IUPS Thermal Commission. 2001. Glossary of terms for thermal physiology. Jpn. J. Physiol. 51:245–280.

Jackson, J.A., Jr., R.W. Hemken, J.A. Boling, R.J. Harmon, R.C. Buckner, and L.P. Bush. 1984. Summer fescue toxicity in dairy steers fed tall fescue seed. J. Anim. Sci. 58:1057–1061.

Kerr, L.A., and W.J. Kelch. 1999. Current veterinary therapy. p. 263-264. *In* J.L. Howard and R.A. Smith (ed.) Food animal practice. Saunders, Philadelphia, PA.

Kerr, L.A., C.P. McCoy, C.R. Boyle, and H.W. Essig. 1990. Effects of ammoniation of endophyte fungus-infested fescue hay on serum prolactin concentration and rectal temperature in beef cattle. Am. J. Vet. Res. 51:76–78.

Lipsey, R.J., D.W. Vogt, G.B. Garner, L.L. Miles, and C.N. Cornell. 1992. Rectal temperature changes of heat and endophyte stressed calves produced by tolerant or susceptible sires. J. Anim. Sci. 70(Suppl. 1):188.

Loew, D.M., E.B. Van Deusen, and W. Meier-Ruge. 1978. Effects on the central nervous system. p. 806-851. *In* B. Berde and H.O. Schild (ed.) Ergot alkaloids and related compounds. Springer-Verlag, New York.

Luker, F.I., D. Mitchell, and H.P. Laburn. 2000. Fever and motor activity in rats following day and night injections of *Staphylococcus aureus* cell walls. Am. J. Physiol. 279:R610–R616.

Mabry, J.W., L.L. Christian, and D.L. Kuhlers. 1981. Inheritance of porcine stress syndrome. J. Hered. 72:429–430.

McCollough, S.F., E.L. Piper, A.S. Moubarak, Z.B. Johnson, R.J. Petroski, and M. Flieger. 1994. Effect of tall fescue ergot alkaloids on peripheral blood flow and serum prolactin in steers. J. Anim. Sci. , Vol. 72(Suppl. 1):144.

McMurphy, W.E., K.S. Lusby, S.C. Smith, S.H. Muntz, and C.A. Strasia. 1990. Steer performance on tall fescue pasture. J. Prod. Agric. 3:100–102.

Murakami, N., Y. Sakai, and S. Ooki. 1980. Behavioral thermoregulation in rats during hyperthermia induced by lysergic acid diethylamide. Neurosci. Lett. 20:105–108.

Neal, W.D., and S.P. Schmidt. 1985. Effects of feeding Kentucky 31 tall fescue seed infected with *Acremonium coenophialum* to laboratory rats. J. Anim. Sci. 61:603–611.

Nesbitt, M.D., and D.E. Spiers. 1995. Thermoregulatory responses of preweanling and adult rats to ergotamine under heat stress and thermoneutral conditions. J. Anim. Sci. 73(Suppl. 1):132.

Nihsen, M.E., E.L. Piper, C.P.West, R.J. Crawford, Jr., T.M. Denard, Z.B. Johnson, C.A. Roberts, D.A. Spiers, and C.F. Rosenkrans, Jr. 2004. Growth rate and physiology of steers grazing tall fescue inoculated with novel endophyte. J. Anim. Sci. 82:878–883.

Oakley, G.A. 1990. Ivermectin—The veterinary handbook. Anpar Books, Hertfordshire, UK.

Osborn, T.G., S.P. Schmidt, D.N. Marple, C.H. Rahe, and J.R. Steenstra. 1992. Effect of consuming fungus-infected and fungus-free tall fescue and ergotamine tartrate on selected physiological variables of cattle in environmentally controlled conditions. J.Anim. Sci. 70:2501–2509.

Paterson, J., C. Forcherio, B. Larson, M. Samford, and M. Kerley. 1995. The effects of fescue toxicosis on beef cattle productivity. J. Anim. Sci. 73:889–898.

Porter J.K., and F.N. Thompson. 1992. Effects of fescue toxicosis on reproduction in livestock. J. Anim. Sci. 70:1594–1603.

Reik, T.R., W.E. Rempel, C.J. McGrath, and P.B. Addis. 1983. Further evidence on the inheritance of halothane reaction in pigs. J. Anim. Sci. 57:826–831.

Rhodes, M.T., J.A. Paterson, M.S. Kerley, H.E. Garner, and M.H. Laughlin. 1991. Reduced blood flow to peripheral and core body tissues in sheep and cattle induced by endophyte-infected tall fescue. J. Anim. Sci. 69:2033–2043.

Roberts, J.E., B.E. Robinson, and A.R. Buchanan. 1949. Oxygen consumption correlated with the thermal reactions of young rats to ergotoxine. Am. J. Physiol. 156:170–176.

Roberts, C.A., D.E. Spiers, A.L. Karr, H.R. Benedict, D.A. Sleper, P.A. Eichen, C.P. West, E.L. Piper, and G.E. Rottinghaus. 2002. Use of a rat model to evaluate tall fescue seed infected with introduced strains of *Neotyphodium coenophialum*. J. Agric. Food Chem. 50:5742–5745.

Sancibrian, M., J.S. Serrano, and F.J. Minano. 1991. Opioid and prostaglandin mechanisms involved in the effects of GABAergic drugs on body temperature. Gen. Pharm. 22:259–262.

Schmidt, S.P., C.S. Hoveland, E.M. Clark, N.D. Davis, L.A. Smith, H.W. Grimes, and J.L. Holliman. 1982. Association of an endophytic fungus with fescue toxicity in steers fed Kentucky 31 tall fescue seed or hay. J. Anim. Sci. 55:1259–1263.

Serrano, J.S., F.J. Minano, and M. Sancibrian. 1986. Gamma aminobutyric-acid-induced hypothermia in rats involvement of serotonergic and cholinergic mechanisms. Gen. Pharm. 17:327–332.

Siegel, M.R. 1993. *Acremonium* endophytes: Our current state of knowledge and future directions for research. Agric. Ecosyst. Environ. 44:301–321.

Snyder, B.L., D.E. Spiers, E.J. Scholljegerdes, S. VanDyke, and G.E. Rottinghaus. 1998. Determination of sensitivity to endophyte-infected tall fescue for beef heifers in a controlled environment. J. Anim. Sci. 76(Suppl. 1):97.

Solomons, R.N., J.W. Oliver, and R.D. Linnabary. 1989. Reactivity of dorsal pedal vein of cattle to selected alkaloids associated with *Acremonium coenophialum*-infected fescue grass. Am. J. Vet. Res. 50:235–238.

Spiers, D.E. 1998. Effect of an anthelmintic in reducing symptoms associated with fescue toxicosis under laboratory conditions. Southern Extension and Research Activity Information Exchange Group-8. Tall Fescue Workshop, Chapel Hill, TN.

Spiers, D.E., and E.R. Adair. 1986. Ontogeny of homeothermy in the immature rat: Metabolic and thermal responses. J. Appl. Physiol. 60:1190–1197.

Spiers, D.E., B.L. Snyder, P.A. Eichen, G.E. Rottinghaus, and G.B. Garner. 1997. Potential benefit of an anthelmintic in reducing hyperthermia associated with fescue toxicosis. J. Anim. Sci. 75(Suppl. 1):212.

Spiers, D.E., Q. Zhang, P.A. Eichen, G.E. Rottinghaus, G.B. Garner, and M.R. Ellersieck. 1995. Temperature-dependent responses of rats to ergovaline derived from endophyte-infected tall fescue. J. Anim. Sci. 73:1954–1961.

Strickland, J.R., E.M. Bailey, L.K. Abney, and J.W. Oliver. 1996. Assessment of the mitogenic potential of the alkaloids produced by endophyte (*Acremonium coenophialum*)-infected tall fescue (*Festuca arundinacea*) on bovine vascular smooth muscle in vitro. J. Anim. Sci. 74:1664–1671.

Thompson, F.N., J.A. Studemann, J.L. Sartin, D.P. Belesky, and O.J. Devine. 1987. Selected hormonal changes with summer fescue toxicosis. J. Anim. Sci. 65:727–733.

Tor-Agbidye, J., L.L. Blythe, and A.M. Craig. 2001. Correlation of endophyte toxins (ergovaline and lolitrem B) with clinical disease: Fescue foot and perennial ryegrass staggers. Vet. Human Toxicol. 43:140–146.

Wiener, N. 1961. Cybernetics or control and communication in the animal and the machine. MIT Press and Wiley & Sons, New York.

Yates, S.G, R.D. Plattner, and G.B. Garner. 1985. Detection of ergopeptine alkaloids in endophyte-infected, toxic KY-31 tall fescue by mass spectroscopy/mass spectroscopy. J. Agric. Food Chem. 33:719–722.

Yousef, M.K. 1985. Thermoneutral Zone. p. 67–74. *In* M. Yousef (ed.) Stress physiology in livestock. Vol. I. Basic principles. CRC Press, Boca Raton, FL.

Zarrindast, M.R., and Y. Oveissi. 1988. GABA A and GABA B receptor sites involvement in rat thermoregulation. Gen. Pharmicol. 19:223–226.

Zhang, Q., D.E. Spiers, G.E. Rottinghaus, and G.B. Garner. 1994. Thermoregulatory effects of ergovaline isolated from endophyte infected tall fescue seed on rats. J. Agric. Food Chem. 42:954–958.

ABSORPTION OF ERGOT ALKALOIDS IN THE RUMINANT

Nicholas S. Hill[1]

The ergopeptine alkaloid, ergovaline, has historically been identified as the putative toxin causing fescue toxicosis in livestock grazing endophyte-infected tall fescue (*Festuca arundinacea*). It is unclear how it was deduced that ergovaline was the toxin, other than it was identified as being the primary ergopeptine alkaloid in endophyte-infected tall fescue (Lyons et al., 1986). However, the ergoline alkaloids make up approximately 50% of the total ergot alkaloid pool in tall fescue, and thus can not be ignored as candidate toxins in addition to, or in place of, ergovaline. The classical literature on the ergot alkaloids may elucidate potential sites and mechanisms of absorption, as well as relative toxicities of specific alkaloids or alkaloid classes within the ergot family.

History suggests ergot toxins have long been problematic in the human food supply. St. Anthony's Fire was a dreaded illness common during the Middle Ages caused by consumption of rye contaminated with ergot. The Salem Witch Trials in the Massachusetts colony were also associated with an affliction of ergotism from bread baked from grain containing a high amount of ergot sclerotia. These historical events provide evidence suggesting ergot alkaloids are indeed toxins, but modern medicine has found uses for ergot alkaloids to control migrane headaches (Diener and McHarg, 2000), deep vein thrombosis (Bongard and Bounameaux, 1991), and to induce uterine contractions during childbirth. Hence, the pharmacological properties, sites of absorption, excretion patterns, and effective and lethal doses of some alkaloids have been determined in monogastric systems and can be used as models for postruminal metabolism in ungulates. Recent investigations into the ruminal metabolism of the ergot alkaloids have provided some interesting and surprising results which illuminate new hypotheses as to the putative toxin(s) for the fescue toxicosis syndrome. The chemical structures of the major classes of ergot alkaloids impact their ability to transport across gastric tissues and, thus, their toxicity to livestock.

---

[1] Department of Crop and Soil Sciences, 3111 Miller Sciences Building, The University of Georgia, Athens, GA

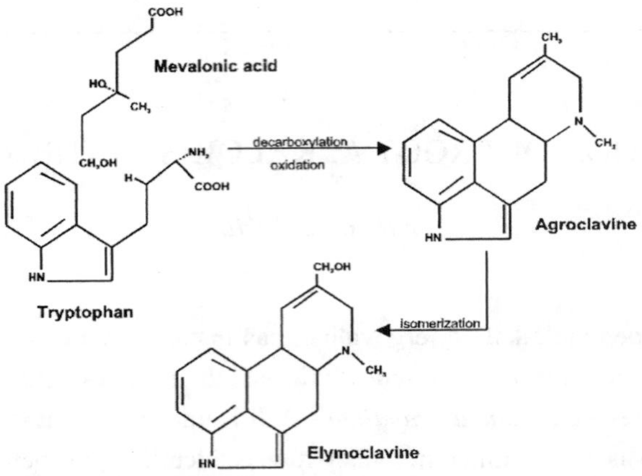

Fig. 12.1. *Synthesis of the clavine alkaloids.*

Thus, understanding ergot alkaloid chemistry is fundamental to a discussion on alkaloid fate and toxicity in livestock.

STRUCTURE OF ERGOT ALKALOIDS

The initiation of biochemical synthesis of the ergoline ring structure common to all ergot alkaloids is via assembly from one molecule of tryptophan and one molecule of mevalonic acid. The resulting intermediate (4-dimethylallyl-*l*-tryptophan) is transformed by various decarboxylation and oxidation steps to close the ergoline ring structure to form agroclavine (Fig. 12.1) (Rutschmann and Stadler, 1978). The 8-methyl group of agroclavine is then oxidized to form the polar form of elymoclavine. These two clavine alkaloids differ from the ergoline ring structure of the ergot alkaloids only by the location of the 8, 9 double bond in the D ring. Subsequent isomerization of the D ring changes the location of the double bond to the 9, 10 position, forming the lysergic moiety common to all ergot alkaloids (Fig. 12.2). The lysergic moiety may have a tricyclic peptide ring attached via a carbonyl at the 8 position in the D ring to form any number of ergopeptine alkaloids. Approximately 50% of the ergot alkaloids produced by *Neotyphodium coenophialum* in planta are the ergopeptine form, but in equal amounts are the ergoline alkaloids as well (Lyons et al., 1986). Ergovaline is the predominant form of the ergopeptine alkaloids and is chemically distinct from other ergopeptine alkaloids with a methyl group at R_1 and an isopropyl group at R_2. Other ergopeptine alkaloids are similar in structure to ergovaline. Ergosine and ergotamine differ in isobutyl or methyl

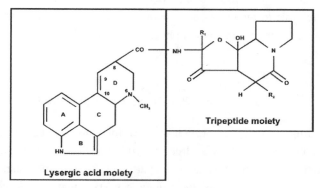

Fig. 12.2. *Structure of the ergot alkaloids.*

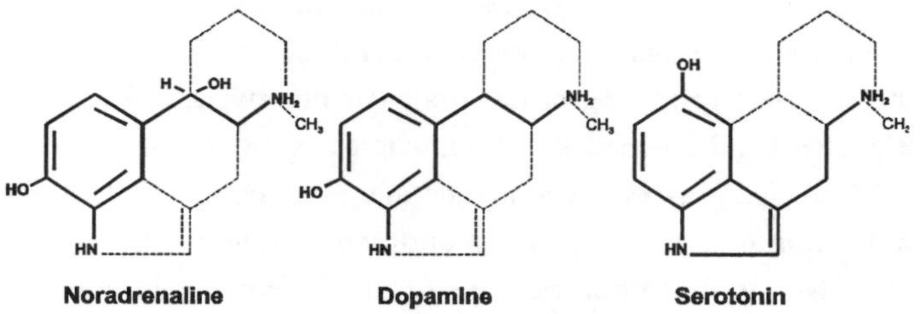

Noradrenaline   **Dopamine**   **Serotonin**

Fig. 12.3. *Structural relationships between the biogenic amines noradrenaline, dopamine, and serotonin reside within the lysergic moiety of the ergot alkaloids and clavine alkaloids.*

benzyl groups at their respective R_2 position, while ergonine and ergocornine differ in an ethyl or isopropyl group at their respective R_1 position (Berde and Sturmer, 1978).

The ergoline alkaloids, on the other hand, have simpler chemical structures attached to the No. 8 carbon in the D ring. Lysergic acid and lysergol, for example, simply have a carboxylic acid or methyl alcohol attached to the 8 carbon, while others may consist of hydrazide, azide, or amide groups (Rutschmann and Stadler, 1978). Of importance to note is the fact that the structural relationships between the biogenic amines noradrenaline, dopamine, and serotonin reside within the lysergic moiety of the ergot alkaloids and clavine alkaloids (Fig. 12.3) (Berde and Sturmer, 1978). Thus, clavine, ergoline, and ergopeptine alkaloids have the chemical structures necessary to elicit a physiological fescue toxicosis type of response in livestock. What is different among these alkaloids is the polarity and relative solubility of the clavine and ergoline alkaloids relative to the ergopeptine alkaloids. Thus, any discussion of toxicity must account for the chemical and physical properties of the respective alkaloid classes.

DRUG ABSORPTION IN MONOGASTRICS

Classical drug or toxin absorption models for ruminant gastric systems are virtually nonexistent. However, drug absorption within monogastric systems is well understood. This permits us to compartmentalize our understanding of drug behavior in the lower tract of the ruminant system that can be coupled with available information about ruminal metabolism of ergot alkaloids to develop a logical scenario of how these toxins are metabolized and absorbed. There are several important considerations that need attention when assessing pharmacological properties of drugs in monogastrics and relating data to ruminants. First, administered drugs in monogastrics, orally or otherwise, are relatively pure in form (Blanchard and Sawchuck, 1979), whereas similar compounds endemic to endophyte-infected grasses are contained in a more complex plant tissue matrix. Thus, simulation of availability in a monogastric system is likely to be a liberal estimation of drug absorption because of the differences in complexity of the two matrices. Second, oral administration of drugs in monogastrics represents a one-time dosed presentation of the drug. On the other hand, oral administration of toxins present in the diet of grazing animals is a continuous exposure because of extended resident time in the digesta and continuous ingestion of fresh herbage. Inasmuch, a single-pass presentation of drug administration in a monogastric is likely to contain higher doses than the continuous flow presentation in ruminant digesta. Third, dissolution of drugs in a monogastric is dependent upon the solubility of the carrier compound (e.g., starch, cellulose, talc) within a tablet or gel cap. Solubility of similar compounds in the ruminant is likely to be time-dependent and related to digestibility of the ingested herbage. A homogeneous carrier compound typically found in drug formulations are likely to make drug presentation more uniform than if presented in plant tissues that vary in fiber and cell solubles. Fourth, solubility of drugs in monogastrics can be a controlled process by predetermining the particle sizes within a formulation, whereas with ruminants, the particle size of the digesta containing similar compounds will vary depending on plant maturity, resident time in the rumen, and extent of digestion. Thus, drug availability from plant tissue is likely to be more complex in ruminant than in monogastric systems. Finally, ruminal microorganisms may alter chemical structures of alkaloids through metabolism. Nonetheless, the abundance of pharmacological data for ergot alkaloids in monogastrics provides excellent insight into the potential for postruminal absorption of putative toxins in endophyte-infected grasses.

Drugs are most commonly absorbed passively via the small intestine of monogastric animals (Muranishi and Yamamoto, 1994; Nelson and Miller, 1979). The structure of the small intestine is comprised of epithelial cells which contain finger-like microvilli projecting from the cell membrane (Anderberg and Artursson, 1994). A glycocalyx barrier is located on the exterior of the micro-villi and functions as a protective barrier to keep the microvilli from being damaged as the digesta passes through the intestinal lumen. Subtending the glycocalyx and surrounding the microvilli is a lipid layer through which all absorbed substances must pass. Once through the lipid matrix, drugs diffuse through the cell membrane, through pores in the cell membrane, or through leaky gaps between adjacent cells (Fig. 12.4).

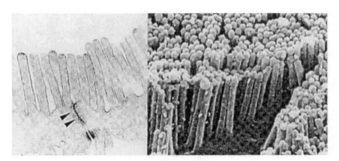

Fig. 12.4. *Transmission electron micrograph of microvilli surfaces of mucosal cells of the intestine. Dark arrowheads point to gaps between cells.*

There are numerous factors affecting drug absorption, and determining how a drug is absorbed is essential to designing drugs and delivery systems to obtain a desired effect. Depending upon the amount of drug passing into the circulatory system, drugs can be present at subefficacious, efficacious, or toxic doses in the blood (Fig. 12.5). Mathematical modeling of drug passage from the intestinal lumen and the microvilli is, therefore, critical to investigative drug therapy. The equation most commonly used is based upon Fick's first law of diffusion as follows (Muranishi and Yamamoto, 1994; Nelson and Miller, 1979):

$$J_s = [D_m \times A_m \times PC_{m/aq} \times (C_L - C_P)]/\Delta X, \tag{1}$$

where J_s is the rate of diffusion of drug across the membrane, D_m is the diffusivity of the drug in the membrane, A_m is the effective surface area of the membrane, $PC_{m/aq}$ is the partition coefficient of drug between the membrane and the aqueous phase of the intestinal lumen, $(C_L - C_P)$ is the drug concentration difference between the intestinal lumen (C_L) and the plasma (C_P), and ΔX is the thickness of the membrane. The A_m and ΔX do not change in a grazing animal, thus, if we want to examine the transport of various ergot alkaloids through the intestine, these variables can be considered constant. Thus, diffusivity of ergot alkaloids in the membrane, their partitioning coefficients between the aqueous and lipid interface in the intestine, and difference in the concentration between the intestine and the plasma can be used to characterize their potential to cause a toxic response.

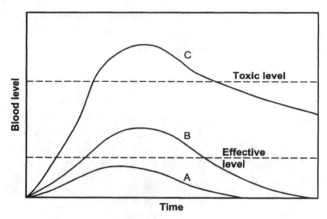

Fig. 12.5. *Hypothetical response curves to a drug administered at* (A) *a subefficacious dose,* (B) *an effective dose, and* (C) *a toxic dose.*

PARTITIONING COEFFICIENTS

Partitioning coefficients were first used to demonstrate that absorption of nonelectrolytes through plant cell membranes was a logarithmic function of their partitioning coefficient (Collander, 1954). Drugs, on the other hand, are either weak acids or bases which exist in the ionized and nonionized forms, the abundance of each dependent upon the ambient pH of the matrix in which they reside. If we assume drug transport can only occur in lipid layers when they are in the nonionized form (Jacobs, 1940), then the relative abundance of each will, in large part, describe their potential as toxins. The relative abundance of the ionized vs. nonionized concentrations of a drug is based upon the dissociation constant of a proton as described by the Henderson-Hasselbach equation (Nelson and Miller, 1979):

$$pKa - pH = \log [HA]/[A^-] \text{ for a weak acid, and} \qquad [2]$$

$$pH - pKa = \log [B]/[BH^+] \text{ for a weak base.}$$

This relationship describes the pH at which half of the conjugate base or acid is in the ionic form, and half the conjugate base or acid is in the nonionic form. With increasing pKa, the basicity of the compound increases, while as pKa decreases the acidity of the compound increases. The relative abundance of the ionized vs. nonionized ergot alkaloids can, therefore, be calculated if we know the pKa of the compound and the pH of the gastric matrix. The proton of concern for the ergot alkaloids is that on the nitrogen at the No. 6 position of the D ring in the ergoline subunit (Eckert et al., 1978) (Fig. 12.2). Lysergic acid, on the other hand, has a second proton of concern at the carboxylic acid attached to the No. 8 position of the D ring in the ergoline subunit. The pKa of the proton associated with the carboxylic acid group of lysergic acid is about 3.4, being a little lower or higher depending upon whether the titration is with acid or base, respectively (Table 12.1) (Craig et al., 1938). However, the dissociation constant

Table 12.1. Dissociation constants of various ergot alkaloids.

Alkaloid	pKa of acid radicle	pKa of basic radicle	Source
Lysergic acid derivatives			
Lysergic acid	3.44	7.70	Craig et al., 1938
Isolysergic acid	3.46	8.61	Craig et al., 1938
Dihydrolysergic acid	3.60	8.45	Craig et al., 1938
Ergoline alkaloids			
Ergonovine		6.70	Craig et al., 1938
Methylergonovine		6.65	Maulding and Zoglio, 1970
Lysergol		8.30	Craig et al., 1938
6-Methyl ergoline		8.80	Maulding and Zoglio, 1970
Methysergide		6.62	Maulding and Zoglio, 1970
Ergopeptine alkaloids			
Ergostine		6.30	Maulding and Zoglio, 1970
Ergotamine		6.25	Maulding and Zoglio, 1970
Dihydroergotamine		6.75	Maulding and Zoglio, 1970
Dihydroergocristine		6.74	Maulding and Zoglio, 1970
Dihydroergocriptine		6.74	Maulding and Zoglio, 1970
Dihydroergocornine		6.75	Maulding and Zoglio, 1970

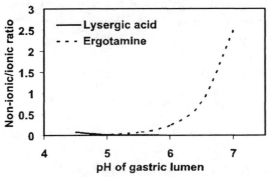

Fig. 12.6. *Calculated ratios of nonionic vs. ionic forms of ergot alkaloids with pKa of 3.4 (lysergic acid) vs. 6.6 (ergotamine).*

for the proton at the No. 6 position of the D ring of ergonovine is 6.7. The dissociation constants for the ergopeptine alkaloids vary from one to another, ranging from 6.25 to 8.8 (Maulding and Zoglio, 1970).

Gastric pH changes among various compartments within the monogastric tract. The stomach has an average pH of about 1.25 (Zoglio et al., 1969), the lumen of the intestine is 6.6, while the pH at the mucosal surface (i.e., the microvilli) is 5.3 (Nelson and Miller, 1979). Hence, the relative abundance of nonionic vs. ionic forms of the alkaloids will vary from one location in the tract to another, and is dependent upon the alkaloid species present in the gastric system. These values were calculated for ergotamine and lysergic acid at various pH values ranging from 4.0 to 7.0, and plotted to provide a visual estimation as to the ionic forms for each (Fig. 12.6). From the graph it is evident that lysergic acid is, for all practical purposes, in the ionic form at pH levels typical in the intestinal lumen or at the mucosal surface of the microvilli. Thus, little or no passive transport of lysergic acid would be expected in the intestines because of the predominance of the ionic form of the alkaloid. Conversely, the basic alkaloids (ergoline and ergopeptine) have near-equimolar concentrations of nonionic and ionic forms when the pH is typical of what might be found in the intestinal lumen, but the proportion of the nonionic form of the basic ergot alkaloids decreases to approximately 0.1 of the total soluble fraction as the pH decreases to that which is typical at the surface of the microvilli (Fig. 12.6). It can be concluded from partitioning coefficients in the intestinal lumen that basic ergot alkaloids (both ergopeptine and ergoline) are more likely to passively transport to the microvilli. However, the pH gradient from the lumen to the microvilli suggests a proportionately low amount of soluble ergoline and ergopeptine alkaloids are in the nonionic form for transport across the mucosal surface and to the blood. Jacobs (1940) described an equation for calculating the equilibrium concentration for nonelectrolytes in two solutions differing in pH and separated by a membrane permeable only to the nonionized species:

$$Ci/Co = [1 + 10^{(pHi-pKa)}]/[1 + 10^{(pHo-pKa)}], \qquad [3]$$

where Ci and Co are the total concentrations of a weak base on the inside (i) and outside (o) of the membrane, pHi is the pH of the matrix inside the membrane, and pHo is the pH on the outside of the membrane. Assuming a dissociation constant of 6.7 for ergot alkaloids and substituting pH values at 6.65 for the intestinal lumen and 5.3 for the lipid matrix surrounding the microvilli suggests a high proportion (55%) of the soluble alkaloids would enter the microvilli.

Table 12.2. Plasma alkaloid concentrations of ergot alkaloids based upon a single oral dose in man (from Eckert et al., 1978).

Alkaloid	Dose amount	Blood concentration
	mg	ng mL^{-1}
Nicergoline	5.00	100.0
LSD	0.16	5.0
	1.00	16.0
	3.00	20.0
Ergotamine tartrate	1.00	1.5
	2.00	1.0
Dihydroergotamine	1.00	0.6
Dihydroergocrytine	1.00	0.6
Dihydroergotoxine	1.00	0.5
Bromoergocryptine	3.00	6.0

Difference in Drug Concentration in Intestinal Lumen vs. the Plasma

Blood and plasma detection remains a primary obstacle to understanding the dynamics of intestinal transport of the ergot alkaloids. Plasma concentrations of alkaloids are often below the detectable limits of fluorometric, photometric, thin layer, high-pressure, or gas chromatographic analytical methods. Immuno-chemical, mass spectrometry, and radioactive methods are necessary to perform plasma alkaloid analysis because these methods can detect alkaloids as low as 0.02 ng mL^{-1} (Eckert et al., 1978). Moubarak et al. (1996) used a solid matrix to concentrate acute intravenously administered ergosine, ergotamine, and ergine in bovine blood and were capable of detecting therapeutic doses ranging from

0.5–24.0 ng mL^{-1} of alkaloids with HPLC methods. However, these alkaloid levels are likely to be artificially high because they were administered intravenously and, thus, assumed 100% transport into the blood. Blood alkaloid values are considerably lower than reported by Moubarak et al. (1996) when purified compounds were orally administered to man via a single dose (Eckert et al., 1978) (Table 12.2). Interestingly, the lysergic amides and nicergoline had at least 20 times higher plasma levels at similar oral doses as the ergopeptine alkaloids. This suggests that ergoline alkaloids have greater potential for absorption and toxicity than the ergopeptine alkaloids.

Table 12.3. Solubility of ergot alkaloids in aqueous solutions.

Alkaloid	Solubility	Reference
Clavine alkaloids		
Agroclavine	slightly	Budavari et al., 1989
Elymoclavine	soluble	Budavari et al., 1989
Ergoline alkaloids		
Lysergic acid	sparingly	Budavari et al., 1989
Lysergol	sparingly	Budavari et al., 1989
LSD	freely	Budavari et al., 1989
Ergonovine	28 mg mL^{-1}	Budavari et al., 1989
Methsergide	7.5 mg mL^{-1}	Zoglio and Maulding, 1970c
Ergopeptine alkaloids		
Ergotamine	0.01 mg mL^{-1}†	Zoglio et al., 1969
Ergosine	insoluble	Budavari et al., 1989
Ergocryptine	insoluble	Budavari et al., 1989
Ergocrystine	insoluble	Budavari et al., 1989
Ergocornine	insoluble	Budavari et al., 1989
Ergotoxine	0.2 mg mL^{-1}†	Zoglio and Maulding, 1970a
Dihydroergocristine	0.4 mg mL^{-1}†	Zoglio and Maulding, 1970b

† Denotes solubility in the presence of xanthines.

Alkaloid Solubility

It is worthy to note that drug solubility is not accounted for in Fick's law of diffusion (Eq. [1]) because drug manufacturers formulate medications that are 100% soluble to minimize dosage costs and provide a more efficacious formulation (Nelson and Miller, 1979). Often, drugs are formulated with inert or slow-release compounds designed to reduce solubility or provide a timed release to prevent medications from entering the blood stream too quickly. Failure to do so can result in a rapid release that will not provide a sustained medicinal effect. Ergot alkaloids, on the other hand, are quite variable and range in solubility from freely soluble to insoluble (Table 12.3). It is difficult to find specific data published on ergot alkaloid solubility, but existing data clearly show that solubility is associated with their chemical structure. The lower molecular weight clavine and ergoline alkaloids are at least 20 times more soluble than the nearly insoluble ergopeptine alkaloids, of which ergovaline is one. The ergopeptine alkaloid which has been most extensively studied is ergotamine, because it mitigates migrane headaches (Diener et al., 2002) and is used to induce contractions during childbirth. However, low solubility of ergotamine has been a hindrance to its utilization for medicinal purposes because of difficulties in obtaining efficacious dosage (Zoglio and Maulding, 1970a, 1970b, 1970c). Solubility and transport of ergotamine can be increased in the presence of xanthine compounds (Fig. 12.7) (Zoglio et al., 1969), and a common formulation for ergotamine administration is to mix it with a 100:1 molar ratio of caffeine to ergotamine (Diener et al., 2002). This formulation lost favor as a migrane medicine because of variability in drug absorption among individuals and cost. A renewed interest has emerged recently because drug formulation has been improved by dissolving the ergotamine in cod liver oil and administering the formulation via buccal pouches placed between the mucosal surfaces of the cheek and gums (Tsutsumi et al., 1998, 2002). The take-home lesson derived from these experiments is that ergotamine, a chemically similar compound to ergovaline, is not absorbed at therapeutic (let alone toxic) levels unless the gastric environment in which the drug resides is chemically altered. Furthermore, a 75-kg human receiving a subtherapeutic dose of 1.5 mg of unadulterated ergotamine per day is dosed with a similar alkaloid dose as a 300-kg steer consuming 2.5% of its body weight of tall fescue forage containing 800 ng g^{-1}

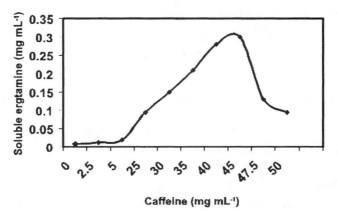

Fig. 12.7. *Effect of caffeine on solubility of ergotamine tartrate in aqueous phosphate buffer (pH 6.65).*

ergovaline (6 mg). This raises doubts as to whether an ergopeptine alkaloid such as ergovaline can be absorbed in sufficient quantity to provide a toxic effect at concentrations found in plant tissues unless some chemical transformation of ergovaline, or adulteration of the absorptive site, occurs.

Relative Toxicities and Excretion Patterns of the Ergot Alkaloids

The toxicological properties of ergot alkaloids are affected by chemical structure and route of administration. Intravenous administration of the alkaloids circumvents the absorption process but, nonetheless, provides insight into the potential toxicity of the compounds. Both the ergoline and ergopeptine alkaloids have similar toxicological properties when administered intravenously (Table 12.4), suggesting both alkaloid classes are equally toxic. However, the ergoline alkaloids are much more potent toxins than the ergopeptine alkaloids when administered orally (Griffith et al., 1978). This suggests the physical and chemical properties of the ergoline alkaloids are more conducive to metabolism and/or transport through monogastric systems and again raises doubts as to whether ergovaline is the toxic entity of *Neotyphodium*-infected grasses or whether other alkaloids in the tall fescue/endophyte complex contribute to animal anomalies.

Table 12.4. Toxicity (LD_{50}) of ergoline and ergopeptine alkaloids when intravenously or orally administered to mice, rats, or rabbits (Adapted from Griffith et al., 1978).

	Intravenous administration		Oral administration	
	Ergoline	Ergopeptine	Ergoline	Ergopeptine
	LD_{50}, mg kg^{-1}			
Mouse	95.0	200.0	240	1942
Rat	53.0	54.0	382	933
Rabbit	1.8	1.4	17	775

RUMINAL METABOLISM AND ABSORPTION OF ERGOT ALKALOIDS

In Vitro and Intravenous Administration of Ergot Alkaloids

Two strategies have been used to determine the effects of specific ergot alkaloids when investigating fescue toxicosis. The first strategy has been to use isolated tissues or cells to investigate alkaloid effects in vitro (Larson et al., 1995; Oliver et al., 1993, 1994; Solomons et al., 1989). These studies indicate the vasoconstrictive and dopamine receptor activity of both the ergoline and ergopeptine alkaloids are similar. A second strategy has been to administer the alkaloids ergonovine and ergotamine intravenously (Browning and Leite-Browning, 1997; Browning and Thompson, 2002; Browning et al., 1997, 1998a, 1998b). Again, both ergoline and ergopeptine alkaloids responded with similar prolactin, growth hormone, and blood pressure. The effect of these alkaloids on respiration rates varied, with ergonovine but not ergotamine elevating respiration in one study (Browning et al., 1997), and ergotamine but not ergonovine elevating respiration in another study (Browning et al., 1998b).

Examination of Excretory Routes of Ergot Alkaloids

On the basis of isotopic experiments using intravenously administered radio-labeled ergot alkaloids in rats, dogs, and monkeys, the route of excretion of the ergot alkaloids was found to be dependent upon molecular weight of the compound investigated (Eckert et al., 1978). With the exception of lysergic acid diethylamide (LSD), ergot alkaloids with a molecular weight < 350 Da were excreted in the renal tubules with the urine, those with molecular weights of 350 to 450 Da often were excreted in roughly equal portions in the urine and the bile, and those above 450 Da were excreted in the bile. Lysergic acid diethylamide is excreted in the urine and bile. Since the lysergic amides have molecular weights < 350 Da and the ergopeptine alkaloids have molecular weights > 450 Da, comparing excretory patterns provides a convenient assessment tool to determine which class of alkaloid(s) are circulating and being excreted. While ruminant metabolism of the ergot alkaloids may differ dramatically from that of monogastric systems, it is highly probable the function of the abomasum, duodenum, liver, and kidneys are highly conserved among mammalian species and can be used to predict postruminal absorption patterns of the alkaloids as well as providing insight as to the toxic entity within the grass.

Stuedemann et al. (1998) used ultrasound instrumentation to locate the gall bladder of steers grazing endophyte-infected (E+) and endophyte-free (E-) tall fescue pastures, and performed a percutaneous puncture to collect bile. Urine was simultaneously collected, and both immunochemically analyzed to determine the excretory route of the ergot alkaloids. Ninety-six percent of the alkaloids excreted were found in the urine. Furthermore, they found accretion of alkaloids in the urine was rapid and measurable within 12 h of introducing steers onto E+ pasture, and by 24 h was equal to steers continuously grazing E+ pasture. Conversely, urinary alkaloid levels were measurably lower 12 h after removing steers from E+ pasture and returned to zero after 48 h. The preponderance of urinary alkaloids and the speed with which they accrue and are voided suggest the primary circulating class of alkaloids is the lysergic amides. In a subsequent study, steers continuously grazed E+ or E− tall fescue, and urine was sampled regularly during the grazing period (Hill et al., 2000). Regression analysis between mean urinary alkaloid concentration and animal performance (average daily gain) established that the two were inversely related. The coefficient of determination (r^2) of the two response variables was 0.86, thus establishing a definitive diagnostic method for fescue toxicosis.

Ruminal Absorption Studies

The end products of microbial fermentation of ingested forage are volatile fatty acids (Van Soest, 1994). The rumen serves a vital function in absorption of these products in order to maintain a microbial population capable of cellulolytic activity. Thus, the fermentation and absorption function of the ruminal ecosystem certainly warrants consideration when contemplating metabolism and transport of ergot alkaloids.

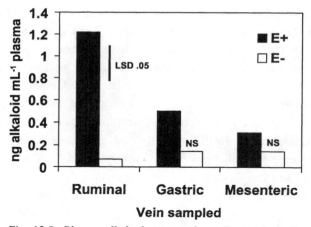

Fig. 12.8. *Plasma alkaloids in samples collected from the right ruminal, right gastric, or cranial mesenteric veins of sheep grazing endophyte-infected (E+) or endophyte-free (E−) tall fescue.*

Absorption of ergot alkaloids was compared for the rumen, omasum, and lower digestive tract by canulating the ruminal, gastric, and mesenteric veins of sheep grazing either E+ or E− tall fescue (A.W. Ayers, N.S. Hill, G.W. Rottinghaus, J.A. Stuedemann, F.N. Thompson, P.T. Purinton, D.H. Seman, D.L. Dawe, A.H. Parks, and D. Ensley, 2004, unpublished data). Plasma was collected from each vein and site of alkaloid absorption postulated by comparing the alkaloid enrichment of each. Only the plasma from ruminal veins of animals grazing E+ tall fescue had alkaloids in greater concentration than in the respective vein from animals grazing E– tall fescue (Fig. 12.8), and thus the rumen is considered the primary site of absorption. Yet, reticular, ruminal, and omasal tissues are all capable of ergot alkaloid transport, thus, the vast majority of the alkaloids must transport across the ruminal/reticular tissues before exiting into the omasum (Hill et al., 2001).

Surprisingly, the posterior dorsal sac had the greater alkaloid transport potential than other ruminal or reticular tissues (Hill et al., 2001) (Table 12.5). It is worthy to note that freshly ingested herbage enters the rumen via the esophageal orifice subtending this area of the rumen, and subsequently rises to the upper strata of the digesta mass (Van Soest, 1994). Thus, the freshly ingested plant tissue resides in proximity to the ruminal tissues with the greatest potential of alkaloid absorption. However, the pH of the ruminal environment is similar to the lumen of the small intestine, suggesting soluble alkaloids would be predominantly in an ionized rather than the nonionized form, and is paradoxical to the passive absorption model presented for monogastrics (Muranishi and Yamamoto, 1994; Nelson and Miller, 1979). Hill et al. (2001) found that ruminal transport was an active process that favors transport of substances in the ionic form, thus resolving the chemical discrepancy of ionic form. It is also noteworthy that ergoline alkaloids have a greater capacity for transport than the ergopeptine alkaloids, regardless of tissue tested (Table 12.5).

Table 12.5. Mole equivalents of ergot alkaloids transported across ruminal or reticular tissues when equimolar concentrations were administered to the mucosal surface in vitro (from Hill et al., 2001).

				Alkaloid				
Tissue	Chamber	Location	Total	Lysergic acid	Lysergol	Ergonovine	Ergotamine	Ergocryptine
					potential transport, nmol			
Reticular	dorsal		1556c†	807c	337b	290b	61cd	71cd
Reticular	ventral		1407c	897c	215b	244b	25d	26d
Rumen	dorsal	posterior	5925a	3441a	578a	1163a	327a	443a
Rumen	dorsal	anterior	3817b	2685ab	318b	622b	83bcd	106bcd
Rumen	ventral	posterior	3733b	2395ab	327b	633b	149ab	229b
Rumen	ventral	anterior	3318b	2145b	305b	622b	129abc	180bc
				12370A‡	2080C	3574B	774D	1055D

† Column means without a common lowercased letter differ ($P < 0.05$).
‡ Row means without a common uppercased letter differ ($P < 0.05$).

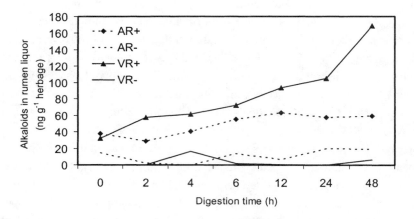

Fig. 12.9. *Total ergot alkaloid concentration in rumen liquor inoculated with auto-claved (AR) or viable (VR) ruminal fluid for both endophyte-infected (+) and endophyte-free (-) tall fescue plant tissue when sampled during a 48-h incubation period.*

Alkaloid Solubility and Fate in the Rumen

Few studies have investigated intraruminal metabolism of the ergot alkaloids, but in those that have, the trends are the same. The aqueous concentration of total ergot alkaloids from E+ tall fescue tissue incubated in autoclaved rumen fluid is consistently 40 to 50 ng mL^{-1} during a 48-hr incubation (Ayers et al., 2004, unpublished data; Stuedemann et al., 1998) (Fig. 12.9) and does not change in the pellet fraction (Stuedemann et al., 1998). However, the aqueous concentration of ergot alkaloids increases with time when viable ruminal micro-organisms decompose the plant tissue (Ayers et al., 2004, unpublished data; Stuedemann et al., 1998). Conversely, the total alkaloid concentration in the E+ tall fescue pellet remains the same in the autoclaved ruminal fluid, but decreases when viable ruminal microorganisms are present (Stuedemann et al., 1998). The concentration of aqueous ergovaline (Ayers et al., 2004, unpublished data; Moyer et al., 1993) and ergonovine (Moyer et al., 1993) decrease in the presence of viable ruminal microorganisms, but do not change in autoclaved aqueous ruminal fluid (Fig. 12.10). It must be noted the concentration of ergovaline in the ruminal fluid, while measurable, was very low compared with that of the total ergot alkaloid concentration. Ruminal fluid from fistulated steers grazing E+ pastures had no ergovaline in the aqueous fraction during a 4-wk period, even when the forage tested 1200 ng g^{-1} in total ergot alkaloids (Ayers et al.,

2004, unpublished data). The logical conclusions from these studies are that ergovaline is not soluble in ruminal fluids, or that ruminal microorganisms metabolize ergovaline, or ergovaline remains in the solid matrix. The final form of this metabolite is unknown, but lysergic acid is the primary alkaloid from in vitro ruminal digestions that transports across ruminal or omasal tissues (Fig. 12.11) (Ayers et al., 2004, unpublished data). Ergonovine is also present in the aqueous fraction, but its concentration is minimal compared with that of lysergic acid, and the concentration of that transported is minor compared with lysergic acid. Immunopurified alkaloids from urine confirmed lysergic acid as the major metabolite.

There is one study in which synthetic ergovaline was fed to lambs consuming a E– tall fescue seed diet, and animal response variables were compared with lambs consuming an E+ tall fescue seed diet, an E+ perennial ryegrass seed diet,

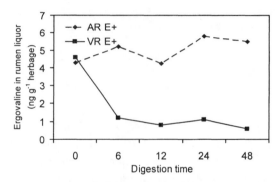

Fig. 12.10. *Ergovaline concentration in rumen liquor inoculated with autoclaved (AR) or viable (VR) ruminal fluid using endophyte-infected (E+) tall fescue plant tissue as substrate when sampled during a 48-h incubation period.*

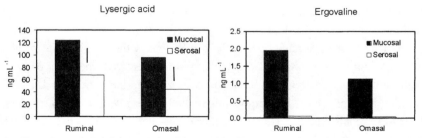

Fig. 12.11. *In vitro transport of lysergic acid and ergovaline across ruminal and omasal tissues. Bars = LSD value.*

or E– (nontoxic) seed (Gadberry et al., 2003). Their results indicated that ergovaline had an effect on some animal response variables, but not as dramatical as those on lambs grazing E+ seed. Lambs consuming E+ perennial ryegrass seed had a lower toxic response than those grazing E+ tall fescue, despite having a higher ergovaline content. Part of their explanation for the discrepancy between ergovaline content and toxic effect was a greater concentration of ergine (lysergic acid amide) in the E+ tall fescue seed as compared with the E+ perennial ryegrass seed. Another likely explanation is the purified ergovaline was not within a plant matrix similar to that of E+ plant tissue, and was partially metabolized by the rumen microflora to a secondary form. Likewise, *Achnatherum inebrians* (drunken horse grass) produces symptoms in sheep and horses typical of ergot alkaloid toxicity when infected with a fungal endophyte (Miles et al., 1996). Examination of the ergot alkaloid fraction indicated ergine and ergonovine were present in high concentrations, but no ergovaline was present. These circumstantial bits of evidence suggest that ergovaline may not be the only toxin responsible for fescue toxicosis, or may not be the toxic component at all.

SUMMARY

The chemical structure of the ergoline ring of the clavine, ergoline, and ergopeptine alkaloids are similar to noradrenaline, dopamine, and serotonin. Therefore, all classes of alkaloids are capable of producing a toxic effect for the associated receptors. The dissociation constants of the ergot alkaloids, with exception to lysergic acid, are also similar for both the ergoline and ergopeptine alkaloids, suggesting they have similar chemical ionization properties and toxicological properties for passive intestinal transport. The solubility of the ergoline alkaloids is greater than that of the ergopeptine alkaloids in aqueous solutions typical of the gastric system and, consequently, the ergoline alkaloids are more toxic when administered orally than the ergopeptine alkaloids even though they have similar toxicological properties (LD_{50}) when administered intravenously. Yet in ruminants, it is the rumen that is the primary site of absorption. Ruminant absorption of ergot alkaloids is an active rather than passive system, and favors the ergoline alkaloids over ergopeptine alkaloid transport, especially lysergic acid. The ergopeptine alkaloid ergovaline is much less soluble in ruminal fluid than lysergic acid or ergonovine, and does not transport across ruminal, reticular, or omasal tissues. Renal tubules are the primary sites of excretion of the ergot alkaloids in ruminants, and lysergic acid is the primary ergot alkaloid found in the urine. These data provide evidence that the toxic entity in endophyte-infected tall fescue may not be ergovaline, and could involve the ergoline alkaloids as well.

REFERENCES

Anderberg, E.K., and P. Artursson. 1994. Cell cultures to assess drug absorption enhancement. p. 101–118. *In* A.G. de Boer (ed.) Drug absorption enhancement: Concepts, possibilities, limitations, and trends. Harwood Academic Publ., Langhorne, PA.

Berde, B., and E. Strumer. 1978. Introduction to the pharmacology of ergot alkaloids and related compounds as a basis of their therapeutic application. p. 1–28. *In* B. Berde and H.O. Schild (ed.) Ergot alkaloids and related compounds. Handbook of experimental pharmacology. Vol. 49. Springer-Verlag, Berlin.

Blanchard, J., and R. J. Sawchuck. 1979. Drug bioavailability: An overview. p. 1–17. *In* J. Blanchard, R.J. Sawchuck, and B.B. Brodie (ed.) Principles and perspectives in drug bioavailability. S. Karger, Basel, Switzerland.

Bongard, O., and H. Bounameaux. 1991. Sever iatrogenic ergotism—Incidence and clinical importance. J. Vasc. Dis. 20:153–156.

Browning, R., and M.L. Leite-Browning. 1997. Effect of ergotamine and ergonovine on thermal regulation and cardiovascular function in cattle. J. Animal Sci. 75:176–181.

Browning, R., M.L. Leite-Browning, H.M. Smith, and T. Wakefield. 1998a. Effect of ergotamine and ergonovine on plasma concentrations of thyroid hormones and cortisol in cattle. J. Anim. Sci. 76:1644–1650.

Browning, R., F.N. Schrick, F.N. Thompson, and T. Wakefield. 1998b. Reproductive hormonal responses to ergotamine and ergonovine in cows during the luteal phase of the estrous cycle. J. Anim. Sci. 76:1448–1454.

Browning, R., and F.N. Thompson. 2000. Endocrine and respiratory responses to ergotamine in Brahman and Hereford steers. Vet. Hum. Toxicol. 44:149–154.

Browning, R., F.N. Thompson, J.L. Sartin, and M.L. Leite-Browning. 1997. Plasma concentrations of prolactin, growth hormone, and luteinizing hormone in steers administered ergotamine or ergonovine. J. Anim. Sci. 75:796–802.

Budavari, S., M.J. O'Neil, A. Smith, and P.E. Heckelman (ed.) 1989. The Merck pharmaceutical index: An encyclopedia of chemicals, drugs, and biologicals. 11th ed. Merck and Co., Rahway, NJ.

Collander, R. 1954. The permeability of Nitella cells to non-electrolytes. Physiol. Plant. 7:420–455.

Craig, L.C., T. Shedlovsky, R.G. Gould, and W.R. Jacobs. 1938. The ergot alkaloids: XIV. The postions of the double bond and the carboxyl group in lysergic acid and its isomer. The structure of the alkaloids. J. Biol. Chem. 125:289–298.

Deiner, H.C., and A. McHarg. 2000. Pharmacology and efficacy of eletriptan for the treatment of migraine attacks. Internat. J. Clinical Pract. 54:670–674.

Diener, H.C., J.P. Jansen, A. Reches, J. Pascual, D. Pitei, and T.J. Steiner. 2002. Efficacy, tolerability and safety, of oral eletriptan and ergotamine plus caffeine (Cafergot®) in the acute treatment of migrane: A multicentre, randomized, double-blind, placebo-controlled comparison. Eur. Neurol. 47:99–107.

Eckert, H., J.R. Kiechel, J. Rosenthaler, R. Schmidt, and E. Schreier. 1978. Biopharmaceutical Aspects: Analytical methods, pharmacokinetics, metabolism and bioavailability. p. 719–803. *In* B. Berde and H.O. Schild (ed.) Ergot alkaloids and related compounds. Handbook of experimental pharmacology. Vol. 49. Springer-Verlag, Berlin.

Gadberry, M.S., T.M. Denard, D.E. Spiers, and E.L Piper. 2003. Effects of feeding ergovaline on lamb performance in a heat stress environment. J. Anim. Sci. 81:1538–1545.

Griffith, R.W., J. Grauwiler, C. Hodel, K.H. Leist, and B. Matter. 1978. Toxicological considerations. p. 805–851 *In* B. Berde and H.O. Schild (ed.) Ergot alkaloids and related compounds. Handbook of experimental pharmacology. Vol. 49. Springer-Verlag, Berlin.

Jacobs, M.H. 1940. Some aspects of cell permeability to weak electrolytes. Cold Spring Harb. Symp. Quant. Biol. 8:30–39.

Hill, N.S., F.N. Thompson, J.A. Stuedemann, D.L. Dawe, and E.E. Hiatt, III. 2000. Urinary alkaloid excretion as a diagnostic tool for fescue toxicosis in cattle. J. Vet. Diag. Invest. 12:210–217.

Hill, N.S., F.N. Thompson, J.A. Stuedemann, G.W. Rottinghaus, H.J. Ju, D.L. Dawe, and E.E. Hiatt III. 2001. Ergot alkaloid transport across ruminant gastric tissues. J. Anim. Sci. 79:542–549.

Larson, B.T., M.D. Samford, J.M. Camden, E.L. Piper, M.S. Kerley, J.A. Patterson, and J.T. Turner. 1995. Ergovaline binding and activation of D2 dopamine-receptors in GH(4)ZR(7) cells. J. Anim. Sci. 73:1396–1400.

Lyons, P.C., R.D. Plattner, and C.W. Bacon. 1986. Occurrence of peptide and clavinet ergot alkaloids in tall fescue. Science (Washington, DC) 232:487–489.

Maulding, H.V., and M.A. Zoglio. 1970. Physical chemistry of ergot alkaloids and derivatives I: Ionization constants of several medicinally active bases. J. Pharm. Sci. 59:700–701.

Miles, C.O., G.A. Lane, M.E. diMenna, I. Garthwaite, E.L. Piper, O.J.P. Ball, G.C.M. Latch, J.M. Allen, M.B. Hunt, L.P. Bush, F.K. Min, L. Fletcher, and P.S. Harris. 1996. High levels of ergonovine and lysergic acid amide in toxic *Achnatherum inebrians* accompany infection by an *Acremonium*-like endophytic fungus. J. Agric. Food Chem. 44:1285–1290.

Moubarak, A.S., E.L. Piper, Z.B. Johnson, and M. Flieger. 1996. HPLC method for detection of ergotamine, ergosine, and ergine after intravenous injection of a single dose. J. Agric. Food Chem. 44:146–148.

Moyer, J.L., N.S. Hill, S.A. Martin, and C.S. Agee. 1993. Degradation of ergoline alkaloids during in vitro ruminal digestion of tall fescue forage. Crop Sci. 33:264–266.

Muranishi, S., and A. Yamamoto. 1994. Mechanisms of absorption enhancement through gastrointestinal epithelium. p. 67–100. *In* A.G. de Boer (ed.) Drug absorption enhancement: Concepts, possibilities, limitations, and trends. Harwood Academic Publ., Langhorne, PA.

Nelson, K.G., and K.W. Miller. 1979. Principles of drug dissolution and absorption related to bioavailability. p. 20–58. *In* J. Blanchard et al. (ed.) Principles and perspectives in drug bioavailability. S. Karger, Basel, Switzerland.

Oliver, J.W., L.K. Abney, J.R. Strickland, and R.D. Linnabary. 1993. Vasoconstriction in bovine vasculature induced by the tall fescue alkaloid lysergamide. J. Anim. Sci. 71:2708–2713.

Oliver, J.W., R.D. Linnabary, L.K. Abney, K.R. van Manen, R. Knoop, and H.S. Adair, III. 1994. Evaluation of a dosing method for studying ergonovine effects in cattle. Am. J. Vet. Res. 55:173–176.

Rutschmann, J., and P.A. Stadler. 1978. Chemical background. p. 29–85. *In* B. Berde and H.O. Schild (ed.) Ergot alkaloids and related compounds. Handbook of experimental pharmacology. Vol. 49. Springer-Verlag, Berlin.

Solomons, R.N., J.W. Oliver, and R.D. Linnabary. 1989. Reactivity of the dorsal pedalvein of cattle to selected alkaloids associated with *Acremonium coenophialum*-infected fescue grass. Am. J. Vet. Res. 50:235–238.

Stuedemann, J.A., N.S. Hill, F.N. Thompson, R.A. Fayrer-Hosken, W.P. Hay, D.L. Dawe, D.H. Seman, and S.A. Marin. 1998. Urinary and biliary excretion of ergot alkaloids from steers that grazed endophyte-infected tall fescue. J. Anim. Sci. 76:2146–2154.

Tsutsumi, K., Y. Obata, T. Nagai, T. Loftsson, and K. Takayama. 2002. Buccal absorption of ergotamine tartrate using the bioadhesive tablet system in guinea-pigs. Int. J. Pharm. 238:161–170.

Tsutsumi, K., Y. Obata, K. Takayama, T. Loftsson, and T. Nagai. 1998. Effect of the cod-liver oil extract on the buccal permeation of ionized and non-ionized forms of ergotamine using the keratinized epithelial-free membrane of hamster cheek pouch mucosa. Int. J. Pharm. 174:151–156.

Van Soest, P.J. 1994. Nutritional ecology of the ruminant. 2nd ed. Comstock, Ithaca, NY.

Zoglio, M.A., and H.V. Maulding. 1970a. Complexes of ergot alkaloids and derivatives II: The interaction of dihydroergotoxine with certain xanthines. J. Pharm. Sci. 59:215–219.

Zoglio, M.A., and H.V. Maulding. 1970b. Complexes of ergot alkaloids and derivatives III: The interaction of dihydroergocristine with xanthine analogs in aqueous media. J. Pharm. Sci. 59:384–386.

Zoglio, M.A., and H.V. Maulding. 1970c. Complexes of ergot alkaloids and derivatives V: Interaction of methysergide maleate and caffeine in aqueous solutions. J. Pharm. Sci. 59:1836–1837.

Zoglio, M.A., H.V. Maulding, and J.J. Windheuser. 1969. Complexes of ergot alkaloids and derivatives I: The interaction of caffeine with ergotamine tartrate in aqueous solution. J. Pharm. Sci. 58:222–225.

PATHOPHYSIOLOGIC RESPONSE TO ENDOPHYTE TOXINS

Jack W. Oliver[1]

Intake by ruminant species of toxic alkaloids from endophyte-infected (E+) tall fescue results in a major economic drain on agribusiness interests world-wide. The annual loss from this E+ tall fescue-associated disease condition is conservatively estimated to exceed $600 million in the USA, with the fescue-toxicosis syndrome considered the major grass-induced toxicity problem in the country (Cheeke, 1995). The estimate of economic loss does not include losses associated with toxicoses from the same alkaloids that are present in ergotized grains and grasses (Bacon et al., 1995). Recommendations for eliminating or reducing fescue-endophyte effects have been documented (Hoveland, 2000). However, at this time, these recommendations (alkaloid-free forage, alkaloid inactivation by vaccines, antagonists of alkaloid uptake, persistent grazing of tolerant perennial clovers for dilution of alkaloid effect, alleviator drug prophy-laxis) remain only partially effective. For additional information on the animal effects and economic concerns from consumption of E+ tall fescue, the reader is directed to the review articles of Cross (2000), Oliver (1997), and Stuedemann and Thompson (1993).

CLINICAL SIGNS/SYNDROMES

Animal disorders (decreased weight gain, decreased milk production, decreased reproductive efficiency) associated with intake of toxic alkaloids in tall fescue by herbivorous animals have been well documented (Cross, 2000; Oliver, 1997). The syndromes have been designated fescue foot, fat necrosis, agalactia, and summer slump—designations that reflect elements of this toxic disorder. Farm-ers and producers are likely to observe some physical signs and behavioral responses in their animals as a result of the consumption of toxic alkaloids in the E+ tall fescue (Aldrich-Markham et al., 2003; Parish et al., 2003). In cooler

---

[1] Department of Comparative Medicine, College of Veterinary Medicine, The University of Tennessee, Knoxville, TN 37996-4543

months, the fescue foot syndrome may be manifested by the following signs, or combinations of them: tenderness and/or swelling around the fetlock and hoof region, lameness, dry gangrene on tips of ears and tails and surrounding tissues, and loss of tail switch. In the most advanced cases, the hooves of affected animals may begin to slough off. These obvious clinical signs of disease are acknowledged to be caused by the alkaloid effects on blood vessels resulting in damage to vessel lining cells, enhanced blood clotting, narrowing (vasoconstriction) of the vessel lumen, and ultimately lack of blood flow with resulting tissue necrosis. Generally, there are no visible signs of fat necrosis apart from poor thrift in cases where the necrotic fat is constricting internal organs. Agalactia results in starvation or poor growth of offspring. *Summer slump* is a term used to describe fescue toxicosis that is less used today, but refers to the poor growth rate of animals in the hot summer months. Animals experiencing fescue toxicosis typically show the following in varying degrees: poor growth rate, rough hair coat, elevated core body temperature, increased respiration rate, and excessive salivation. Aldrich-Markham et al. (2003) list the threshold levels of ergovaline ($\mu g\ kg^{-1}$) in the diet that produce clinical disease as: horses (300–500), cattle (400–750), and sheep (500–800). Affected animals are likely to seek shade, form wallows around water troughs and in popular shade areas, and spend less time grazing than their unaffected counterparts (Parish et al., 2003). Ryegrass infected with a similar endophyte (*Neotyphodium lolii*) also produces the ergot alkaloid ergovaline. Consumption of the infected ryegrass results in a similar, but milder, syndrome in grazing livestock. These clinical signs are often incorrectly included in the ryegrass staggers syndrome. Ryegrass staggers is caused by a tremorgenic mycotoxin, lolitrem B, which is produced by the same ryegrass–endophyte association. However, this disease syndrome is not related to the ergot alkaloids that are responsible for fescue toxicosis signs, although both are produced by the same ryegrass–endophyte association. Aldrich-Markham et al. (2003) can be consulted for lolitrem B levels that produce clinical disease. In horses, the major problems associated with intake of E+ tall fescue occur in the periparturient period (Cross, 2000). Problems noted include agalactia, prolonged gestation, foaling difficulties, and sometimes mare and foal deaths due to dystocia. Recent work by Christiansen et al. (2003) indicates that midgestation mares grazing E+ tall fescue are at risk for poor pregnancy outcome as well.

PHYSIOLOGICAL MANIFESTATIONS OF ENDOPHYTE TOXICOSIS

Biogenic Amine Receptor Effects

Various biogenic amine receptors are known to be impacted by the ergopeptide-alkaloids that are present in E+ tall fescue (Dyer, 2000; Larson et al., 1999; Oliver, 1997; Oliver et al., 1998; Schoning et al., 2001). Stimulation of these receptors by the alkaloids of E+ tall fescue contributes to the well-known vasoconstrictive effects (increased contractility) of the alkaloids that are known to deprive tissues of proper blood flow and nutrition. Stimulation of α-2 adrenergic receptors by the alkaloids (Oliver et al., 1998) results in enhancement of blood platelet aggregation that likely is involved in the coagulopathies and tissue necrosis (fescue foot) that occur with severe toxicity (Oliver, 1997). Stimulation of the dopamine-2 receptor by alkaloids found in E+ tall fescue is known to cause a decrease in prolactin secretion, which is the cause of the decreased milk secretion (agalactia) in animals that graze endophyte-infected tall fescue (Larson et al., 1999). The drugs domperidone and metoclopramide, which are dopamine-2 receptor antagonists, are known to alleviate the toxic signs of tall fescue in horses (Cross, 2000) and cattle (Jones et al., 2003; Lipham et al., 1989), particularly those associated with prolactin suppression. Ergot alkaloids (ergine, ergovaline) also have important effects on serotonin-2 receptors (Dyer, 2000; Oliver, 1997; Schoning et al., 2001), and serotonin-induced mitogenesis of vascular smooth muscle is associated with serotonergic-2 receptor activity. Further, serotonin is known to have effects on the hypothalamic thermoregulatory and satiety centers, with increased levels resulting in increased body temperatures and suppression of appetite. Porter et al. (1990) have shown that cattle that graze E+ tall fescue have increased serotonin metabolites in the central nervous system tissues, and Oliver et al. (2000) have shown increased tryptophan levels in sera of cattle that graze E+ tall fescue, a condition that results in decreased feed intake (Rossi-Fanelli and Cangiano, 1991). One major problem in cattle exposed to E+ tall fescue toxins is lack of feed intake (Parish et al., 2003) that results in severe economic losses from decreased weight gain.

General Metabolism-Related Manifestations from Exposure to E+ Tall Fescue

Increased core body temperatures can be expected in animals that consume E+ tall fescue (Spiers et al., 1999). At the same time, because of the α-adrenergic vasoconstrictor effect of alkaloids in subcutaneous areas, skin temperatures are decreased. Respiration rates are variable (Oliver, 1997), with central nervous system effects of alkaloids decreasing rates, while direct effects on receptors in lung tissues (and blood platelets) appear to cause hypoxemia (decreased oxygenation of tissues) and subsequent reflex increase in respiration rates. Heart rate is usually unaffected, or reflexively decreased, in response to peripheral vasoconstriction and increased impedance to tissue blood flow that results in increased blood pressure in animals (Oliver, 1997).

Serum Enzyme/Biochemical/Immunologic Changes

Enzymatic activity, in general, and hepatic enzyme activity, in particular, are decreased by intake of E+ tall fescue (Oliver et al., 2000; Schultze et al., 1999; Table 13.1). Multiple factors are likely responsible for this lowered enzymatic activity, but decreased feed intake may be the most causative factor (Oliver et al., 2000; Parish et al., 2003). Serum cholesterol levels are consistently lowered (Table 13.1), as are serum globulin levels (Oliver et al., 2000; Tables 13.1 and 13.2). Specifically, alpha and gamma globulin levels are lowered (Schultze et al., 1999; Table 13.2). Most mineral levels are unaffected by E+ tall fescue intake (Oliver et al., 2000), but serum copper levels are decreased (Oliver et al., 2000; Saker et al., 1998; Table 13.2). Dennis et al. (1998) have documented lowered copper levels in E+ tall fescue that may account for the lowered serum copper levels in cattle that graze this forage. Saker et al. (1998) have found that steers that grazed E+ tall fescue had lowered immune function that was associated with copper deficiency. Thus, compromise in immune status does occur in animals that consume E+ tall fescue.

Blood Cellular Parameters

Blood cellular parameters are largely spared by E+ tall fescue intake, but erythrocyte numbers are increased. However, erythrocytes are smaller in size (microcytic) and have less hemoglobin content (hypochromic) as shown by mean corpuscular volume (MCV) and mean corpuscular hemoglobin (MCH) values (Oliver et al., 2000; Table 13.2). The reduced copper levels likely contribute to the decreased MCV and MCH values due to effect on hemoglobin synthesis.

Table 13.1. Mean serum chemistry values of significance for steers grazing endophyte-free (E–) and endophyte-infected (E+) tall fescue.†

Item	Steers grazing E– ($N = 8$)	Steers grazing E+ ($N = 8$)	Reference range‡
Albumin to globulin ratio	1.10 ± 0.04	1.32 ± 0.04***	0.84–0.94
Alanine aminotransferase, U L⁻¹	37.8 ± 1.8	31.3 ± 1.8**	11–40
Cholesterol, mg 100 mL⁻¹	107.0 ± 4.2	91.0 ± 4.2**	80–120
Creatinine, mg 100 mL⁻¹	1.38 ± 0.04	1.58 ± 0.04***	1.0–2.0
Globulin, g 100 mL⁻¹	3.37 ± 0.07	2.97 ± 0.07***	3.00–3.48
Prolactin, ng mL⁻¹	214 ± 57	13.2 ± 57.0*	–
Total bilirubin, mg 100 mL⁻¹	1.63 ± 0.04	1.74 ± 0.04*	0.01–0.50
Total protein, g 100 mL⁻¹	6.93 ± 0.09	6.68 ± 0.08*	6.74–7.46

* $P < 0.05$.
** $P < 0.01$.
*** $P < 0.001$.
† Repeated measures ANOVA. Twenty different bleed dates, April–August 1996–1998. Least square means (± SEM). From: Oliver et al. (2000).
‡ Kaneko et al. (1997) and Clinical Chemistry Service, College of Veterinary Medicine, University of Tennessee.

Table 13.2. Mean serum analyte values of significance for steers grazing endophyte-free (E–) and endophyte-infected (E+) tall fescue.

Item	Steers grazing E– ($N = 8$)	Steers grazing E+ ($N = 8$)	Reference range‡
Globulin electrophoresis†			
Alpha globulins, g dL⁻¹	1.57 ± 0.05	1.32 ± 0.01*	–
Gamma globulins, g dL⁻¹	2.30 ± 0.14	1.40 ± 0.51*	–
Serum minerals‡			
Copper, mg L⁻¹	0.72 ± 0.02	0.62 ± 0.02**	0.72 ± 0.02§
Hemogram values¶			
Erythrocytes, 1×10^6, μL⁻¹	8.33 ± 0.15	8.83 ± 0.15*	5–10#
MCV, fL	41.4 ± 0.46	39.5 ± 0.46**	40–60
MCH, pg	15.1 ± 0.18	14.3 ± 0.18**	11–17
Eosinophils 1×10^3, μL⁻¹	0.68 ± 0.06	0.48 ± 0.06*	0.0–2.4

* $P < 0.05$.
** $P < 0.01$.
*** $P < 0.001$.
† Data are presented as mean ± standard error. $N = 3$/group. From: Schultze et al. (1999).
‡ Repeated measures ANOVA. Fourteen different bleed dates, April–August, 1998–1999. Least square means (± SEM). From: Oliver et al. (2000).
§ Saker et al. (1998).
¶ Repeated measures ANOVA. Twenty different bleed dates, April–August 1996–1998. Least square means (± SEM). From: Oliver et al. (2000).
# Clinical Chemistry Service, College of Veterinary Medicine, University of Tennessee.

Haircoat Changes

The reason for the rough and shaggy-looking haircoats that cattle have when they consume E+ tall fescue remains to be defined. Speculation on cause could be directed at nutrient deficiency due to vasoconstriction and peripheral ischemia, or due to altered hormonal levels at hair follicles, but the exact cause remains to be delineated. The known copper deficiency that accompanies E+ tall fescue consumption (Table 13.2) likely contributes to the change in hair coat color that occurs in cattle with fescue toxicosis.

Growth Hormone

No consistent effect of E+ tall fescue on growth hormone has been documented (Lipham et al., 1989; Oliver, 1997). However, Lipham et al. (1989) found decreased levels of insulin-like growth factor in cattle that consumed E+ vs. E– tall fescue that was attributed to decreased feed intake. Thus, growth hormone secretory changes do not appear to be a contributing cause to the decreased weight gains associated with E+ tall fescue intake.

Biotransformation

Administration of ergot and loline alkaloids to ruminants resulted in 50-60% recovery in abomasal contents, while very little reached the ileum, and only 5% was recovered in fecal collections, implicating extensive absorption from the gastrointestinal tract (Westendorf et al., 1993). Pharmacokinetic studies of bromocriptine have been conducted in a number of species (Oliver, 1997), and indicate significant first-pass biotransformation of this alkaloid in liver. When alkaloids are injected intravenously, they are rapidly cleared from the circulation (Moubarak et al., 1994). Stuedemann et al. (1998) have developed an ELISA technique that detects ergot alkaloids in the urine (major route of excretion) and bile (minor route of excretion) of cattle that has been used as a way to detect alkaloid exposure. In recent studies, Ayers et al. (2003) collected plasma from the ruminal, gastric, and mesenteric veins of sheep grazing E– and E+ tall fescue and determined that only ruminal vein plasma had ergot alkaloid content; a finding they suggest implicates the rumen as the site of alkaloid absorption. In vitro studies of rumen digesta by these researchers have determined that the digesta of cattle on E+ tall fescue contains ergovaline, ergonovine, lysergic acid, and lysergol, but that only the lysergic acid was able to cross from the mucosal to the serosal side of ruminal tissues in their in vitro model. Ayers et al. (2003) also conducted a grazing study with rumen fistulated steers to examine ruminal alkaloids in vivo and urinary alkaloid excretion. No ergovaline was found in the ruminal fluid of either E− or E+ steers. High performance liquid chroma- tographic (HPLC) analysis of urine was also negative for ergovaline, but did reveal lysergic acid isomers. Ayers et al. (2003) and Hill (Chapter 12, this volume) implicate lysergic acid as the toxin causing fescue toxicosis. Much more definition of alkaloid pharmacokinetics in animals is needed.

Temperature-Related Effects

Several investigators have reported on the relationship between ergot alkaloid intake and core body temperature in both laboratory and large animal species (Oliver, 1997). From these reports, it is obvious that animals exposed to ergot alkaloids do not handle heat stress well because of the associated peripheral vasoconstriction and reduced blood flow in skin, and decreased water vaporiza- tion. Additional heat loss compromise is likely due to vasoconstrictor and bronchoconstrictor effects of α-2 adrenergic and serotonergic receptor stimula- tion in lungs (Oliver, 1997). Aldrich-Markham et al. (2003) have defined the levels of both ergovaline and lolitrem B that are needed to cause toxicity in cattle and sheep, and Spiers et al. (1999) have identified biological markers of cattle response to heat stress in association with endophyte-infected tall fescue.

Ergotism/Fescue Toxicity

For a comprehensive review of the pathologic changes in animals that are associated with endophyte-infected tall fescue intake, see the review of Oliver (1997). Similar lesions are known to occur in both cattle and people with ergo- tism (Oliver, 1997). In general, fescue toxicosis has major impact on the cardiovascular system, with vasoconstriction occurring along with a thickened medial layer of blood vessels, endothelial cell damage, vascular stasis and thrombosis, ischemia, and finally gangrene (Garner and Cornell, 1978; Coppock et al., 1989; Oliver and Schultz, 1997). There is also evidence of decreased nitric oxide (NO) synthesis in animals that have grazed E+ tall fescue (Oliver et al., 2001), which facilitates the vasoconstrictive and coagulopathy conditions. If blood flow is severely compromised, then animals manifest with avascular necrosis (gangrene) of the extremities (feet, ears, and tail). These severe mani- festations of fescue toxicosis occur infrequently in comparison with the much more common changes of adverse haircoat quality and suppressed daily gains (Bouton et al., 2002; Parish et al., 2003), that result in profound economic losses to producers.

Serum Inflammatory Response

The vascular endothelium has been described as the cellular target for initial injury by the ergot alkaloids (Thompson et al., 1950). The serum inflammatory response (SIRS) is a general term that infers that a number of bioactive compounds have been released from the cardiovascular tissues, with associated pathological response. In recent studies, isolated bovine endothelial cells were exposed to ergot alkaloids (ergovaline, ergine) in vitro. The ergine treatment has been described (Schultze and Oliver, 1999), and treatment of endothelial cells at 10^{-5} M concentration had minimal effect. In further studies with this system, ergine continued to have minimal effect at 10^{-5} M concentration, while ergovaline at this dose level gave cytotoxic effects (1999, unpublished data). At 10^{-4} M concentration, ergovaline caused marked cytotoxicity, while ergine only caused a slight toxic effect. These studies are supportive of the improved response in herbivores when ergovaline is removed from tall fescue (Bouton et al. 2002; Parish et al., 2003). Thus, the presence of ergot-alkaloid-induced toxicity to endothelial cells in vivo (possibly as ergovaline effect alone, or maybe as a synergistic response with other alkaloids present in E+ tall fescue) would result in elaboration of inflammatory mediators, as well as change in the clotting factor profile of the animal (SIRS). Strickland et al. (1996) demonstrated a mitogenic effect of alkaloids (ergonovine, α-ergocryptine, ergovaline, N-acetyl loline) on isolated bovine vascular smooth muscle cells from the dorsal metatarsal artery, indicating that these alkaloids have an effect on blood vessel reactivity in cattle; and further, that synergistic effect of the alkaloids may be possible. The mitogenic effect likely contributes to the well-described thickening of blood vessels that widely occurs in cattle that graze E+ tall fescue (Garner and Cornell, 1978; Oliver and Schultze, 1997), or that suffer from ergotism (Coppock et al., 1989). Endothelial cells damaged from exposure to the alkaloids present in E+ tall fescue thus allow access of various mediator agents to the underlying smooth muscle cell layer of blood vessels that contributes to the thickened vessels seen at necropsy (Oliver and Schultz, 1997). The mitogenesis, with resulting narrowing of lumen size and potential for increased vasoactivity from increased smooth muscle component, contributes to the vascular complications that arise in a dose-dependent manner (Aldrich-Markham et al., 2003). Indeed, the pathological changes that occur from ergot alkaloid exposure in tall fescue toxicosis have inflammation as an important etiologic factor. Inflammation (SIRS) can be linked to many of the clinical signs of fescue toxicosis, including febrile response, hyperalgesia and lameness, and stasis of blood in peripheral vessels that leads to coagulation defects. Biologically active substances in the sera of cattle that are produced by endothelial cells are impacted by exposure of the endothelium to ergot alkaloids. Included is increased angiotensin converting enzyme activity that leads to increased angiotensin II levels (Oliver et al., 1998). Angiotensin II is a potent vasoconstrictor substance that also impacts fluid and electrolyte balance through release of antidiuretic hormone and aldosterone. In addition, activation of angiotensin II receptors on smooth muscle cells is known to induce mitogenesis (Oliver, 1997). There is a trend for increase in von Willibrand factor (vWF) levels in cattle exposed to endophyte-infected tall fescue (Oliver et al., 1998), with vWF being a factor that functions as an adhesion molecule in the subendothelium for blood platelets during vascular injury (Oliver, 1997). Increased vWF levels are associated with increased risk for hypercoagulability and thrombosis; events prominent in severe fescue toxicosis. Additionally, thromboxane A_2 (TXA_2) levels are increased in cattle that graze endophyte-infected tall fescue (Oliver et al., 1998). Thromboxane A_2 is a potent vasoconstrictor substance that is formed in blood platelets, and increased blood levels of TXA_2 indicates that thrombotic events are occurring. Thromboxane A_2 is also known to have broncho-constrictive effects in the lungs (Oliver, 1997). Studies by Filipov et al. (1999) documented increased tumor necrosis factor ($TNF\alpha$) in steers exposed to E+ tall fescue that were challenged with lipopolysaccharide (LPS). Overall, the cattle exposed to E+ tall fescue had a greater host response to LPS challenge (Filipov et al., 1999). Filipov et al. (1999) also determined that mice treated with ergotamine for 10 d had greater production of the proinflammatory cytokine IL-6 (compared with nontreated controls). More recently, Oliver and co-workers (2001) have shown that arginine levels are decreased in cattle that graze E+ tall fescue, resulting in NO deficiency that further contributes to increased vasoactivity and blood platelet activation. Results of studies reported by these workers also revealed that NO synthase activity was similar in lateral saphenous veins of control and E+ cattle, an enzyme that converts arginine to citrulline and frees NO for vasodilatory effect. However, total nitrite and nitrate levels of E+ cattle were significantly less, reflecting on the arginine-deficient condition of E+ cattle (Oliver et al., 2001). In support of NO being an important factor in toxic effects of alkaloids in tissues, Al-Tamimi et al. (2001, 2002) have shown relief from heat stress in rats on E+

diets by two different NO donor agents. Further, these workers demonstrated that treatment with NO donor agents overrides peripheral vasoconstrictive effects of E+ diet. In studies with cattle fed E+ diet, Al-Tamimi et al. (2002) showed that nitroglycerine (NO donor) enhanced peripheral circulation, as indicated by significantly reduced differences between body core and skin temperatures. In another study with cattle under heat stress conditions, and control group vs. ergovaline-treated group, nitroglycerine reversed the heat stress in ergovaline-treated cattle but not in the control group. Finally, these researchers showed that nitroglycerine was effective in preventing ergotamine-induced rise in both intragastric and intraperitoneal temperatures. These studies are corroborative of cases in humans that have become toxic to dihydroergotamine treatment (intense vasospasm), with the most effective of various treatments to relieve the debilitating effect of vasospasm on tissues being NO donor agents (nitroglycerine, nitroprusside; Ashenburg and Phillips, 1989). Pharmacologic treatments reported in the literature for reversal of the severe vasospastic disease induced by ergot alkaloids have also been successful when α-adrenergic receptor antagonists have been used (Oliver, 1997).

SUMMARY

The pathogenesis of fescue toxicosis in animals is due in part to effects of alkaloids (ergovaline is prominent) on the endothelium of blood vessels, with resultant effects leading to thickened blood vessels, increased vascular tone, hypercoagulability, sludging of blood, and depending on severity, ischemia, and gangrenous necrosis of tissues. That ergovaline is causally dominant in these effects is shown by the relief afforded animals by the removal of ergovaline from new fescue cultivars (Bouton et al., 2002; Parish et al., 2003), and by studies where vaccine removal of the alkaloid in animals gives relief from fescue toxicosis (Filipov et al., 1998; Hill et al., 1994). Ischemic response of blood vessels in the skin and lungs contributes to heat stress problems (Lakritz et al., 2002), and also likely influences reproductive performance and nutrient flux in the gastrointestinal system. The well-known effect of alkaloids on reducing feed intake (Parish et al., 2003) may have an associated effect on important nutrient factors. Oliver et al. (2000) demonstrated a trend for increased tryptophan levels to occur in the blood of cattle that had grazed E+ tall fescue, which may impact serotonin levels in the hypothalamus and affect feeding behavior. Porter et al. (1990) have documented that serotonin metabolites increase in the hypothalamus of cattle on E+ tall fescue. Similarly, Oliver et al. (2001) demonstrated suppressive effect of E+ tall fescue intake on serum arginine levels, which were associated with NO deficiency (decreased total nitrate levels), a condition that would enhance vasoconstrictive and hypercoagulability effects. In addition, Al-Tamimi et al. (2001, 2002) have demonstrated the protective effects against heat stress from NO donor agents. Thus, it is desirable to minimize exposure to the ergot alkaloids of E+ tall fescue. Progress has been made in this area by forage manipulation using two approaches: (i) selection of fescue genotypes which suppress the expression of ergovaline in the wild-type endophyte–tall fescue association (Bouton, 1996), and (ii) inoculation of elite endophyte-free tall fescue germplasm with selected endophytes which do not produce ergot alkaloids (Bouton et al., 2002). Cultivars of tall fescue with a nontoxic endophyte have been developed, tested, and marketed (Bouton et al., 2002; Parish et al., 2003). No signs of fescue toxicosis have been reported in recent years in these nontoxic endophyte associations in trials, nor in the multitude of acres that have been sown on commercial and research farms throughout the USA (Bouton et al., 2002; Parish et al., 2003). While there has been success in reducing the toxic nature of the grass with selected endophytes, adoption of this technology has been slow. The latter is due to the widespread, existing pasturelands that are populated with tall fescue infected with wild-type endophyte, and the cost of renovating these pastures with the improved fescue–endophyte associations. Recently, there has been some success with alleviator drug treatments (domperidone, metoclopramide, NO donor agents) that have been given to minimize the impact of alkaloids on animals grazing E+ tall fescue. Seaweed extract products have been shown to improve immunologic status of cattle on E+ fescue (Saker et al., 2001), and ammoniation of toxic fescue forage reduces toxic alkaloid content (Roberts et al., 2002). Development of rumen microbes that are capable of enhanced degradation of ergot alkaloids has been shown to be feasible (Craig and Blythe, 2002). Dietary manipulation with introduction of clovers into pastures is a proven method to reduce toxin effects (Andrae et al., 2003), and there is promise that supplementation with arginine may provide relief from the vasoactive aspects of tall fescue toxicosis (Oliver et al., 2001). Thus, there are methods now available for agriproducers to minimize the heretofore considerable losses associated with fescue toxin effects on cattle, which still allow the usage of this desirable forage that has so many positive agronomic traits.

REFERENCES

Aldrich-Markham, S., G. Pirelli, and A.M. Craig. 2003. Endophyte toxins in grass seed fields and straw; effects on livestock. Publication EM 8598. Report of Oregon State Univ. Ext. Serv., Corvallis, OR.

Al-Tamimi, H., and D. Spiers. 2002. Relationship between nitric oxide activity and fescue toxicosis-induced hyperthermia. p. 77–79. *In* Proc. Tall Fescue Toxicosis Workshop, SERAIEG-8, Wildersville, TN.

Al-Tamimi, H., D.E. Spiers, and G.E. Rottinghaus. 2001. Refinement of a rat model to study fescue toxicosis. Nitric oxide studies. p. 33. In Proc. Tall Fescue Toxicosis Workshop, SERAIEG-8, Chapel Hill, TN.

Andrae, J., N. Hill, and J. Bouton. 2003. Effect of tall fescue endophyte status and white clover addition on ergot alkaloid urinary excretion and performance of stocker cattle grazing tall fescue. p. 18–19. *In* Proc. Tall Fescue Toxicosis Workshop, SERAIEG-8, Wildersville, TN.

Ashenburg, R.J., and D.A. Phillips. 1989. Ergotism as a consequence of thromboembolic prophylaxis. Radiology 170:375–376.

Ayers, A.W., G.E. Rottinghaus, J.A. Stuedemann, F.N. Thompson, P.T. Purinton, D.L. Dawe, D. Ensley, and N.S. Hill. 2003. Transport of tall fescue alkaloids across gastric tissues. p. 32–33. *In* Proc. Tall Fescue Toxicosis Workshop, SERAIEG-8, Wildersville, TN.

Bacon, C.W. 1995. Toxic endophyte-infected tall fescue and range grasses: Historic perspectives. J. Anim. Sci. 73:861–870.

Bouton, J.H. 1996. Breeding tall fescue cultivars for the Southeast. p. 27–30. *In* R.R. Duncan (ed.) Proc. Grass Breeders Work Planning Conf., 34th, Griffin, GA. 15–17 Sept. 1996. Nat'l. Forage Seed Prod. Res. Center, Corvallis, OR.

Bouton, J.H., G.C.M. Latch, N.S. Hill, C.S. Hoveland, M.A. McCann, R.H. Watson, J.A. Parish, L.L. Hawkins, and F.N. Thompson. 2002. Reinfection of tall fescue cultivars with non-ergot alkaloid producing endophytes. Agron. J. 94:567–574.

Cheeke, P.R. 1995. Endogenous toxins and mycotoxins in forage grasses and their effects on livestock. J. Anim. Sci. 73:909–918.

Christiansen, D., G. Olsen, R. Hopper, N. Filipov, D. Lang, N. Hill, B. Fitzgerald, and P. Ryan. 2003. Pregnancy outcome in mares grazing endophyte-infected tall fescue in the fall. p. 36–37. *In* Proc. Tall Fescue Toxicosis Workshop, SERAIEG-8, Wildersville, TN.

Coppock, R.W., M.S. Mostrom, J. Simon, D.J. McKenna, B. Jacobsen, and H.L. Szlachta. 1989. Cutaneous ergotism in a herd of dairy calves. JAVMA 194:549–551.

Craig, A.M., and L.L. Blythe. 2002. Oregon report. p. 87. *In* Proc. Tall Fescue Toxicosis Workshop, SERAIEG-8, Wildersville, TN.

Cross, D.L. 2000. Toxic effects of *Neotyphodium coenophialum* in cattle and horses. p. 219–235. *In* V.H. Paul and P.D. Dapprich (ed.) Proc. 4th Int. *Neotyphodium*/Grass Interactions Symp., Soest, Germany. Univ. Paderborn, Germany.

Dennis, S.B., V.G. Allen, K.E. Saker, J.P. Fontenot, J.Y.M. Ayad, and C.P. Brown. 1998. Influence of *Neotyphodium coenophialum* on copper concentration in tall fescue. J. Anim. Sci. 76:2687–2693.

Dyer, D.C. 2000. Additional evidence for the antagonism of ergovaline-induced vasoconstriction by 5-hydroxytryptamine-2 (5-HT^2) antagonists. p. 505–513. *In* V.H. Paul and P.D. Dapprich (ed.) Proc. 4th Int. *Neotyphodium*/Grass Interactions Symp., Soest, Germany. Univ. Paderborn, Germany.

Filipov, N.M., F.N. Thompson, N.S. Hill, D.L. Dawe, J.A. Stuedemann, J.C. Price, and C.K. Smith. 1998. Vaccination against ergot alkaloids and the effect of endophyte-infected fescue seed-based diets on rabbits. J. Anim. Sci. 76:2456–2463.

Filipov, N.M., F.N. Thompson, R.P. Sharma, and R.R. Dugyala. 1999. Increased pro-inflammatory cytokine production by ergotamine in male BALB/c mice. J. Toxicol. Envir. Health 58:145–155.

Filipov, N.M., F.N. Thompson, J.A. Stuedemann, T.H. Elsasser, S. Kahl, R.P. Sharma, C.R. Young, L.H. Stanker, and C.K. Smith. 1999. Increased responsiveness to intravenous lipopolysaccharide challenge in steers grazing endophyte-infected tall fescue compared with steers grazing endophyte-free tall fescue. J. Endocrinol. 163:213–220.

Garner, G.B., and C.N. Cornell. 1978. Fescue foot in cattle. p. 45–62. In T.D. Wylie and L.G. Morehouse (ed.) Mycotoxic fungi, mycotoxins and mycotoxicoses. Vol. 2. Marcel Dekker, New York.

Hill, N.S., F.N. Thompson, D.L. Dawe, and J.A. Stuedemann. 1994. Antibody binding of circulating ergot alkaloids in cattle grazing tall fescue. Am. J. Vet. Res. 55:419–424.

Hoveland, C. 2000. Endophytes—Research and impact. p. 1–8. In V.H. Paul and P.D. Dapprich (ed.) Proc. 4th Int. Neotyphodium/Grass Interactions Symp., Soest, Germany. 27–29 Sept. 2000. Univ. Paderborn, Germany.

Jones, K.L., S.S. King, K.E. Griswold, D. Cazac, and D. Cross. 2003. Domperidone can ameliorate deleterious reproductive effects and reduced weight gain associated with fescue toxicosis in heifers. J. Anim. Sci. 81:2568–2574.

Lakritz, J., M.J. Leonard, P.A. Eichen, G.E. Rottinghaus, G.C. Johnson, and D.E. Spiers. 2002. Whole-blood concentrations of glutathione in cattle exposed to heat stress or a combination of heat stress and endophyte-infected tall fescue toxins in controlled environmental conditions. Am. J. Vet. Res. 63:799–803.

Larson, B.T., D.L. Harmon, E.L. Piper, L.M. Griffis, and L.P. Bush. 1999. Alkaloid binding and activation of D_2 dopamine receptors in cell cultue. J. Anim. Sci. 77:942–947.

Lipham, L.B., F.N. Thompson, J.A. Stuedemann, and J.L. Sartin. 1989. Effects of metoclopramide on steers grazing endophyte-infected fescue. J. Anim. Sci. 67:1090–1097.

Moubarak, A.S., E.L. Piper, and Z. Johnson. 1994. Detection of ergotamine, ergosine and ergine after intravenous injection of a single dose. p. 11–12. In Proc. Tall Fescue Toxicosis Workshop, SERAIEG-8, Atlanta, GA.

Oliver, J.W. 1997. Physiological manifestations of endophyte toxicosis in ruminant and laboratory species. p. 311–346. In C.W. Bacon and N.S. Hill (ed.) Neotyphodium/Grass Interactions. Proc. Int. Symp. Acremonium/Grass Interactions, 3rd, Athens, GA, USA. 28–31 May 1997. Plenum Press, New York.

Oliver, J.W., H. Al-Tamimi, J. Waller, H. Fribourg, K. Gwinn, L. Abney, and R. Linnabary. 2001. Effect of chronic exposure of beef steers to the endophytic fungus of tall fescue; comparative effects on nitric oxide synthase activity and nitrate/nitrite levels in lateral saphenous veins. p. 55–56. In Proc. Tall Fescue Toxicosis Workshop, SERAIEG-8, Chapel Hill, TN.

Oliver, J.W., S.K. Cox, J.C. Waller, H.A. Fribourg, K.D. Gwinn, B.W. Rohrbach, and R.D. Linnabary. 2001. Effect of chronic exposure of beef steers to the endophytic fungus of tall fescue; comparative effects on serum arginine levels. p. 56–57. In Proc. Tall Fescue Toxicosis Workshop, SERAIEG-8, Chapel Hill, TN.

Oliver, J.W., R.D. Linnabary, A.E. Schultze, J.C. Waller, H.A. Fribourg, B.W. Rohrbach, L.K. Abney, and E.M. Bailey. 1998. Characterization of inflammatory response to ergot alkaloid presence in toxic tall fescue. p. 36–38. In Proc. Tall Fescue Toxicosis Workshop, SERAIEG-8, Nashville, TN.

Oliver, J.W., and A.E. Schultze. 1997. Histologic lesions in cattle fed toxic tall fescue grass. The Toxicologist 36:46.

Oliver, J.W., A.E. Schultze, B.W. Rohrbach, H.A. Fribourgh, T. Ingle, and J.C. Waller. 2000. Alterations in hemograms and serum biochemical analytes of steers after prolonged consumption of endophyte-infected tall fescue. J. Anim. Sci. 78:1029–1035.

Oliver, J.W., J.R. Strickland, J.C. Waller, H.A. Fribourg, R.D. Linnabary, and L.K. Abney. 1998. Endophytic fungal toxin effects on adrenergic receptors in bovine lateral saphenous veins (cranial branch) of cattle grazing tall fescue. J. Anim. Sci. 76:2853–2856.

Oliver, J.W., J. Waller, H. Fribourg, K. Gwinn, M. Cottrell, and S. Cox. 2000. Aminoacidemia in cattle grazed on endophyte-infected tall fescue. p. 241–245. *In* P.H. Volker and P.D. Dapprich (ed.) Proc. 4th Int. *Neotyphodium*/Grass Interactions Symp., Soest, Germany. Univ. Paderborn, Germany.

Parish, J.A., M.A. McCann, R.H. Watson, N.N. Paiva, C.S. Hoveland, A.H. Parks, B.L. Upchurch, N.S. Hill, and J.H. Bouton. 2003. Use of non-ergot alkaloid-producing endophytes for alleviating tall fescue toxicosis in stocker cattle. J. Anim. Sci. 81:2856–2868.

Porter, J.K., J.A. Stuedemann, F.N. Thompson, Jr., and L.B. Lipham. 1990. Neuroendocrine measurements in steers grazed on endophyte-infected fescue. J. Anim. Sci. 68:3285–3292.

Roberts, C.A., R.L. Kallenbach, and N.S. Hill. 2002. Harvest and storage methods affect ergot alkaloid concentration in tall fescue [Online]. Available at www.plantmanagementnetwork.org/cm/. Crop Manage. DOI 10.1094/cm-2002-0917-01-BR.

Rossi-Fanelli, F., and C. Cangiano. 1991. Increased availability of tryptophan in brain as common pathogenic mechanism for anorexia associated with different diseases. Nutrition 7:364–367.

Saker, K.E., V.G. Allen, J.P. Fontenot, C.P. Bagley, R.L. Ivy, R.R. Evans, and D.B. Wester. 2001. Tasco-forage: II. Monocyte immune cell response and performance of beef steers grazing tall fescue treated with a seaweed extract. J. Anim. Sci. 79:1022–1031.

Saker, K.E., V.G. Allen, J. Kalnitsky, C.D. Thatcher, W.S. Swecker, Jr., and J.P. Fontenot. 1998. Monocyte immune cell response and copper status in beef steers that grazed endophyte-infected tall fescue. J. Anim. Sci. 76:2694–2700.

Schoning, C., M. Flieger, and H.H. Pertz. 2001. Complex interaction of ergovaline with 5-HT2A, 5 HT1B/1D, and alpha-1 receptors in isolated arteries of rat and guinea pig. J. Anim. Sci. 79:2202–2209.

Schultze, A.E., and J.W. Oliver. 1999. Cytotoxicity of ergine to bovine endothelial cells in culture. The Toxicologist 48:298.

Schultze, A.E., B.W. Rohrbach, H.A. Fribourg, J.C. Waller, and J.W. Oliver. 1999. Alterations in bovine serum biochemistry profiles associated with prolonged consumption of endophyte infected tall fescue. Vet. Human Toxicol. 41:133–139.

Spiers, D.E., J. Lakritz, P.A. Eichen, G.E. Rottinghaus, H.J. Al-Tamimi, J.R. Dodam, and J. Underwood. 1999. Biological markers of cattle response to heat stress on endophyte-infected tall fescue. J. Anim. Sci. (Suppl. 1) 77:146.

Strickland, J.R., E.M. Bailey, L.K. Abney, and J.W. Oliver. 1996. Assessment of the mitogenic potential of the alkaloids produced by endophyte *Acremonium coenophialum*-infected tall fescue (*Festuca arundinacea*) on bovine vascular smooth muscle in vitro. J. Anim. Sci. 74:1664–1671.

Stuedemann, J.A., N.S. Hill, F.N. Thompson, R.A. Fayrer-Hosken, W.P. Hay, D.L. Dawe, D.H. Seman, and S.A. Martin. 1998. Urinary and biliary excretion of ergot alkaloids from steers that grazed endophyte-infected tall fescue. J. Anim. Sci. 76:2146–2154.

Stuedemann, J.A., and F.N. Thompson. 1993. Management strategies and potential opportunities to reduce the effects of endophyte-infected tall fescue on animal performance. p. 103–114. *In* D.E. Hume et al. (ed.) Proc. Int. Symp. on *Acremonium*/Grass Interactions, 2nd, Palmerston North, New Zealand. 4–6 Feb. 1993. AgResearch Grasslands, Palmerston North.

Thompson, W.S., W.W. McClure, and M. Landowne. 1950. Prolonged vasoconstriction due to ergotamine tartrate. Archiv. Intern. Med. 85:691–698.

Westendorf, M.L., G.E. Mitchell, Jr., R.E, Tucker, L.P. Bush, R.J. Petroski, and R.G.Powell. 1993. *In vitro* and *in vivo* ruminal and physiological responses to endophyte-infected tall fescue. J. Dairy Sci. 76:555–563.

INTEGRATING GENETICS, ENVIRONMENT, AND MANAGEMENT TO MINIMIZE ANIMAL TOXICOSES

John A. Stuedemann and Dwight H. Seman[1]

The role of tall fescue (*Festuca arundinacea*) as a valuable pasture grass is well established. In the USA, it is grown on more than 20 million ha, primarily in the humid areas of the East and South (Bouton, 2001). It has been estimated that more than 8.5 million beef cows and 700 000 horses are maintained on tall fescue in the USA (Ball et al., 2002, p. 198–205). It is also well established that its contribution to agricultural production and environmental quality is dependant upon the mutualistic relationship between the endophyte (*Neotyphodium coenphialum*) and the plant. Presence of the wild-type endophyte (E+) expanded the area of adaptation for tall fescue beyond what would have been possible as an endophyte-free plant. Ironically, much of this expansion occurred before the discovery of the endophyte (Bacon et al., 1977) and long after initial observations suggested that the plant could reduce animal production (Pratt and Davis, 1954; Pratt and Haynes, 1950). Producers and agricultural researchers placed a greater emphasis on the agronomic attributes of the plant than on the negative effects on animal production. In other words, plant survival with reduced animal production was more important than no plant or a weak plant.

The lesson is, that to overcome the negative effects of E+ tall fescue, we must not decrease the survivability of the plant. There may be a region of adaptation where the endophyte is not necessary for survival and excellent plant growth. In this region, endophyte-free (E-) tall fescue is a management option.

Our objective is to review and summarize recent advances in mechanisms that reduce negative impacts of wild-type, endophyte-infested tall fescue on animal production and advances in endophyte and/or plant genetics that result in improved animal production.

---

[1] J. Phil Campbell, Sr. Natural Resource Conservation Center, USDA, Agricultural Research Service, Watkinsville, GA

IMPACT OF ENDOPHYTE-INFESTED TALL FESCUE ON ANIMAL RESPONSE

Extensive reviews concerning the effects of consuming E+ tall fescue on grazing animals (Bush et al., 1979; Cross, 2001; Stuedemann and Hoveland, 1988; Stuedemann and Thompson, 1993; Thompson et al., 2001) have categorized the effects as follows: (i) performance and production, (ii) behavioral, (iii) physiological (e.g., respiration rate, heart rate, rectal or core body temperature, etc.), and (iv) tissue concentrations (e.g., sera or plasma concentrations of minerals, enzymes, hormones, etc.). These effects have traditionally been referred to collectively as fescue toxicosis. Cross (2001) reviewed the effects of consuming E+ tall fescue in cattle and horses with particular emphasis on horses. In the horse, the most noted effects appear to be associated with the mare, where consumption of E+ tall fescue results in increased gestation lengths, agalactia, foal and mare mortality, tough and thickened placentas, weak and dysmature foals, and reduced serum prolactin and progesterone levels (Monroe et al., 1988).

Fescue foot and fat necrosis are two conditions associated with consuming E+ tall fescue. Fescue foot is a gangrenous condition of the animal's extremities such as hooves, tail, and occasionally ears, that is often observed during the winter months (Garner and Cornell, 1978) and is a result of vasoconstriction (Solomons et al., 1989; Rhodes et al., 1991). In the USA, the condition rarely occurs in the lower South. Consequently, there is an interaction between environment, endophyte, and the occurrence of fescue foot. Fat necrosis is typically expressed by masses of hard fat primarily located in the abdominal cavity of mature cows grazing tall fescue (Stuedemann et al., 1975; Williams et al., 1969). Stuedemann et al. (1985) reported that steers grazing E+ tall fescue had reduced serum cholesterol levels. In a separate experiment, they observed that the incidence of fat necrosis was highest in cow herds with the lowest serum cholesterol levels, implicating altered lipid metabolism in cows grazing E+ tall fescue. Thompson et al. (2001) concluded that fat necrosis and/or altered lipid metabolism, as well as fescue foot, are signs of fescue toxicosis.

Although cattle grazing E+ tall fescue may have an unthrifty appearance with a rough hair coat throughout the year, it was obvious signs of reduced tolerance to heat, such as elevated respiration rates and altered grazing behavior, that lead to the use of the terms *summer slump* or *summer fescue toxicosis* (Garner and Cornell, 1985). These terms are used less frequently today because it is well documented that fescue toxicosis can occur throughout the year. However, it is important to note that there is an interaction between the magnitude of the negative response and ambient temperature, making environmental conditions important when assessing the magnitude of the impact of E+ tall fescue on animal response.

Heat tolerance can be increased in *Bos taurus* cattle by crossing *Bos indicus* and *Bos taurus* cattle (Turner, 1980). This information, coupled with the interaction between animal response and ambient temperature, prompted research to investigate whether Brahman × Angus and Angus × Brahman crossbred cows and their calves might be more tolerant of E+ tall fescue. Brown et al. (1997) reported that, although crossbred cows and their calves were more tolerant of E+ tall fescue, the use of such a cross would not overcome all production losses as a result of grazing E+ tall fescue. In addition, some of these differences could be due to heterotic effects. Just because crossbred cattle outperformed the straight-bred cattle does not mean that they are able to overcome fescue toxicosis. Possibly they are simply less sensitive to heat and the interaction that exists between the consumption of E+ tall fescue and ambient temperature.

The role of pasture fertilization rate and the use of poultry litter have long been implicated as factors related to the development of fescue toxicosis. Williams et al. (1969) suggested that fertilization with high rates of poultry litter increased the occurrence of fat necrosis in cows. Stuedemann et al. (1975) suggested that it was the high level of nitrogen fertilization, and not poultry litter, that increased symptoms of fescue toxicosis. Missouri researchers found a positive correlation between the occurrence of fescue foot and N application to E+ tall fescue (Garner and Cornell, 1978, 1985). Belesky et al. (1988) reported that increased N fertilization of E+ tall fescue increased ergopeptine alkaloid concentrations in the forage. The accumulation of ergot alkaloids in E+ tall fescue was shown to be affected by nitrogen source (Arechavaleta et al., 1992), with the NH_4^+ form having a greater effect on accumulation of ergot alkaloids than the NO_3^- form. Azevedo et al. (1993) studied the role of P in growth and ergot alkaloid production in E+ tall fescue. They found that ergovaline production responded to increased P availability and later reported that endophyte hyphae accumulated inorganic P (Azevedo and Welty, 1995). Although not consistent across all cultivars, Malinowski et al. (1998) found that increased P fertilization of E+ tall fescue, particularly under P-deficient soil conditions, resulted in elevated ergot alkaloid concentrations in plant tissue. The authors did not determine hyphae concentration; therefore, it was not known whether the increase in alkaloid concentration was a result of increased hyphae or increased biosynthesis. In a subsequent review, Malinowski and Belesky (2000) provided an extensive discussion of N and P involvement in E+ tall fescue and mechanisms affecting abiotic (drought, mineral) stresses and tolerance of selected biotic stresses in cool-season grasses. Because poultry litter typically provides high levels of N and P, it may be effective at increasing the negative effects of E+ tall fescue, and the effect may be enhanced under P-deficient soil conditions.

Residual effects on the animal, of having grazed E+ tall fescue, are less clearly delineated than other responses to E+ tall fescue. Hill et al. (2000) reported that urinary alkaloid analysis was a useful tool in determining whether animals were consuming toxic fescue. They found that there was a significant quadratic relationship between average daily weight gain of steers and the amount of ergot alkaloids excreted via the urine. They also found that within 96 h after removal from E+ tall fescue, urinary levels of ergot alkaloids were the same as for animals that were grazing E− tall fescue. The results of studies of steer performance in drylot following grazing of E+ tall fescue have been inconsistent. Some researchers have reported compensatory effects in performance even though there were some relatively short-term effects on certain response variables (Beconi et al., 1995; Cole et al., 2001). Others have reported reduced feedlot performance (Hancock et al., 1987). The results could be impacted by interactions with ambient temperature. If cattle are removed from E+ tall fescue during the hot time of the year and placed in drylot, the results may differ from those received in drylot during the cool time of the year.

The effects of grazing E+ tall fescue cover a wide range of response criteria, all the way from performance and production variables to physiological and tissue effects. The importance of some of the effects in a production setting is debatable; however, the effects on performance and production directly translate to economic loss. In the horse, effects on the gravid mare appear to be the most important in terms of economic significance. As we review and discuss strategies for minimizing or alleviating the effects of consuming E+ tall fescue in cattle, emphasis will be given to options that improve performance and production rather than variables that do not directly affect economic returns. We understand the importance of these response variables as they relate to the mechanisms involved in the development of fescue toxicosis, but we will focus on variables that producers equate with economic return.

STRATEGIES FOR REDUCING OR ALLEVIATING THE EFFECTS OF ENDOPHYTE-INFESTED FESCUE

Options for reducing or alleviating the effects of E+ tall fescue were extensively reviewed (Schmidt and Osborn, 1993; Stuedemann and Thompson, 1993) about

10 yr ago. Since then, significant advances have occurred in our understanding of the endophyte and fescue toxicosis. Even though it appears that the ergot alkaloids are the major animal toxins, the contributions of the various ergot alkaloids to the development of clinical signs of fescue toxicosis are unclear. This uncertainty has contributed to the difficulty in making progress in developing methods for overcoming fescue toxicosis, and particularly in developing a vaccine.

Pasture Management

Planting endophyte-free seed. Reseeding of E+ tall fescue pastures to other grasses has been an option available to farmers since the discovery of the endophyte (Bacon et al., 1977) and its association with fescue toxicosis (Hoveland et al., 1983). The establishment of laboratories for testing seed and pasture samples aided in assessing infection rates and verified that a high percentage of pastures were highly infested with the endophyte. In a survey of more than 1500 pasture samples from throughout the USA, >70% of the samples had 60% or more infection rates (Shelby and Dalrymple, 1987). It has generally been assumed that existing tall fescue stands, particularly if they had been established for more than 10 yr, were highly infested.

The major problems with reseeding include: out-of-pocket expense, loss or reduced use of the pasture during establishment, and the question of what to plant. Endophyte-free tall fescue seed became available soon after the discovery that the endophyte was transmitted through the seed and that it would die if seed were stored at ambient temperature for one year. Unfortunately, many producers soon learned that if they established E– tall fescue, the stands were soon thinned and eventually lost (Lacefield et al., 1993). Ball (1997) reviewed many of the aspects of producer attitudes and management that are critical for establishment and maintenance of E– tall fescue. In much of the lower southern portions of the USA, an alternative persistent, cool-season perennial grass was not available, so many producers switched to warm-season grasses such as bermudagrass (*Cynodon dactylon*) and overseeded with a cool-season annual in order to extend grazing and more fully utilize the climatic resource.

Incorporating non-ergot alkaloid-producing endophytes in fescue. During the past 10 yr, there have been comparatively few innovations for reducing or alleviating fescue toxicosis. However, a potentially important breakthrough in overcoming fescue toxicosis has been the development of non-ergot alkaloid-producing endophyte-infested tall fescue. This has been accomplished by reinfecting E– tall fescue cultivars with non-ergot alkaloid-producing endophytes (Bouton et al., 2002; West et al., 1998).

At the present time, development of these cultivars has centered around two groups: the Georgia, USA–New Zealand group (Bouton–Latch) and the Arkansas–Missouri, USA group (West–Sleper). Research with these cultivars has and is being executed at various research institutions. However, it must be emphasized that two key questions about these cultivars are still unanswered. First, how long will they persist when grazed? Second, are there critical management inputs necessary for persistence?

The Bouton–Latch group developed non-ergot alkaloid-producing endophyte-infected tall fescue by reinfecting E– tall fescue cultivars, Jesup and Georgia 5, with non-ergot alkaloid-producing endophytes (Bouton et al., 2002). Bouton et al. (2002) reported that the cultivars with non-ergot alkaloid-producing endophytes (AR542 and 502) had survival rates similar to their E+ controls when close grazed in bermudagrass sod. Subsequent studies with lambs (Parish et al., 2003a) and stocker cattle (Parish et al., 2003b) revealed that, in general, average daily gain (ADG) and gain per hectare were higher on AR542 (MaxQ™), AR502, and E– than on E+ tall fescue. Similar findings were reported with cow-calf units grazing wild-type E+ Georgia 5 tall fescue or non-ergot alkaloid-producing endophyte-infested Georgia 5 (Watson et al., 2001, 2004). Cow-body weights and condition scores were higher on the non-ergot alkaloid-producing endophyte-infested tall fescue as compared with the wild-type infested tall fescue. Weaning weights of steer calves were higher on the non-ergot alkaloid-producing endophyte-infested tall fescue (260 kg) as compared with those on wild-type E+ tall fescue (226 kg). Timper et al. (2001) conducted studies to determine if the non-ergot alkaloid strain AR542 conferred the same level of resistance to the lesion nematode (*Pratylenchus scribneri*) as the wild-type strain. Tall fescue with the wild-type endophyte had fewer lesion nematodes in their roots than the E– plants or those with the AR542 strain. Therefore, if nematodes contribute to poor growth and persistence of tall fescue, particularly on sandy soils, there may be reduced survival of the AR542-infected tall fescues as compared with those infected with the wild-type endophyte.

The West–Sleper group incorporated non-ergot alkaloid-producing endophytes developed by West et al. (1998) into an endophyte-free cultivar of tall fescue called 'HiMag', a cultivar with a low risk of causing grass tetany in cows (Sleper et al., 2002). Two different non-ergot alkaloid-producing endophytes were incorporated into HiMag, resulting in HiMag4 and HiMag9. When grazing

'Kentucky 31' E+ (K31 E+), endophyte-free HiMag (HiMag–), HiMag 4, or HiMag9, steer ADG on the HiMag–, HiMag4, and HiMag9 was nearly double the ADG on K31 E+ (Nihsen et al., 2004).

Dilution with other grasses or legumes. Pasture reseeding to overcome fescue toxicosis would typically require large commitments of resources and time. A less costly method of pasture management includes the dilution of E+ tall fescue with other grasses or legumes such as bermudagrass (Chestnut et al., 1991) or clovers (*Trifolium* spp.) (Chestnut et al., 1991; Coffey et al., 1990; Hoveland et al., 1981; McMurphy et al., 1990). Chestnut et al. (1991) observed greater ADG and gain per unit of land area when including bermudagrass in both E+ and E– tall fescue pastures. Although the dilution of E+ tall fescue pastures has resulted in improved animal performance, results have been inconsistent. This is undoubtedly partly due to the diversity of environmental conditions and pasture management across the fescue growing area. Thompson et al. (1993) combined data from 12 independent studies conducted in seven states in the eastern USA, and revealed that the addition of clover to pastures with high levels of endophyte infection may have little positive effect on ADG of steers in the spring. In contrast, when clover was present in pastures with low to moderate endophyte infestation, ADG was improved. Potential for growing clover varies greatly among regions. Clovers are sensitive to viruses and other diseases, as well as being shallow rooted and subject to frequent drought stress during summer months. Bouton (2003) investigated the inclusion of 'Durana' white clover (*T. repens*) and 'Cinnamon Plus' red clover (*T. pratense*), both developed for grazing, in Jesup E+, E–, or MaxQ-infested tall fescue, and found that Durana improved steer ADG during autumn and spring in all treatments compared with grass alone.

Although the effects of diluting E+ tall fescue with other grasses or legumes is not well defined scientifically, it appears to be one of the major techniques used by producers to minimize the effects of E+ tall fescue in cattle. In the southern and western USA where tall fescue is adapted, this is often accomplished with other grasses. In northern and eastern regions, this may be accomplished with a combination of legumes and grasses.

According to Cross (2001), dilution with other grasses or legumes is typically not an option for owners of mares because it appears that horses exhibit signs of toxicosis even when consuming only small quantities in hay or when grazing small patches of E+ tall fescue. Therefore, diluting E+ tall fescue with other forages does not appear to be a viable option for horse owners.

Use of plant growth regulators, mowing, or clipping. Mefluidide is a plant growth regulator that was used to suppress heading and growth in E+ tall fescue in early spring, resulting in 47% greater organic matter intake in July and a 17-kg weight gain improvement in steers as compared with those grazing untreated E+ tall fescue (Turner et al., 1990). In a greenhouse experiment, Salminen et al. (2003) found that increasing clipping height from 2.5 to 7.5 cm increased concentrations of ergonovine, ergocryptine, perloline methyl ether, and an unidentified alkaloid. Concentrations varied with time and clipping height.

Stockpiling of forage. Kallenbach et al. (2003) reported that herbage mass of stockpiled K31 E+ tall fescue did not change from mid-December to mid-March in Missouri. However, the concentration of ergovaline decreased to relatively low levels by mid- to late winter. Pattern and rate of loss differed in the 2-yr study, suggesting that weather conditions may be an important factor. Results under Missouri conditions suggest that producers can minimize fescue toxicosis problems by delaying the use of stockpiled E+ tall fescue until midwinter.

Application of Tasco-Forage. The influence of Tasco™-Forage, an *Ascophyllum nodosum* seaweed-based product prepared by a proprietary process, on animals grazing E+ tall fescue has been reviewed by Allen et al. (2001a). Tasco application to pastures in Virginia and Mississippi did not influence steer gain on either E+ or E– tall fescue (Fike et al., 2001), nor did it influence subsequent gain in the feedlot (Allen et al., 2001b). A short-term study with sheep indicated an improvement of sheep gains with the application of Tasco (Fike et al., 2001). Steers that had grazed Tasco-treated pastures had higher marbling scores regardless of the endophyte. Tasco increased monocyte phagocytic activity and major histocompatibility complex (MHC) class II expression in steers that grazed either E+ or E– pastures (Allen et al., 2001b; Saker et al., 2001). Steers that grazed E+ tall fescue had depressed monocyte immune cell function throughout the feedlot-finishing period, which was reversed by Tasco. There were indications that Tasco applied to tall fescue during the grazing period improved color stability and extended beef shelf-life (Montgomery et al., 2001). Although results are very interesting, including the effects on marbling and shelf-life of meat, and it appears that Tasco affects some responses influenced by the endophyte, in particular, E+ effects on the immune system. The application of Tasco to pastures does not appear to influence performance of steers grazing E+ tall fescue.

Animal Management

Stocking rate and differential forage mass available. Because the concentration of ergot alkaloids is typically highest in the seed (Rottinghaus et al., 1991), possibly clipping or close grazing would reduce the toxicity of E+ tall fescue pastures. Bransby et al. (1988) observed that during a comparatively dry 84-d spring period, the ADG and gain per ha of steers grazing E+ pastures were improved by increased stocking rates.

From a different perspective, Gwinn et al. (1998) reported that endophyte infection levels increased when grazing at moderate to low forage mass, but did not change under high forage mass. The authors suggested that stress caused by grazing for low forage mass and not environmental stress or contamination resulted in the increase in endophyte infestation levels. Therefore, if endophyte infection levels are intermediate, close grazing or grazing for low forage mass could ultimately result in higher infection levels and potentially increase pasture toxicity.

Animal removal from E+ tall fescue. Considerable research has been done on removing (often referred to as rotating) animals from E+ tall fescue pastures during summer months to other forages, both annual and perennial. For example, after 84 d of grazing either E+ or E− tall fescue, Aldrich et al. (1990) rotated steers to sorghum-sudangrass (*Sorghum* × *drummondii*) pasture for an additional 56 d. Other researchers have investigated moving steers to caucasian bluestem (*Bothriochloa bladhii*) in early summer (Forcherio et al., 1992).

Evidence suggests that removal of mares at least 30 d prior to expected foaling greatly reduces the impact of E+ tall fescue (Cross, 2001). However, recommendations by veterinarians for removal of mares from E+ pastures ranges from 30 and up to 90 d prior to expected foaling date in order to minimize chances of possible abortion and *red bagging* or premature placental separations (Cross, 2001).

Rotational grazing. Extensive research has been conducted on rotational grazing; however, little data is available for comparing effects of endophyte infestation. Davenport et al. (1993) grazed steers on E+ and E− tall fescue on a 14-d alternate basis vs. continuously grazing E+ and E− tall fescue. Discontinuous grazing of E+ tall fescue at 14-d intervals did not improve ADG over the E+ tall fescue. Coffey et al. (2003) observed little difference in cow/calf performance of autumn-calving cows between rotation frequencies of two times per month vs. two times per week on E+ tall fescue that had been overseeded with crabgrass (*Digitaria* spp.), lespedeza (*Lespedeza* spp.), and red and white clover.

Season of calving. Although no research exists comparing season of calving on E+ tall fescue, the interactions that exist between time of year and ergot alkaloid production and the interaction between ambient temperature and effect of the ergot alkaloids suggest that season of calving on E+ tall fescue could have an impact on toxicosis. One could speculate that autumn calving with rebreeding during the late autumn and early winter months could result in improved calf performance and cow reproductive efficiency.

Creep grazing or creep feeding of calves. Vicini et al. (1982a, 1982b) compared presumably E+ tall fescue pastures with and without creep grazing of calves on bluegrass (*Poa* spp.)–white clover pastures. Creep grazing enhanced calf production by increasing forage consumption of the calves. The concept of creep grazing presumes that calves will have access to pasture that is superior in quality to that in which the cows and calves commingle. This places logistic constraints on grazing system design, time of year that these pastures may be available, and would need to be coordinated with the calving season. Creep feeding of calves suckling cows grazing E+ tall fescue has economic limitations (Stricker et al., 1979).

Use of Feed Treatments and Dietary Additives

Ammoniation of hay. Ammoniation of E+ tall fescue hay has resulted in consistent improvement of animal performance (Newsome et al., 1989; Essig and Armstrong, 1990). In studies that contrasted E– and E+ tall fescue hay, ammoniation resulted in similar serum prolactin concentrations in steers (Kerr et al., 1990) and sheep (Chestnut et al., 1991). In these studies, animals that were consuming untreated E+ hay had greatly reduced prolactin concentrations. Weight gains were similar among lambs fed either ammoniated E+ or E– tall fescue hay. Anhydrous ammonia was effective for detoxifying E+ tall fescue hay.

Ensiling. Research is limited, but Jackson et al. (1988) reported that ensiling fall accumulated E+ tall fescue hay produced toxicosis in calves as indicated by hay intake, respiration rates, and rectal temperatures.

Activated carbon. Supplementing cattle with either of two types of activated carbon had no effect on gain when grazing either E- or E+ tall fescue pastures (Geotsch et al., 1988).

Aluminosilicate. In vitro studies with hydrated sodium calcium aluminosilicate (HSCAS) revealed that it could remove >90 % of the ergotamine from aqueous solutions at pH 7.8 or lower (Chestnut et al., 1992). However, rats receiving E+ tall fescue seed and supplemented with HSCAS still exhibited the negative effects of E+ tall fescue on intake and prolactin concentrations. When E+ and E- tall fescue hay diets were fed to sheep in a digestion study, HSCAS did not change organic matter and N digestion coefficients, nor did it affect apparent absorption of organic matter, N, Ca, P, K, Na, or Cu, but it did reduce absorption of Mg, Mn, and Zn.

Anthelmintic compounds. Results suggest that some anthelmintic compounds may influence animal response to E+ tall fescue in ways beyond that of eliminating or reducing the incidence of parasites. The compound receiving the most interest has been ivermectin, which is known to be a modest gamma-amino butyric acid (GABA) agonist (Turner and Shaeffer, 1989). Receptors for GABA are widely distributed throughout the central nervous system and peripheral organs (Erdo, 1990). Unfortunately, the effects of ivermectin treatment of animals grazing E+ tall fescue have not been consistent. Goetsch et al. (1988) reported that treatment of steers with ivermectin or fenbendazole did not improve gain of steers on E- or E+ pastures. In contrast, Crawford and Garner (1991) found that treatment with ivermectin at 14-d intervals improved gains of steers grazing E+ pastures. Bransby (1997) also reported improved gains when steers were given ivermectin on E+ tall fescue. Rosenkrans et al. (2001) reported no effect of ivermectin slow release boluses on steer gain while fed E+ tall fescue hay but observed physiologic responses suggesting that ivermectin could have a positive effect on animals grazing E+ tall fescue. Conclusive evidence that ivermectin could be used to overcome fescue toxicosis on a routine basis is lacking.

Protein and energy supplementation. When beef cows wintered on either hay or hay plus grain were placed on either low E (21%) or high E (77%) tall fescue, Tucker et al. (1989) reported that winter supplementation with grain resulted in similar pregnancy rates among cows grazing either E level. When comparing supplementation of cows nursing calves from late May to late July while grazing

E+ tall fescue, Forcherio et al. (1995) reported that there was little difference in calf ADG among supplements and between supplemented and nonsupplemented groups when cows were supplemented with either cracked corn or soybean hulls as a source of energy and 100 vs. 200 g d^{-1} of undegraded intake protein.

Thiamin supplementation. Lauriault et al. (1990) suggested that reduced performance of cattle grazing E+ tall fescue may involve thiamin deficiencies caused by rumen thiaminases that use alkaloids found in E+ tall fescue as cosubstrates. Using the tethered grazing technique, they reported that supplementing cows exposed to E+ tall fescue with 1 g of thiamin d^{-1} increased intake compared with non-thiamin-supplemented cows. When adult cows were fed E+ (70%) tall fescue seed at 0 or 1 kg d^{-1} and were supplemented with 0 or 1 g of thiamin d^{-1}, Dougherty et al. (1991) reported that thiamin overcame decreased intake by extending grazing time.

Zeranol. Brazle and Coffey (1991) observed an interactive effect of the hormone implant zeranol when given to steers grazing low E (20%) and high E (82%) tall fescue. They found a 12% improvement in ADG on low E and a 37% improvement in ADG on high E with a 36-mg implant. Zeranol did not influence serum prolactin. Goetsch et al. (1988) found no gain improvement with zeranol treatment as well as with progesterone plus estradiol benzoate.

Copper, selenium, vitamin E, and others. Fike et al. (2001) reported lower whole blood Se and serum vitamin E concentrations in cattle grazing E+ tall fescue. The endophyte also appears to reduce plant Cu concentrations (Dennis et al., 1998), and serum Cu concentrations in cattle grazing E+ tall fescue (Saker et al. (1998). However, at the present time, supplementation with these elements has not resulted in an improvement in animal performance. Coffey et al. (1992) observed no advantage when steers grazing E+ tall fescue were supplemented with Cu. Similarly, no advantage was reported for Se supplementation of mares grazing E+ tall fescue (Monroe et al., 1988). In studies that only included E+ hay, Essig and Armstrong (1990) reported no improvement in steer gains when diets were supplemented with lasalocid, salicylic acid, Se, Zn, or vitamin E.

Use of pharmacologic agents. Lipham et al. (1989) reported that the oral administration of metoclopramide, a D2 dopamine receptor blocker, increased ADG and grazing time of steers grazing E+ tall fescue compared with untreated controls. In sheep, Aldrich et al. (1993) reported that administration of metoclo-pramide increased intake, but did not reduce core body temperature.

Redmond et al. (1994) orally administered domperidone, which is also a D2 dopamine receptor blocker, to gravid mares grazing E+ tall fescue. Domperi-done resulted in what appeared to be nearly complete recovery from having grazed E+ tall fescue without neuroleptic side effects. Treated mares had milk, live, healthy foals, and gestation lengths similar to calculated gestation lengths. An extensive field study by Cross et al. (1999) revealed that domperidone was 94.5% effective when administered to 1423 periparturient mares. Jones et al. (2003) reported that administration of domperidone to cycling heifers fed an E+ diet resulted in midcycle blood plasma progesterone levels that were similar to heifers fed an E– diet. Although the study was only 21 d in length, domperidone administration did increase weight gain of the heifers on the E+ diet. This needs to be tested under practical grazing conditions.

Use of immunologic protection. Both active and passive immunization are well-established methods of protecting animals against the action of toxins (LaGrange and Capron, 1978). Although vaccination against ergot alkaloids has been investigated (Filipov et al., 1998), at present there are no practical vaccines available to protect against fescue toxicosis.

THE ROLE OF PRODUCER EXPECTATIONS OR GOALS

Many livestock producers are aware of the impact of E+ tall fescue on animal production. In spite of this knowledge, many producers feel that there are few options that will yield a practical solution to the problem.

The southeastern USA is the major tall fescue-growing region of the USA, with a number of distinctive features that set it apart from many other areas of the country and world. These features have a direct impact on the approach that producers take in attempting to mitigate the effects of endophyte-infested tall fescue. Some of these features include: (i) the climatic resource enables both cool- and warm-season grass production; (ii) much of the land resource is composed of easily eroded soil; (iii) much of the area receives comparatively high rainfall, increasing environmental concerns about erosion; (iv) the majority of the farms in the region are comparatively small, in the 10-state area of the middle and upper South, the average herd size is fewer than 30 cows; (v) many farmers have off-farm jobs to supplement their incomes, for many the lifestyle, rather than economic efficiency, often impacts decisions; and (vi) often the livestock are secondary to other enterprises such as poultry or row-crops (Stuedemann et al., 1993).

It is obvious that producers want forages that persist. This was evident after it became apparent that E− tall fescue would not persist under many environmental and management constraints present in much of the southeastern USA (Ball, 1997). Tall fescue cultivars containing non-ergot alkaloid-producing endophytes must persist in order to be successful. Even though the development of the non-ergot alkaloid-producing endophytes and their insertion into productive tall fescue cultivars is a major breakthrough, it is estimated that <10% of the traditional E+ tall fescue will be reestablished during the next 10 yr. As a result, many producers will be seeking a practical way to manage wild-type E+ tall fescue. Unfortunately, as was written more than 10 yr ago, although there are numerous pasture management options as well as dietary additives and supplements that may have value in reducing the impact of endophyte-infested tall fescue on animals, none appear to have universal application (Stuedemann and Thompson, 1993).

SUMMARY

Since the discovery of the wild-type endophyte and the accumulation of evidence associating its presence in tall fescue with reduced animal productivity, much research has been devoted to developing methods aimed at reducing the impact of endophyte-infested tall fescue on animal production. Our objective was to summarize this body of research. We briefly assessed the impact of the endophyte on animal performance and production. We have summarized opportunities for eliminating, reducing, or minimizing negative effects on animal production by pasture and/or animal management. These opportunities include reseeding with endophyte-free fescues, or fescues containing non-ergot alkaloid-producing endophytes, dilution with other grasses or legumes, differential stocking rate, differential forage mass available, stockpiling of forage, and mechanical management of the pasture such as mowing. We reviewed the impact of a host of feed treatments and dietary additives on the symptoms of fescue toxicosis. The role of pharmacologic agents and the potential for immunologic protection have been included. Lastly, we have coupled the role of producer expectations with management time and expertise in making decisions to minimize the effects of fescue toxicosis.

REFERENCES

Aldrich, C.G., K.N. Grigsby, J.A. Paterson, and M.S. Kerley. 1990. Performance, OM intake, and digestibility by steers when rotationally grazed from tall fescue pastures to warm-season annual grass pastures. J. Anim. Sci. 68 (Suppl. 1):559.

Aldrich, C.G., M.T. Rhodes, J.L. Miner, M.S. Kerley, and J.A. Paterson. 1993. The effects of endophyte-infected tall fescue consumption and use of a dopamine antagonist on intake, digestibility, body temperature, and blood constituents in sheep. J. Anim. Sci. 71:158–163.

Allen, V.G., K.R. Pond, M.F. Miller, J.L. Montgomery, J.P. Fontenot, K.E. Saker, J.H. Fike, C.P. Bagley, R.R. Evans, and R.L. Ivy. 2001a. Influence of Tasco-Forage on immune response and carcass characteristics of steers grazing endophyte-infected tall fescue—A review. p. 245–253. In V.H. Paul and P.D. Dapprich (ed.) The Grassland Conference 2000, Proc. 4th Int. Neotyphodium/Grass Interactions Symp. Soest, Germany. 27–29 Sept. 2000. Univ. Paderborn, Germany.

Allen, V.G., K.R. Pond, K.E. Saker, J.P. Fontenot, C.P. Bagley, R.L. Ivy, R.R. Evans, C.P. Brown, M.F. Miller, J.L. Montgomery, T.M. Dettle, and D.B. Wester. 2001b. Tasco-Forage: III. Influence of a seaweed extract on performance, monocyte immune cell response, and carcass characteristics in feedlot-finished steers. J. Anim. Sci. 79:1032–1040.

Arechavaleta, M., C.W. Bacon, R.D. Plattner, C.S. Hoveland, and D.E. Radcliffe. 1992. Accumulation of ergopeptide alkaloids in symbiotic tall fescue grown under deficits of soil water and nitrogen fertilizer. Appl. Environ. Microbiol. 58:857–861.

Azevedo, M.D., and R.E. Welty. 1995. A study of the fungal endophyte Acremonium coenophialum in the roots of tall fescue seedlings. Mycologia 87:289–297.

Azevedo, M.D., R.E. Welty, A.M. Creaig, and J. Bartlett. 1993. Ergovaline distribution, total nitrogen and phosphorus content of two endophyte-infected tall fescue clones. p. 59–62. In D.E. Hume et al. (ed.) Proc. Int. Symposium on Acremonium/Grass Interactions, 2nd, Palmerston North, New Zealand. 4–6 Feb. 1993. AgResearch, Grassland Research Centre, Palmerston North.

Bacon, C.W., J.K. Porter, J.D. Robbins, and E. S. Luttrell. 1977. Epichloe typhina from toxic tall fescue grasses. Applied Environ. Microbiol. 34(5):576–581.

Ball, D.M. 1997. Significance of endophyte toxicosis and current practices in dealing with the problem in the United States. p. 395–410. In C.W. Bacon and N.S. Hill (ed.) Neotyphodium/Grass Interactions. Proc. Int. Symp. Acremonium/Grass Interactions, 3rd, Athens, GA. 28–31 May 1997. Plenum Press, New York.

Ball, D.M., C.S. Hoveland, and G.D. Lacefield. 2002. Southern forages. Potash and Phosphate Institute, Norcross, GA.

Beconi, M.G., M.D. Howard, T.D.A. Forbes, R.B Muntifering, N.W. Bradley, and M.J. Ford. 1995. Growth and subsequent feedlot performance of estradiol-implanted vs. nonimplanted steers grazing fall-accumulated endophyte-infested or low-endophyte tall fescue. J. Anim. Sci. 73:1576–1584.

Belesky, D.P., J.A. Stuedemann, R.D. Plattner, and S.R. Wilkinson. 1988. Ergopeptine alkaloids in grazed tall fescue. Agron. J. 80:209–212.

Bouton, J. 2001. The use of endophytic fungi for pasture improvement in the USA. p. 163–168. In V.H. Paul and P.D. Dapprich (ed.) The Grassland Conference 2000. Proc. Int. Neotyphodium/Grass Interactions Symposium, 4th, Soest, Germany. 27–29 Sept. 2000. Univ. Paderborn, Germany.

Bouton, J.H. 2003. New forage cultivars developed with the farmer in mind. p. 167–171. In Proc. Am. Forage Grassl. Counc., Lafayette, LA. 26–30 Apr. 2003. AFGC, Georgetown, TX.

Bouton, J.H., G.C.M. Latch, N.S. Hill, C.S. Hoveland, M.A. McCann, R.H. Watson, J.A. Parish, L.L. Hawkins, and F.N. Thompson. 2002. Reinfection of tall fescue cultivars with non-ergot alkaloid-producing endophytes. Agron. J. 94:567–574.

Bransby, D.I. 1997. Steer weight gain responses to ivermectin when grazing fescue. Large Anim. Pract. 18(3):16–19.

Bransby, D.I., S.P. Schmidt, W. Griffey, and J.T. Eason. 1988. Heavy grazing is best for infected fescue. Alabama Agric. Exp. Stn. Highlights Agric. Res. 35(4):12.

Brazle, F.K., and K.P. Coffey. 1991. Effect of zeranol on performance of steers grazing high- and low-endophyte tall fescue pastures. Prof. Anim. Sci. 7:39–42.

Brown, M.A., A.H. Brown, Jr., W.G. Jackson, and J.R. Miesner. 1997. Genotype × environment interactions in Angus, Brahman, and reciprocal cross cows and their calves grazing common bermudagrass and endophyte-infected tall fescue pastures. J. Anim. Sci. 75:920–925.

Bush, L.P., J. Boling, and S. Yates. 1979. Animal disorders. p. 247–292. In R.C. Buckner and L.P. Bush (ed.) Tall fescue. Agron. Monogr. No. 20. ASA, CSSA, and SSSA, Madison, WI.

Chestnut, A.B., P.D. Anderson, M.A. Cochran, H.A. Fribourg, and K.D. Gwinn. 1992. Effects of hydrated sodium calcium aluminosilicate on fescue toxicosis and mineral absorption. J. Anim. Sci. 70:2838–2846.

Chestnut, A. B., H.A. Fribourg, J.B. McLaren, D.G. Keltner, B.B. Reddick, R.J. Carlisle, and M.C. Smith. 1991. Effects of Acremonium coenophialum infestation, bermudagrass, and nitrogen or clover on steers grazing tall fescue pastures. J. Prod. Agric. 4:208–213.

Coffey, K., W. Coblentz, T. Smith, I.D. Hubbell, D. Scarbrough, B. Humphry, and C. Rosenkrans. 2003. Impact of rotation frequency and weaning date on performance by fall-calving cow-calf pairs grazing endophyte-infected tall fescue pastures. Arkansas Agric. Exp. Stn. Res. Ser. 509:91–94.

Coffey, K.P., L.W. Lomas, and J.L. Moyer. 1990. Grazing and subsequent feedlot performance by steers that grazed different types of fescue pasture. J. Prod. Agric. 3:415–420.

Coffey, K.P., J.L. Moyer, L.W. Lomas, J.E. Smith, D.C. La Rue, and F.K. Brazle. 1992. Implant and copper oxide needles for steers grazing Acremonium coenophialum-infected tall fescue pastures: Effects on grazing and subsequent feedlot performance and serum constituents. J. Anim. Sci. 70:3203–3214.

Cole, N.A., J.A. Stuedemann, and F.N. Thompson. 2001. Influence of both endophyte infestation in fescue pastures and calf genotype on subsequent feedlot performance of steers. Prof. Anim. Sci. 17:174–182.

Crawford, R.J., and G.B. Garner. 1991. Evaluation of corn gluten feed and fermenting to ameliorate the negative impact of endophyte-infected tall fescue on cattle performance. p. 255–258. In Proc. Am. Grassl. Counc., Columbia, MO. 1-4 Apr. 1991. AFGC, Georgetown, TX.

Cross, D.L. 2001. Toxic effects of Neotyphodium coenophialum in cattle and horses. p. 219–235. In V.H. Paul and P.D. Dapprich (ed.) The Grassland Conf. 2000. Proc. Int. Neotyphodium/Grass Interactions Symp, 4th, Soest, Germany. 27–29 Sept. 2000. Univ. Paderborn, Germany.

Cross, D.L., K. Anas, W.C. Bridges, and J.H. Chappell. 1999. Clinical effects of domperidone on fescue toxicosis in pregnant mares. Proc. Am. Assoc. Equine Pract. 45:203–206.

Davenport, G.M., J.A. Boling, and C.H. Rahe. 1993. Growth and endocrine responses of cattle to implantation of estradiol-17B during continuous or discontinuous grazing of high- and low-endophyte-infected tall fescue. J. Anim. Sci. 71:757–764.

Dennis, S.B., V.G. Allen, K.E. Saker, J.P. Fontenot, J.Y.M. Ayad, and C.P. Brown. 1998. Influence of Neotyphodium coenophialum on copper concentration in tall fescue. J. Anim. Sci. 76:2687–2693.

Dougherty, C.T., L.M. Lauriault, N.W. Bradley, N. Gay, and P.L. Cornelius. 1991. Induction of tall fescue toxicosis in heat-stressed cattle and its alleviation with thiamin. J. Anim. Sci. 69:1008–1018.

Erdo, S.L. 1990. GABA outside the CNS. Springer-Verlag, New York.

Essig, H.W., and T.W. Armstrong. 1990. Evaluation of additives to endophyte-infected fescue hay on growth performance and body temperature of steers. Prof. Anim. Sci. 6:11–14.

Fike, J.H., V.G. Allen, R.E. Schmidt, X. Zhang, J.P. Fontenot, C.P. Bagley, R.L. Ivy, R.R. Evans, R.W. Coelhio, and D.B. Wester. 2001. Tasco-Forage: I. Influence of a seaweed extract on antioxidant activity in tall fescue and in ruminants. J. Anim. Sci. 79:1011–1021.

Filipov, N.M., F.N. Thompson, N.S. Hill, D.L. Dawe, J.A. Stuedemann, J.C. Price, and C.K. Smith. 1998. Vaccination against ergot alkaloids and the effect of endophyte-infected fescue seed-based diets on rabbits. J. Anim. Sci. 76:2456–2463.

Forcherio, J.C., G.E. Carlett, J.A. Paterson, M.S. Kerley, and M.R. Ellersieck. 1995. Supplemental protein and energy for beef cows consuming endophyte-infected tall fescue. J. Anim. Sci. 73:3427–3436.

Forcherio, J.C., M.S. Kerley, and J.A. Paterson. 1992. Performance of steers grazing tall fescue during the spring followed by warm-season grasses during the summer. J. Anim. Sci. 70(Suppl. 1):182.

Garner, G.B., and C.N. Cornell. 1978. Fescue foot in cattle. p. 45–62. In T.D. Wylie and L.G. Morehouse (ed.) Mycotoxic fungi, mycotoxins, mycotoxicoses. An encyclopedic handbook. Vol. 2. Marcel Dekker, New York.

Garner, G.B., and C.N. Cornell. 1985. Cattle response to the endophyte of tall fescue. Am. Assoc. Vet. Laboratory Diagnosticians 28:145–154.

Goetsch, A.L., K.L. Landis, G.E. Murphy, B.L. Morrison, Z.B. Johnson, E.L. Piper, A.C. Hardin, and K.L. Hall. 1988. Supplements, parasite treatments, and growth implants for Brahman or English crossbred steers grazing endophyte-infected or noninfected fescue in the spring and fall. Prof. Anim. Sci. 4:32–38.

Gwinn, K.D., H.A. Fribourg, J.C. Waller, A.M. Saxton, and M.C. Smith. 1998. Changes in Neotyphodium coenophialum infestation levels in tall fescue pastures due to different grazing pressures. Crop Sci. 38:201–204.

Hancock, D.L., J.E. Williams, H.B. Hedrick, E.E. Beaver, D.K. Larrick, M.R. Ellersieck, G.B. Garner, R.E. Morrow, J.A. Paterson, and J.R. Gerrish. 1987. Performance, body composition and carcass characteristics of finishing steers as influenced by previous forage systems. J. Anim. Sci. 65:1381–1391.

Hill, N.S., F.N. Thompson, J.A. Stuedemann, D.L. Dawe, and E.E. Hiatt III. 2000. Urinary alkaloid excretion as a diagnostic tool for fescue toxicosis in cattle. J. Vet. Diag. Invest. 12:210–217.

Hoveland, C.S., R.R. Harris, E.E. Thomas, E.M. Clark, J.A. McGuire, J.T. Eason, and M.E. Ruf. 1981. Tall fescue with ladino clover or birdsfoot trefoil as pasture for steers in northern Alabama. Bull. 530. Alabama Agric. Exp. Stn., Auburn.

Hoveland, C.S., S.P. Schmidt, C.C. King, Jr., J.W. Odom, E.M. Clark, J.A. McGuire, L.A. Smith, H.W. Grimes, and J.L. Holliman. 1983. Steer performance and association of Acremonium coenophialum fungal endophyte on tall fescue pasture. Agron. J. 75:821–824.

Jackson, J.A., Jr, Z. Sorgho, and R.H. Hatton. 1988. Effect of nitrogen fertilization or urea addition and ensiling as large round bales of endophyte infected tall fescue on fescue toxicosis when fed to dairy calves. Nutr. Rep. Int. 37:335–345.

Jones, K.L., S.S. King, K.E. Griswold, D. Cazac, and D.L. Cross. 2003. Domperidone can ameliorate deleterious reproductive effects and reduced weight gain associated with fescue toxicosis in heifers. J. Anim. Sci. 81:2568–2574.

Kallenbach, R.L., G.J. Bishop-Hurley, M.D. Massie, G.E. Rottinghaus, and C.P. West. 2003. Herbage mass, nutritive value, and ergovaline concentration of stockpiled tall fescue. Crop Sci. 43:1001–1005.

Kerr, L.A., C.P. McCoy, C.R. Boyle, and H.W. Essig. 1990. Effects of ammoniation of endophyte fungus-infected fescue hay on serum prolactin concentration and rectal temperature in beef cattle. Am. J. Vet. Res. 51:76–78.

Lacefield, G., D. Ball, and J. Henning. 1993. Results of two-state survey of county agents regarding status of endophyte-infected and endophyte-free tall fescue. p. 22–25. Proc. Southern Pasture and Forage Crop Improvement Conf., 49th, Longboat Key, FL. 14–16 June 1993.

LaGrange, P.H., and A. Capron. 1978. Immune response directed against infections and parasitic agents. p. 410–444. In J.F. Bach (ed.) Immunology. John Wiley & Sons, New York.

Lauriault, L.M., C.T. Dougherty, N.W. Bradley, and P.L. Cornelius. 1990. Thiamin supplementation and the ingestive behavior of beef cattle grazing endophyte-infected tall fescue. J. Anim. Sci. 68:1245–1253.

Lipham, L.B., F.N. Thompson, J.A. Stuedemann, and J.L. Sartin. 1989. Effects of metoclopramide on steers grazing endophyte-infected fescue. J. Anim. Sci. 67:1090–1097.

Malinowski, D.P., and D.P. Belesky. 2000. Adaptations of endophyte-infected cool-season grasses to environmental stresses: Mechanisms of drought and mineral stress tolerance. Crop Sci. 40:923–940.

Malinowski, D.P., D. P Belesky, N.S. Hill, V.C. Baligar, and J.M. Fedders. 1998. Influence of phosphorus on the growth and ergot alkaloid content of *Neotyphodium coenophialum*-infected tall fescue (*Festuca arundinacea* Schreb.). Plant Soil 198:53–61.

McMurphy, W.E., K.S. Lusby, S.C. Smith, S.H. Muntz, and C.A. Strasia. 1990. Steer performance on tall fescue pasture. J. Prod. Agric. 3:100–102.

Monroe, J.L., D.L. Cross, L.W. Hudson, D.M. Henricks, S.W. Kennedy, and W.C. Bridges, Jr. 1988. Effect of selenium and endophyte-contaminated fescue on performance and reproduction in mares. J. Equine Vet. Sci. 8:148–154.

Montgomery, J.L., V.G. Allen, K.R. Pond, M.F. Miller, D.B. Wester, C.P. Brown, R. Evans, C.P. Bagley, R.L. Ivy, and J.P. Fontenot. 2001. Tasco-Forage: IV. Influence of a seaweed extract applied to tall fescue pastures on sensory characteristics, shelf-life, and vitamin E status in feedlot-finished steers. J. Anim. Sci. 79:884–894.

Newsome, J.E., H.W. Essig, K.P. Boykin, and S.L. Hamlin. 1989. Performance of steers fed heated and (or) ammoniated endophyte infected fescue hay. Prof. Anim. Sci. 5:36–38.

Nihsen, M.E., E.L. Piper, C.P. West, R.J. Crawford, Jr., T.M. Denard, Z.B. Johnson, C.A. Roberts, D.A. Spiers, and C.P. Rosenkrans, Jr. 2004. Growth rate and physiology of steers grazing tall fescue inoculated with novel endophytes. J. Anim. Sci. 82:878–883.

Parish, J.A., M.A. McCann, R.H. Watson, C.S. Hoveland, L.L. Hawkins, N.S. Hill, and J.H. Bouton. 2003a. Use of nonergot alkaloid-producing endophytes for alleviating tall fescue toxicosis in sheep. J. Anim. Sci. 81:1316–1322.

Parish, J.A., M.A. McCann, R.H. Watson, N.N. Paiva, C.S. Hoveland, A.H. Parks, B.L. Upchurch, N.S. Hill, and J.H. Bouton. 2003b. Use of nonergot alkaloid-producing endophytes for alleviating tall fescue toxicosis in stocker cattle. J. Anim. Sci. 81:2856–2868.

Pratt, A.D., and R.R. Davis. 1954. Kentucky-31 fescue. Ohio Farm Home Res. 39:93–94.

Pratt, A.D., and J.L. Haynes. 1950. Herd performance on Kentucky-31 fescue. Ohio Farm Home Res. 35:10–11.

Redmond, L.M., D.L. Cross, J.R. Strickland, and S.W. Kennedy. 1994. Efficacy of domperidone and sulpiride as treatment for fescue toxicosis in horses. Am. J. Vet. Res. 55:722–729.

Rhodes, M.T., J.A. Paterson, M.S. Kerley, H.E. Garner, and M.H. Laughlin. 1991. Reduced blood flow to peripheral and core body tissues in sheep and cattle induced by endophyte-infected tall fescue. J. Anim. Sci. 69:2033–2043.

Rosenkrans, Jr., C., T. Bedingfield, and E. Piper. 2001. Physiological responses of steers to ingestion of endophyte-infected fescue hay, ivermectin treatment and immune challenge.

p. 255–260. *In* V.H. Paul and P.D. Dapprich (ed.) The Grassland Conference 2000. Proc. Int. *Neotyphodium*/Grass Interactions Symposium, 4th, Soest, Germany. 27–29 Sept. 2000. Univ. Paderborn, Germany.

Rottinghaus, G.E., G.B. Garner, C.N. Cornell, and J.L. Ellis. 1991. HPLC method for quantitating ergovaline in endophyte-infested tall fescue: Seasonal variation of ergovaline levels in stems with leaf sheaths, leaf blades, and seed heads. J. Agric. Food Chem. 39:112–125.

Saker, K.E., V.G. Allen, J.P. Fontenot, C.P. Bagley, R.L. Ivy, R.R. Evans, and D.B. Wester. 2001. Tasco-Forage: II. Monocyte immune cell response and performance of beef steers grazing tall fescue treated with a seaweed extract. J. Anim. Sci. 79:1022–1031.

Saker, K.E., V.G. Allen, J. Kalnitsky, C.D. Thatcher, W.S. Swecker, Jr., and J.P. Fontenot. 1998. Monocyte immune cell response and copper status in beef steers that grazed endophyte-infected tall fescue. J. Anim. Sci. 76:2694–2700.

Salminen, S.O., P.S. Grewal, and M.F. Quigley. 2003. Does mowing height influence alkaloid production in endophytic tall fescue and perennial ryegrass? J. Chem. Ecol. 29:1319–1328.

Schmidt, S.P. and T.G. Osborn. 1993. Effects of endophyte-infected tall fescue on animal performance. Agric. Ecosyst. Environ. 44:233–262.

Shelby, R.A., and L.W. Dalrymple. 1987. Incidence and distribution of the tall fescue endophyte in the United States. Plant Dis. 71:783–786.

Sleper, D.A., H.F. Mayland, R.J. Crawford, Jr., G.E. Shewmaker, and M.D. Massie. 2002. Registration of HiMag tall fescue germplasm. Crop Sci. 42:318.

Solomons, R.N., J.W. Oliver, and R.D. Linnabary. 1989. Reactivity of dorsal pedal vein of cattle to selected alkaloids associated with *Acremonium coenophialum* infected fescue grass. Am. J. Vet. Res. 50:235–238.

Stricker, J.A., A.G. Matches, G.B. Thompson, V.E. Jacobs, F.A. Martz, H.N. Wheaton, H.D. Currence, and G.F. Krause. 1979. Cow-calf production on tall fescue-ladino clover pastures with and without nitrogen fertilization or creep-feeding: Spring calves. J. Anim. Sci. 48:13–25.

Stuedemann, J.A., D.L. Breedlove, and D.H. Seman. 1993. Raising replacement beef heifers in the South—Birth to parturition: A review of current practices. J. Anim. Sci. 71:3131–3137.

Stuedemann, J.A., and C.S. Hoveland. 1988. Fescue endophyte: History and impact on animal agriculture. J. Prod. Agric. 1:39–44.

Stuedemann, J.A., T.S. Rumsey, J. Bond, S.R. Wilkinson, L.P. Bush, D.J. Williams, and A.B. Caudle. 1985. Association of blood cholesterol with the occurrence of fat necrosis in cows and tall fescue summer toxicosis in steers. Am. J. Vet. Res. 46:1990–1995.

Stuedemann, J.A., and F.N. Thompson. 1993. Management strategies and potential opportunities to reduce the effects of endophyte-infested tall fescue on animal performance. p. 103–114. *In* D.E. Hume et al. (ed.) Proc. Int. Symposium on *Acremonium*/Grass Interactions, 2nd, Palmerston North, New Zealand. 4–6 Feb. 1993. AgResearch, Grassland Research Centre, Palmerston North.

Stuedemann, J.A., S.R. Wilkinson, D.J. Williams, H. Ciordia, J.V. Ernst, W.A. Jackson, and J.B. Jones, Jr. 1975. Long-Term Broiler Litter Fertilization of Tall Fescue Pastures and Health and Performance of Beef Cows. Managing Livestock Wastes. p. 264–268. *In* Proc. Int. Symp. Livestock Wastes, 3rd, Champaign, IL. 21–25 Apr. 1975. ASAE, St. Joseph, MI.

Thompson, R.W., H.A. Fribourg, J.C. Waller, W.L. Sanders, J.H. Reynolds, J.M. Phillips, S.P. Schmidt, R.J. Crawford, Jr., V.G. Allen, D.B. Faulkner, C.S. Hoveland, J.P. Fontenot, R.J. Carlisle, and P.P. Hunter. 1993. Combined analysis of tall fescue steer grazing studies in the eastern United States. J. Anim. Sci. 71:1940–1946.

Thompson, F.N., J.A. Stuedemann, and N.S. Hill. 2001. Anti-quality factors associated with alkaloids in eastern temperate pasture. J. Range Manage. 54:474–489.

Timper, P., R.N. Gates, and J.H. Bouton. 2001. Nematode reproduction in tall fescue infected with different endophyte strains. J. Nematol. 33:280.

Tucker, C.A., R.E. Morrow, J.R. Gerrish, C.J. Nelson, G.B. Garner, V.E. Jacobs, W.G. Hires, J.J. Shinkle, and J.R. Forwood. 1989. Forage systems for beef cattle: Effect of winter supplementation and forage system on reproductive performance of cows. J. Prod. Agric. 2:217–221.

Turner, J.W. 1980. Genetic and biological aspects of Zebu adaptability. J. Anim. Sci. 50:1201–1205.

Turner, K.E., J.A. Paterson, M.S. Kerley, and J.R. Forwood. 1990. Mefluidide treatment of tall fescue pastures: Intake and animal performance. J. Anim. Sci. 68:3399–3405.

Turner, M.J., and J.M. Schaeffer. 1989. Mode of action of ivermectin. p. 73. *In* W.C. Campbell (ed.) Ivermectin and abamectin. Springer-Verlag, New York.

Vicini, J.L., E.C. Prigge, W.B. Bryan, and G.A. Varga. 1982a. Influence of forage species and creep grazing on a cow-calf system. I. Intake and digestibility of forages. J. Anim. Sci. 55:752–758.

Vicini, J.L., E.C. Prigge, W.B. Bryan, and G.A. Varga. 1982b. Influence of forage species and creep grazing on a cow-calf system. II. Calf production. J. Anim. Sci. 55:759–764.

Watson, R.H., M.A. McCann, J.A. Bondurant, J.H. Bouton, C.S. Hoveland, and F.N. Thompson. 2001. Liveweight and growth rate of cow-calf pairs grazing tall fescue pastures infected with either non-toxic (MaxQ^{TM}) or toxic endophyte strains. J. Anim. Sci. 79 (Suppl. 1):220.

Watson, R.H., M.A. McCann, J.A. Parish, C.S. Hoveland, F.N. Thompson, J.H. Bouton. 2004. Productivity of cow-calf pairs grazing tall fescue pastures infected with either the wild-type endophyte or a non-ergot alkaloid-producing endophyte strain, AR542. J. Anim. Sci. 82:(In Press).

West, C.P., M.L. Marlett, M.E. McConnell, E.L. Piper, and T.J. Kring. 1998. Novel endophyte technology: Selection for the fungus. p. 105–115. *In* E.C. Brummer et al. (ed.) Molecular and cellular technologies for forage improvement. CSSA Spec. Publ. 26. CSSA, Madison, WI.

Williams, D.J., D.E. Tyler, and E. Papp. 1969. Abdominal fat necrosis as a herd problem in Georgia cattle. J. Am. Vet. Med. Assoc. 154:1017–1026.

Section V

Technology Transfer and Quality Assurance

ENDOPHYTES IN FORAGE CULTIVARS

Joe Bouton[1] and Syd Easton[2]

The agricultural importance of *Neotyphodium* endophytes has been recognized for many years. Associations of these endophytes with the widely planted pasture and turf species tall fescue (*Festuca arundinacea*) and perennial ryegrass (*Lolium perenne*) have also formed the basis for much of the information presented at international symposia on *Neotyphodium* endophytes. Since commercialization of endophytes occurred mainly in cultivars of these two grass species (Bouton and Hopkins, 2003), this current discussion will concentrate on them.

ENDEMIC SITUATION IN TALL FESCUE

Tall fescue is an important pasture species in all temperate regions of the world. In the USA, pastures are highly infected and the agronomic benefits of using endemic, endophyte-infected (E+) versions of this grass are recognized due mainly to the introduction and use of the E+ cultivar 'Kentucky 31' (Bouton, 2000). In Europe, pastures are low to moderately infected, but the commercial seed supply is highly infected (Lewis, 1997). In contrast, South American (DeBattista et al., 1997) and New Zealand (Hume, 1993) tall fescue cultivars were introduced mainly as endophyte-free (E–).

In a survey by Lacefield et al. (1993), 80% of U.S. tall fescue acreage was infected with *N. coenophialum* at a mean infection rate of 76%, and 50% of the respondents considered their new plantings with endophyte-free (E–) cultivars to be failures. In the same survey, it was found that the main reasons farmers do not convert to E– pastures were lack of confidence in E– cultivars and a perception that the benefits do not outweigh costs. From these results, it was concluded that *N. coenophialum* is critical for the persistence and increased *ecological fitness* of this grass species (Bouton, 2000).

---

[1] Forage Improvement Division, Noble Foundation, Ardmore, OK 73402
[2] AgResearch Grasslands, Private Bag 11008, Palmerston North, New Zealand

Infection of Kentucky 31 with its endemic *N. coenophialum* in most cases leads to a condition called *fescue toxicosis*. Fescue toxicosis is generally characterized by poor weight gain and reproduction in ruminant animals (Stuedemann and Thompson, 1993) ingesting ergot alkaloids derived from the endophyte association (Hill et al., 1994). Livestock grazing tall fescue pastures in the USA, therefore, probably suffer from some degree of fescue toxicity.

Although released in 1943 (Buckner, 1985), Kentucky 31 is still the most widely planted U.S. cultivar today and has apparently been highly infected since its release (Bouton, 2000). Even though other cultivars were released during the four decades after Kentucky 31, none assumed a dominant role in the commercial seed trade. This was probably because Kentucky 31 had already been established on millions of hectares and was doing a good job as a conservation crop in preventing erosion in these major land areas, but also probably because of the confounding effects of the endophyte. A good example of this may be speculated with the cultivar Kenwell. Although it demonstrated good palatability during selection and initial testing, there were third-party trials that put the cultivar in a bad light. An excerpt from the Virginia Forage Research Station report (Anonymous, 1969) demonstrates this:

> Kenwell, a new and apparently palatable fescue variety, proved to be inferior to Ky. 31. During a four-year comparison, the 14 percent higher milk yield per cow grazing Ky. 31 fescue pastures was attributed to less pasture consumption for Kenwell. During the last two years, the cows grazing Kenwell had severe fescue foot, and some of the cows died.

From this description, it is easy to conclude that the Kenwell pastures were highly infected, while the seed source that was used to plant the Kentucky 31 pastures was lightly infected or even E−. Kenwell never assumed a dominant role in the commercial seed trade probably due to this confounding.

By 1985, and surely after 1985, with the clear association of fescue toxicosis with *Neotyphodium* endophytes in Kentucky 31, the trend among breeders was to develop E− cultivars for pastures, but as mentioned above, this approach was limited from a commercial point of view by the poor agronomic performance of the E− cultivars (Bouton, 2000). In the coastal plain region of the southern USA, however, where tall fescue is not currently used due to poor summer survival, E+ germplasms selected for persistence demonstrated increased summer survival (Bouton et al., 1993a). These findings resulted in the release of 'Georgia-5' tall fescue as an E+ cultivar for use in this region of the country (Bouton et al. 1993b). When grown in mixtures with bermudagrass (*Cynodon dactylon*) and bahiagrass (*Paspalum notatum*), E+ Georgia 5 gave good winter gains on beef heifers with no hay feeding; animal toxicities were acceptable under these management conditions (Gates et al., 1999).

ENDEMIC SITUATION IN PERENNIAL RYEGRASS

Perennial ryegrass is naturally infected with *Neotyphodium lolii*, giving the plant several agronomic advantages, but the presence of ergot and tremorgenic alkaloids can cause severe problems in both cattle and sheep. In most of New Zealand, endophyte-free ryegrass pastures do not persist, and almost all old pastures show high infection frequency (Easton, 1999), but this is not the case everywhere. In particular, perennial ryegrass in commerce in the European Community is free of endophyte, many ryegrass pastures are uninfected (Lewis, 1997), and even wild populations may be E– (Lewis et al., 1997). The most obvious symptom of endophyte toxicosis in stock grazing infected perennial ryegrass is ryegrass staggers (RGS) (Fletcher and Harvey, 1981). The active factor produced by endophyte causing RGS is lolitrem B (Gallagher et al., 1981), and ergovaline has been shown to act synergistically with it (Fletcher and Easton, 1997). Livestock are affected in other ways, and an inverse linear relation has been shown between lamb growth rates and percentage infection of a sward with endemic endophyte (Fletcher and Easton, 1997).

Some perennial ryegrass cultivars have a reputation as unusually toxic. Lean (2001) criticized 'Grasslands Impact' as particularly likely to cause RGS. However, analysis of samples from several plot trials has shown ergovaline and lolitrem B levels comparable with those of other cultivars. This cultivar is later flowering than most available in Australia and New Zealand, and it forms a dense sward with a very high ryegrass content, so that management factors may contribute to the incidence of RGS. Alternatively, there may be a factor exacerbating RGS that is not being analyzed.

A clearly documented difference between cultivars arose from a trial evaluating 'Aries HD'. This cultivar was released with a claim for higher digestibility, and a trial to verify this found that sheep suffered significantly less RGS than those grazing the control cultivar, 'Yatsyn 1' (Bluett et al., 1999). Chemical analysis found no difference in herbage concentrations of lolitrem B, the compound directly responsible for RGS, but Aries herbage contained less ergovaline.

Old populations of perennial ryegrass taken from fields that have not been cultivated for several decades sometimes prove to have very high levels of fungal toxins. A survey of farms following up sowings of a selected endophyte with higher concentrations of ergovaline than expected (see below) found that fields sown with that product did have higher concentrations than fields recently sown with proprietary cultivars infected with endemic endophyte (Easton et al., 1993). However, old fields on the same farms had herbage ergovaline concentrations significantly higher than what was considered acceptable in the seed market.

Cultivars and old populations are genetically different, but also they are usually independently derived, and so their endemic endophyte may not be the same. There is no information on whether or not endophyte strains coexist in a host population. Certainly, comparisons between cultivars or populations with endemic endophyte cannot distinguish between effects due to host genetics and effects due to differences in endophyte strain.

REINFECTION OF ELITE CULTIVARS WITH NOVEL ENDOPHYTES

Currently, research programs are under way to either improve cultivars and (or) endophytes in order to keep the positive agronomic aspects of infection while reducing toxicity. These include more persistent E− cultivars, reduction of alkaloid levels via selection within current E+ plant–endemic strains populations, isolation and genetic manipulation of E+ strains to eliminate toxic alkaloid pathways with reinfection later into elite cultivars, and selection and reinfection of naturally occurring, nontoxic strains into elite cultivars (Bouton and Hopkins, 2003). In Fig. 15.1, an overview of various strategies for development of endophyte–cultivar combinations is shown. The approach of selection and reinfection of nontoxic strains into elite cultivars will be the focus of our remaining discussion, as it has achieved the most recent commercial success. Since the term nontoxic actually means nil levels of a specific toxin (e.g., ergot alkaloids, lolitrem B, etc.), and does not account for other unidentified toxins, the more popular current term for these strains is *novel*.

Selected Endophytes in Tall Fescue

In the first proof of concept studies, reinfection of two tall fescue cultivars, 'Jesup' (Bouton et al., 1997) and Georgia 5, with a patented but naturally occurring non-ergot alkaloid-producing endophyte strain, AR542, was found to eliminate production of these known toxic alkaloids and provide better animal

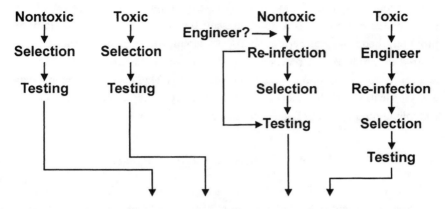

New Cultivar/Novel Endophyte Combination

Fig. 15.1. Strategies for cultivar development with fungal endophytes. In all cases, "nontoxic" means incapable of producing a specific toxic alkaloid(s) and "selection" designates plant selection.

performance and pasture stand survival (Bouton et al., 2000, 2002). In this strategy, endophyte strains with nil production of total ergot alkaloids, yet capable of producing pest-deterring peramine and loline alkaloids, were isolated from field grown plants collected throughout the world (Latch, 1993, 1997). Different cultivar/non-ergot-producing strain combinations were tested against the E+ and E– versions of the same cultivars for stand survival and dry matter yield at multiple locations; while in separate experiments, they were assessed for toxicity in lambs (Bouton et al., 2000, 2002). The best combination, Jesup (AR542), produced no toxic ergot alkaloids and possessed stand survival better than the E– checks and equal to the E+ check. Gains for lambs on both cultivars containing AR542 were equivalent to those from E– forage, but much greater than animals on E+ forage (Parish et al., 2003a). Lambs consuming E– forage or forage from AR542 also did not exhibit the depressed serum prolactin or elevated body temperatures of animals consuming E+ forage. Further testing with beef cattle, the primary commercial livestock group in the USA, showed gains with AR542 to be equivalent to gains on E– pastures, and much higher than animals on E+ forage (Parish et al., 2003b). The AR542 strain was therefore successful in making safe substitution for the endemic toxic strain of both cultivars and is marketed under the trade name of "MaxQ" (Bouton et al., 2000, 2002).

Overall, these studies demonstrated that reinfection of naturally occurring, novel endophyte strains into elite tall fescue cultivars after removal of their toxic, endemic strain(s) is a good strategy for eliminating fescue toxicity and providing better animal performance yet keeping the agronomic characteristics desired by producers. A recent report of successfully isolating nontoxic strains and reinfecting them into the tall fescue cultivar HiMag also provides independent verification of the success of this strategy (Nihsen et al., 2004). This endophyte product in HiMag is sold under the trade name ArkPlus[TM].

Selected Endophytes in Perennial Ryegrass

Screening in perennial ryegrass has sought strains that do not produce ergovaline or lolitrem B. In the majority of cases, strains producing no ergovaline or lolitrem B in a single sampling in a glasshouse experiment with their natural host plant have proved to not produce the alkaloid in other host plants and in other environments (B.A. Tapper and G.C.M. Latch, 2003, personal communication). The capacity to produce the molecules is not present. In one case, however, the production of ergovaline and peramine, and the amount of mycelia in the host as determined by semiquantitative ELISA, were shown to be subject to host genetic control (Easton et al., 2002).

Endophyte strains may differ quantitatively in the concentrations of alkaloid accumulated in host herbage. The first strain commercialized in New Zealand as free of lolitrem B proved to produce ergovaline in proprietary cultivars at about twice the concentrations found in those same cultivars with their endemic endophyte strains (Easton et al., 1993). As described above, many old fields with endemic endophytes had similar elevated levels, with the extra disadvantage of also producing lolitrem B.

In all perennial ryegrass cultivars studied, the same elevated concentrations of ergovaline were measured. However, in 'Grasslands Greenstone', a tetraploid hybrid cultivar [Lolium (L. perenne × L. multiflorum) × L. perenne], ergovaline concentrations were not unusually high, even though peramine concentrations were similar to those in the perennial ryegrasses.

The endophyte strain for which we have the most information is AR1. After intensive testing of agronomic performance (Popay et al., 2000) and health of livestock grazing swards (Fletcher and Easton, 2000), AR1 was commercialized in significant quantities from early 2003, and immediately captured a major share of the proprietary ryegrass seed market in New Zealand (Easton and Tapper, Chapter 1.3, this volume). Every cultivar association with AR1 has been individually tested with sheep to verify that the nontoxic properties of AR1 are recovered in the association. Many cultivars and breeding populations have been inoculated. In 2003, nearly 8000 seedlings of perennial ryegrass were inoculated with AR1. From this, 1700 successfully infected plants were recovered, giving a success rate of 22% (A. de Bonthe and W.R. Simpson, 2003, personal communication). However, rates of success vary (Easton et al., 2000), with results for individual seed lines varying from 3.5 to above 50%. Potentially, this could lead to the newly infected subpopulation not being an unbiased representation of the original cultivar. There have been unpublished reports of drifts in susceptibility to crown rust (*Puccinia coronata*) and in seedling fluorescence, a standard test for contamination with *L. multiflorum*. Again, where there have been large experiments, they have shown little evidence for this (Easton, 1993), but a strategy for reselection may be necessary to overcome these genetic shifts (Fig. 15.1).

However, a study of half-sibling families in a breeding population did suggest that the subpopulation infected with AR1 had a weak tail in the distribution. Programs of selection within populations infected with AR1 are now pursued by several organizations in New Zealand (Easton et al., 2000).

TECHNICAL REQUIREMENTS DURING DEVELOPMENT AND COMMERCIALIZATION OF ENDOPHYTES

Common technical issues during commercialization of endophytes include screening methods to assess both amount and type of infection, maintaining endophyte viability during seed increase and dissemination, conducting the requisite agronomic and animal testing before actual release of new cultivars, and intellectual property issues and overall costs of commercialization (Bouton and Hopkins, 2003).

Systems of determination of endophyte presence and alkaloid concentration have been regularly refined, allowing efficient handling of large sample sets. Use of ELISA to detect endophyte was an early development. While identification of toxin-free endophyte strains was achieved using HPLC, development of immunoblot procedures for detecting endophyte presence (Hiatt et al., 1997) and ELISA methods for assessment of ergot alkaloids (Adcock et al., 1997) were very important during refinement and evaluation of MaxQ. Use of detection techniques based on molecular markers such as microsatellites, due to their specificity, should also be considered when considering the proper detection method.

Maintaining endophyte viability during the breeding, seed production, marketing, and on-farm establishment is a problem due to endophytes dying in the seed quicker than the embryo itself. Rolston et al. (1993) reported low seed moisture (10% seed moisture content) and low temperatures (5°C) were effective in maintaining endophyte viability during seed storage. Paul et al. (2000) found three main causes for loss of endophyte infection in commercial cultivars. First, there is a 5.5% per generation cycle loss during the generative or seed increase phase. Second, phenotypic selection for *general appearance* desired by most European breeders resulted in loss of colonization density of 13% per cycle of selection. Third, the common practice of seed storage in open bins without air conditioning led to significant loss in endophyte infection and viability.

Other common practices in the commercial seed industry may also result in poor levels of endophyte infection or viability. In one instance (J.H. Bouton, 1995, unpublished data), two seed lots of E+ Georgia 5 tall fescue varying in time of swathing (early or late) were followed for endophyte viability. Although the fields providing seed for these two lots were from the same Oregon location, and were nearly identical in infection levels, the early swathed lot had less initial endophyte and subsequently loss viable infection at a faster rate than the later swathed seed lot. Similar problems have occurred with some systemic fungicide use to control common rust diseases in the seed fields (N.S. Hill, 2002, personal communication). The best approach for the seed industry, therefore, is to view the endophyte, while in the seed, as a very fragile organism that can be lost or injured when mistreated. This will require completely different seed production practices from those in current use.

Since the main basis for any endophyte commercialization will be through a plant cultivar, the testing phase before actual release into the commercial seed trade is the most time- and resource-consuming process. For commercial forage cultivars with novel endophytes, there is an added requirement of both standard agronomic testing combined with trials of animal performance and (or) toxicity.

Animal responses to tall fescue endophytic toxins can be grouped into four categories (Stuedemann and Thompson, 1993): (i) decreased productivity (weight gain, milk production, and pregnancy rate), (ii) behavioral criteria demonstrated with decreased feed but increased water intake, (iii) physiological responses such as increased respiration and elevated rectal and core body temperatures, and (iv) sera or plasma levels of constituents such as decreased serum prolactin and cholesterol. One or more of these responses are usually measured when assessing fescue toxicity (Stuedemann and Thompson, 1993). Similar types of studies are needed for perennial ryegrass infected with nontoxic *N. lolii* strains (Fletcher and Easton, 2000). Toxicity trials to measure these responses are therefore expensive and resource consuming, but they must be conducted if the commercial product is to be sold as nontoxic.

From an agronomic point of view, the most successful combinations should provide the advantages of E+ cultivars. Agronomic testing for new cultivar–reinfected strain combinations therefore must be rigorous, examine the main environmental stresses for the region of interest (biotic and abiotic), contain the best checks (e.g., E+ and E– entries), and be realistically applied and proven. In addition to measuring yield and persistence across broadly based environments, there is also a need to conduct trials that assess response to specific stresses, such as insects, diseases, and nematodes, that can individually lead to poor plant survival and performance. For example, although the AR1 endophyte when placed in commercial perennial ryegrass cultivars has met the general criteria for success, such as acceptable animal and agronomic performance, it has shown some susceptibility to African black beetle (Popay and Baltus, 2001). This has caused any commercialization effort to be combined with an educational effort so that farmers know they may have to control African black beetle in areas where this pest is a problem.

Such rigorous testing and associated quality control are expensive, and follow an expensive development process. Patent protection is also being sought for endophytes. This patenting process could potentially impact farmers in two ways. First, it will result in higher seed costs. Second, if intellectual property protection is granted and enforced on endophytes, it will eliminate a farmer's ability to harvest and sell seed from his own fields unless he has clearance from the patent holders. However, in the current market, most grass seed is sold as a cheap commodity product and farmer-produced seed is common. Recent attempts to market newly released forage cultivars as value-added, proprietary cultivars have had mixed success due to market pressures to lower prices to meet competition from the large quantities of inexpensive commodity and farmer-produced seed. To now move toward high-cost technology products will surely require a reeducational effort for both the seed marketers and farmers.

FUTURE

It is envisioned that the approach of using reinfection into commercially viable cultivars with naturally occurring novel strains will continue in the near term.

The basis of this strategy is to keep the main agronomic traits of the current cultivar intact, but to add the value of the endophyte. It relies on the best strain–cultivar niche to be successful. However, all infections at the cultivar level were not found to be successful from an agronomic point of view (Bouton et al., 2002). Reselection within these types of populations (Fig. 15.1) for better survival is an alternate strategy that has shown success in improving survival (J.H. Bouton, 2003, unpublished data). Finally, once elite breeding populations with novel endophytes are commonly found in most cultivar development programs, reselection within them of new cultivars with improved agronomic and nutritional traits will also be pursued.

The use of biotechnology to improve endophyte strains and (or) the endophyte/plant association is also an area of interest (Bacon and White, 1994). With the negative reactions in some quarters to biotech traits in important food crops, any work in this area must continue to monitor factors such as public perception, government regulation, and intellectual property issues. However, biotechnologies are too useful not to be part of any future commercialization efforts. For example, the strategy of removing the toxic strains from successful toxic cultivars, using genetic engineering techniques to eliminate their toxic ergot alkaloid pathways, and reinserting them into the same cultivar (Fig. 15.1) was reported previously (Wilkinson and Schardl, 1997). Recent work on cloning genes responsible for the endophyte toxic pathways should allow for more efficient screening of natural nontoxic strains or development of strategies for eliminating these genes in toxic associations (Scott, 2000). The ability to improve host-plant cross compatibility is also an important area of future investigation.

Fungal endophytes, because of their unique properties of living nonpathogenically throughout the grass host, are a potential vector for carrying value-added genes. They are also more easily manipulated than their genetically complex plant hosts. Therefore, genetically engineering the strain to contain other genes is also a viable strategy (Fig. 15.1). In this regard, Turner et al. (1993) proposed engineering the endophyte instead of the plant with genes that produce insect Bt toxins. With the growth of the mycelium throughout the plant, the plant would have insect deterrence capability without having these same genes inserted directly into its own genome. This approach would also ensure pollen containment, an important problem in obtaining regulatory approval on any biotech trait, as the endophyte itself has never been reported transmitted through the seed.

SUMMARY

The endophytes with value in forage systems are currently in the *Neotyphodium* genus with use and development focused on the tall fescue–*N. coenophialum* and the perennial ryegrass–*N. lolii* associations. The successful reinfection of novel endophytes such as AR1 into perennial ryegrass cultivars and MaxQ and ArkPlus into tall fescue cultivars demonstrates the commercial importance of this approach to improve both agronomic and animal performance. This reinfection approach will probably continue for the near future. Reselection of improved cultivars from elite breeding populations infected with novel endophytes is also the logical next step in cultivar development, as are biotechnological approaches for improving strains or strain–cultivar interactions. Technical requirements during commercialization of any endophytes, especially these commercially important novel strains, include devising screening methods to assess amount and type of infection, maintaining endophyte viability during seed increase and dissemination, and conducting the requisite agronomic and animal testing before actual release of new cultivars. Finally, novel endophytes will become patented in elite grass cultivars, but the current commodity approach of the grass seed industry, coupled with the common practice of farmer-produced seed, seems ill equipped to deal with intellectual property protection issues and high seed prices. Seed marketers also need price discipline currently not found in the market, and farmers need to demand guarantees on the product, especially endophyte viability.

REFERENCES

Adcock, R.A., N.S. Hill, J.H. Bouton, H.R. Boerma, and G.O. Ware. 1997. Symbiont regulation and reducing ergot alkloid concentration by breeding endophyte-infected tall fescue. J. Chem. Ecol. 23:691–704.

Anonymous. 1969. Managing forages for animal production. Res. Bull. 45. Virginia Polytechnic Inst., Blacksburg.

Bacon, C.W., and J.F. White, Jr. 1994. Biotechnology of endophytic fungi of grasses. CRC Press, Boca Raton, FL, USA.

Bluett, S. J., J. Hodgson, P.D. Kemp, and T.N. Barry. 1999. Performance of lambs and the incidence of staggers and heat stress on two perennial ryegrass cultivars over three summers in the Manawatu. p. 143–150. *In* D.R. Woodfield and C. Matthew (ed.) Ryegrass endophyte: An essential New Zealand symbiosis. Grassland Research and Practice Series, Vol. 7. New Zealand Grassland Assoc., Palmerston North.

Bouton, J. 2000. The use of endophytic fungi for pasture improvement in the USA. p. 163–168. *In* V.H. Paul and P.D. Dapprich (ed.) Proc. Int. *Neotyphodium*/Grass Interactions Symp., 4th, Soest, Germany. 27–29 Sept. 2000. Univ. Paderborn, Germany.

Bouton, J.H., R.R. Duncan, R.N. Gates, C.S. Hoveland, and D.T. Wood. 1997. Registration of 'Jesup' tall fescue. Crop Sci. 37:1011–1012.

Bouton, J.H., R.N. Gates, D.P. Belesky, and M. Owsley. 1993a. Yield and persistence of tall fescue in the southeastern coastal plain after removal of its endophyte. Agron J. 85:52–55.

Bouton, J.H., R.N. Gates, G.M. Hill, M. Owsley, and D.T. Wood. 1993b. Registration of 'Georgia 5' tall fescue. Crop Sci. 33:1405.

Bouton J., N. Hill, C. Hoveland, M. McCann, F. Thompson, L. Hawkins, and G. Latch. 2000. Performance of tall fescue cultivars infected with non-toxic endophytes. p. 179–185. *In* V.H. Paul and P.D. Dapprich (ed.) Proc. 4th Int. *Neotyphodium*/Grass Interactions Symp., Soest, Germany. 27–29 Sept. 2000. Univ. Paderborn, Germany.

Bouton, J.H., and A.A. Hopkins. 2003. Commercial applications of endophytic fungi. p. 495–516. *In* J.F. White et al. (ed.) Clavicipitalean fungi: Evoluntionary biology, chemistry, biocontrol, and cultural impacts. Marcel Dekker, New York.

Bouton, J.H., G.C.M. Latch, N.S. Hill, C.S. Hovland, M.A. McCann, R.H. Watson, J.A. Parish, L.L. Hawkins, and F.N. Thompson. 2002. Reinfection of tall fescue cultivars with non-ergot alkaloid-producing endophytes. Agron. J. 94:567–574.

Buckner, R.C. 1985. The fescues. p. 233–240. *In* M.E. Heath et al. (ed.) Forages, the science of grassland agriculture. 4th ed. Iowa State Univ. Press, Ames.

DeBattista, J., N. Altier, D.R. Galdames, and M. Dall'Agnol. 1997. Significance of endophyte toxicosis and current practices in dealing with the problem in South America. p. 383–388. *In* C.W. Bacon and N.S. Hill (ed.) *Neotyphodium*/Grass Interactions. Proc. Int. Symp. *Acremonium*/Grass Interactions, 3rd, Athens, GA, USA. 28–31 May 1997. Plenum Press, New York.

Easton, H.S. 1999. Endophyte in New Zealand ryegrass pastures, an overview. p. 1–9. *In* D.R. Woodfield and C. Matthew (ed.) Ryegrass endophyte: An essential New Zealand symbiosis. Grassland Research and Practice Series, Vol. 7. New Zealand Grassland Assoc., Palmerston North.

Easton, H.S. 1993. Will endophyte strain affect variety performance? p. 195–197. *In* D.E. Hume et al. (ed.) Proc. 2nd Int. Symp. on *Acremonium*/Grass Interactions, Palmerston North, New Zealand. 4–6 Feb. 1993. AgResearch Grasslands, Palmerston North.

Easton, H. S., B.M. Cooper, T.B. Lyons, C.G.L. Pennell, A.J. Popay, B.A. Tapper, and W.R. Simpson. 2000. Selected endophyte and plant variation. p. 351–356. *In* V.H. Paul and P.D. Dapprich (ed.) Proc. 4th Int. *Neotyphodium*/Grass Interactions Symp., Soest, Germany. 27–29 Sept. 2000. Univ. Paderborn, Germany.

Easton, H. S., G.A. Lane, and B.A. Tapper. 1993. Ergovaline in endophyte-infected ryegrass pastures. N. Z. Vet. J. 41(4):214.

Easton, H.S., G.C.M. Latch, B.A. Tapper, and O.J.P. Ball. 2002. Ryegrass host genetic control of concentrations of endophyte-derived alkaloids. Crop Sci. 42:51–57.

Fletcher, L.R., and H.S. Easton. 1997. The evaluation and use of endophytes for pasture improvement. *Neotyphodium*/Grass Interactions. p. 209–227. *In* C.W. Bacon and N.S. Hill (ed.) *Neotyphodium*/Grass Interactions. Proc. Int. Symp. *Acremonium*/Grass Interactions, 3rd, Athens, GA, USA. 28–31 May 1997. Plenum Press, New York.

Fletcher, L.R., and H.S. Easton. 2000. Using endophytes for pasture improvement in New Zealand. p. 149–162. *In* V.H. Paul and P.D. Dapprich (ed.) Proc. 4th Int. *Neotyphodium*/ Grass Interactions Symp., Soest, Germany. 27–29 Sept. 2000. Univ. Paderborn, Germany.

Fletcher, L.R., and I.C. Harvey. 1981. An association of a *Lolium* endophyte with ryegrass staggers. N. Z. Vet. J. 29(10):185–186.

Gallagher, R.T., E.P. White, and P.H. Mortimer. 1981. Ryegrass staggers: Isolation of potent neurotoxins lolitrem A and lolitrem B from staggers producing pastures. N. Z. Vet. J. 29:189–190.

Gates, R.N., G.M. Hill, and J.H. Bouton. 1999. Wintering beef cattle on mixtures of 'Georgia 5' tall fescue and warm-season perennial grasses on coastal plain soils. J. Prod. Agric. 12:581–587.

Hiatt, E.E. III, N.S. Hill, J.H. Bouton, and C.W. Mims. 1997. Monoclonal antibodies for detection of *Neotyphodium coenophialum*. Crop Sci. 37:1265–1269.

Hill, N.S., F.N. Thompson, D.L. Dawe, and J.A. Stuedemann. 1994. Antibody binding of circulating ergopeptine alkaloids in cattle grazing tall fescue. Am. J. Vet. Res. 55:419–424.

Hume, D.E. 1993. Agronomic performance of New Zealand pastures: Implications of *Acremonium* presence. p. 31–38. *In* D.E. Hume et al. (ed.) Proc. 2nd Int. Symp. on *Acremonium*/grass interactions: Plenary papers. Palmerston North, New Zealand. 4–6 Feb. 1993. AgResearch Grasslands, Palmerston North.

Lacefield, G., D. Ball, and J. Henning. 1993. Results of two-state survey of county agents regarding status of endophyte-infected and endophyte-free tall fescue. p. 22–25. *In* Proc. Southern Pasture and Forage Crop Improvement Conf., 49th, Sarasota, FL, USA. 14–16 June 1993.

Latch, G.C.M. 1997. An overview of *Neotyphodium*–grass interactions. p. 1–11. *In* C.W. Bacon and N.S. Hill (ed.) *Neotyphodium*/Grass Interactions. Proc. Int. Symp. *Acremonium*/Grass Interactions, 3rd, Athens, GA, USA. 28–31 May 1997. Plenum Press, New York.

Latch G.C.M. 1993. Physiological interactions of endophytic fungi and their hosts. Biotic stress tolerance imparted to grasses by endophytes. p 143–156. *In* R. Joost and S. Quisenberry (ed.) *Acremonium*/Grass Interactions. Elsevier Science Publ., Amsterdam.

Lean, I.J. 2001. Association between feeding perennial ryegrass (*Lolium perenne* cultivar Grasslands Impact) containing high concentrations of ergovaline, and health and productivity in a herd of lactating dairy cows. Aust. Vet. J. 79(4):262–264.

Lewis, G.C. 1997. Significance of endophyte toxicosis and current practices in dealing with the problem in Europe. p. 377–382. *In* C. W. Bacon and N. S. Hill (ed.) *Neotyphodium*/ Grass Interactions. Proc. Int. Symp. *Acremonium*/Grass Interactions, 3rd, Athens, GA, USA. 28–31 May 1997. Plenum Press, New York.

Lewis, G.C., C. Ravel, W. Naffaa, C. Astier, and G. Charmet. 1997. Occurrence of *Acremonium* endophytes in wild populations of *Lolium* spp. in European countries and a relationship between level of infection and climate in France. Ann. Appl. Biol. 130:227–238.

Nihsen, N.E., E.L. Piper, C.P. West, R.J. Crawford, Jr., T.M. Denard, Z.B. Johnson, C.A. Roberts, D.A. Spiers, and C.F. Rosenkrans, Jr. 2004. Growth rate and physiology of steers grazing tall fescue inoculated with novel endophytes. J. Anim. Sci. 82:878–883.

Parish, J.A., M.A. McCann, R.H. Watson, C.S. Hoveland, L.L. Hawkins, N.S. Hill, and J.H. Bouton. 2003a. Use of nonergot alkaloid-producing endophytes for alleviating tall fescue toxicosis in sheep. J. Anim. Sci. 81:1316–1322.

Parish, J.A., M.A. McCann, R.H. Watson, N.N. Paiva, C.S. Hoveland, A.H. Parks, B.L. Upchurch, N.S. Hill, and J.H. Bouton. 2003b. Use of nonergot alkaloid-producing endophytes for alleviating tall fescue toxicosis in stocker cattle. J. Anim. Sci. 81:2856–2868.

Paul, V.H., H. Ostbohmke, and U. Feuerstein. 2000. Studies on the dynamics of colonization of the endophytic fungus *Neotyphodium lolii* ((Latch, Christensen & Samuels) Glenn, Bacon & Hanlin comb. Nov.) (syn. *Acremonium lolii* Latch, Christensen & Samuels) in perennial ryegrass (*Lolium perenne* L.) in an European breeding process. p. 71–77. *In* V.H. Paul and P.D. Dapprich (ed.) Proc. 4th Int. *Neotyphodium*/Grass Interactions Symp., Soest, Germany. 27–29 Sept. 2000. Univ. Paderborn, Germany.

Popay, A.J., and J.G. Baltus. 2001. Black beetle damage to perennial ryegrass infected with AR1 endophyte. Proc. N. Z. Grassl. Assoc. 63:267–271.

Popay, A.J., J.G. Baltus, and C.G.L. Pennell. 2000. Insect resistance in perennial ryegrass infected with toxin-free *Neotyphodium* endophytes. p. 187–193. *In* V.H. Paul and P.D. Dapprich (ed.) Proc. 4th Int. *Neotyphodium*/Grass Interactions Symp., Soest, Germany. 27–29 Sept. 2000. Univ. Paderborn, Germany.

Rolston, M.P., J.R. Crush, M.D. Hare, and K.K. Moore. 1993. *Lolium* endophyte viability: Effect of seed storage. p. 1876–1877. *In* Proc. 17th Int. Grassland Congr., Palmerston North, New Zealand.

Scott, B. 2000. Molecular methods for the identification and modification of *Epichloë*-grass endophytes. p. 273–282. *In* V.H. Paul and P.D. Dapprich (ed.) Proc. 4th Int. *Neotyphodium*/Grass Interactions Symp., Soest, Germany. 27–29 Sept. 2000. Univ. Paderborn, Germany.

Stuedemann, J.A., and F.N. Thompson. 1993. Management strategies and potential opportunities to reduce the effects of endophyte-infected tall fescue on animal performance. p. 103–114. *In* D.E. Hume et al. (ed.) Proc. 2nd Int. Symp. on *Acremonium*/Grass Interactions: Plenary papers. 4–6 Feb. 1993. Palmerston North, New Zealand. AgResearch Grasslands, Palmerston North.

Turner, J.T., J.L. Kelly, and P.S. Carlson. 1993. Endophytes: An alternative genome for crop improvement. p. 555–560. *In* D.R. Buxton et al. (ed.) International Crop Science I. CSSA, Madison, WI, USA.

Wilkinson, H.H., and C.L. Schardl. 1997. The evolution of mutualism in grass–endophyte associations. p. 13–25. *In* C.W. Bacon and N.S. Hill (ed.) *Neotyphodium*/Grass Interactions. Proc. Int. Symp. *Acremonium*/Grass Interactions, 3rd, Athens, GA, USA. 28–31 May 1997. Plenum Press, New York.

ENDOPHYTES IN TURFGRASS CULTIVARS

Leah A. Brilman[1]

Utilization of turfgrass containing endophytes is an important part of best management practices to reduce inputs. The presence of endophytes in turfgrass has been demonstrated to provide many benefits, including resistance to surface-feeding insects, increased disease resistance, and increased stress tolerance (Funk and White, 1997). However, significant challenges still remain before seed companies and consumers realize the true potential of this important asset.

Currently the species of turfgrass with endophytes that are available on the market include perennial ryegrass (*Lolium perenne*), tall fescue (*Festuca arundinacea*), Chewings fescue (*F. rubra* subsp. *commutata*), strong creeping red fescue (*F. rubra* subsp. *rubra*), slender creeping red fescue (*F. rubra* subsp. *litoralis*), and hard fescue (*F. trachyphylla*). In some of these species, most cultivars available will have high levels of viable endophyte, while in other species there may be only a few cultivars with high endophyte levels. The species of endophytes that infect the turf species include *Neotyphodium lolii* in perennial ryegrass, *N. coenophialum* in tall fescues, and strains of *Epichloë festucae* in the fine fescues. Attempts have been made to find or introduce endophytes into other turf species such as Kentucky bluegrass (*Poa pratensis*), but so far these associations have not been stable and have not led to marketable cultivars.

Little research has been done to establish the percentage of endophyte needed in a turfgrass stand to provide the benefits to the end user. The required levels may depend on the species of grass, the desired benefit, and the strain of endophyte in the cultivar. It is the lack of an established level for customer benefit, the difficulties in supplying a product with high levels throughout the supply chain, and the lack of a well-documented test to rapidly establish viable endophyte levels for all species and endophytes that have retarded obtaining a market benefit for endophyte enhancement.

---

[1] Seed Research of Oregon, Corvallis, OR

ENDOPHYTE BENEFITS FOR TURF

Initial reports on endophytes in grasses documented their detrimental effects on grazing animals. It was not until endophytes were removed from forage ryegrass in New Zealand (Prestidge et al.,1982) and an association of the ryegrass endophyte with resistance to Argentine stem weevil (*Listronotus bonariensis*) was discovered that the additional aspects of this association were explored, including the benefits to turfgrass species. A series of papers by C. R. Funk and his associates at Rutgers University documented many of these turfgrass benefits. In perennial ryegrass turf, resistance to sod webworms (*Pyralidae* spp.) was associated with the presence of the endophyte (Funk et al., 1984), as well as resistance to bluegrass billbug (*Sphenophorus parvulus*) (Ahmed et al., 1986), and antibiosis to larvae of the southern armyworm (*Spodoptera eridania*) (Ahmed et al., 1987). Breen (1993a) found fall armyworm (*S. frugiperda*) development varied on different genotypes of endophyte-infected perennial ryegrass. It was also found that all genotypes of infected perennial ryegrass caused feeding deterrence, antibiosis, and subsequent death in greenbug (*Schizaphis graminum*) and yellow sugarcane aphid (*Sipha flava*). Resistance to bird cherry-oat aphid (*Rhopalosiphum padi*) depended on the concentration of endophyte hyphae in the perennial ryegrass host (Breen, 1993b). It appeared many of the original ryegrass plants used to develop turf varieties had originally contained endophytes, but in many cases these were later lost due to lack of knowledge of the endophytes presence, and thus the importance of storage conditions. In many cases this led to inconsistent performance of varieties once they reached the marketplace.

A survey of the 1986 National Turfgrass Evaluation Program (NTEP) Perennial Ryegrass Trial revealed that 12 cultivars had endophyte levels above 75%, the level at which Saha et al. (1988) demonstrated benefits to the host, with an additional four cultivars noted to have levels above 60%. Many additional cultivars had some level of endophytes, and without established standards could also state they were endophyte enhanced. It was also noted by Ahmed et al. (1986) that, with time, endophyte levels in perennial ryegrass turf trials increased from 50 and 25% infection in the seeds to 84 and 64% infected tillers from 1978 to 1984. Additional advantages were documented in endophyte-enhanced perennial ryegrasses under low maintenance conditions or during summer stress. These included drought resistance and improved summer stress recovery. Funk et al. (1984) also reported improved persistence and less crabgrass invasion in endophyte-infected perennial ryegrass, but these benefits were only observed under less-than-optimal maintenance.

More recent work has shown further benefits of endophyte-infected perennial ryegrass and has helped elucidate the levels of infection needed in a mixed turfgrass community. Carriere et al. (1998) found that hairy chinch bugs (*Blissus leucopterus hirstus*) avoided toxic endophyte-infected perennial ryegrass when mixed with Kentucky bluegrass and the presence of the bluegrass as a food source enabled their survival. However, Richmond and Shetlar (1999) found that in mixed turfgrass stands of perennial ryegrass and Kentucky bluegrass, as little as 40% endophyte infection in the ryegrass resulted in significant reductions in bluegrass webworm (*Parapediasia teterrella*) through a combination of emigration and reduced survival. Richmond and Shetlar (2000) also found that hairy chinch bug damage and population density decreased linearly as the percentage of endophyte infection increased in perennial ryegrass, nymphs emigrated more from high endophyte stands, and adult emigration was unaffected. This increase in movement of the insects makes them more vulnerable to predators and may be enough to cause death of newly hatched larvae or nymphs (Grewal and Richmond, 2004).

Many of the studies in tall fescue with endophyte have been primarily in forage varieties, and much of the work is explained elsewhere in this volume. Initial work on turf varieties included documentation of aphid resistance by Breen (1993b). Crutchfield and Potter (1995) studied endophyte-free and infected tall fescue in turf plots infested with root-feeding grubs of the southern masked chafer (*Cyclocephala lurida*) and Japanese beetle (*Popillia japonica*). They found greater damage by the chafer on endophyte-free compared with endophyte-infected tall fescue. Grewal and Richmond (2004) reported that ergot alkaloids strongly deterred feeding by Japanese beetle grubs and reduced their survival and gain in endophyte-containing tall fescues, although Richmond et al. (2004) found no benefit of endophyte against Japanese beetle grub when the tall fescue was in competition with dandelion (*Taraxacum officinale* aggr.). This study demonstrated that endophyte-infected tall fescue had larger tillers and increased below- and aboveground biomass, which would make them more competitive across time. The influence of turf management practices on the alkaloids produced by tall fescue and perennial ryegrass was demonstrated in that decreasing the mowing frequency from weekly to biweekly increased alkaloid levels in both tall fescue and perennial ryegrass (Salminen and Grewal, 2002). It was also demonstrated that increasing the mowing height in both species tended to increase alkaloid levels (Salminen et al., 2003). More studies of the complex interactions in a turf stand with and without endophytes need to be done, and in mixed communities of different turf species and weeds.

Fine fescues utilized for turf have also been shown to gain a significant advantage from the presence of endophytes. The initial documentation of occurrence and significance in the fine fescues by Saha et al. (1987) also documented chinch bug resistance from the presence of the endophyte in hard and Chewings fescue and improved performance during drought and summer heat. Further studies by Yue et al. (2000) found various endophyte and inoculated fine fescue combinations showed varying levels of different alkaloids, but all were toxic to chinch bugs. Breen (1993a, 1993b) demonstrated that fall armyworms did not survive to pupation when fed infected Chewings and hard fescues, and three species of aphids had greatly reduced survival on infected fine fescues. Endophyte-enhanced fine fescues have also shown improved dollar spot (*Lanzia* and *Moellerodiscus*) resistance (Clark et al., 1994). In certain endophyte–host combinations, it was observed that infected fine fescue had improved aluminum tolerance compared with endophyte-free plants (Zaurov et al., 2001).

CHALLENGES IN IMPLEMENTING ENDOPHYTIC BENEFITS IN TURF

The problems in developing and maintaining high-endophyte turfgrass varieties depend partially on the turf species involved. Although benefits have been documented for perennial ryegrass, tall fescue, and fine fescues, they are often subtle benefits that do not show up rapidly or under high maintenance conditions unless an attack by a insect pest is deterred. The end user is not aware that less insects are present in his lawn unless his neighbor suffers a serious attack. Most individuals supply excessive irrigation and will only notice improved drought resistance if their lawn survives a drought period. Endophytes supply an insurance policy and the customer must be willing to pay for the benefit.

Most perennial ryegrasses are currently developed to have high endophyte levels. In the 1999 NTEP Perennial Ryegrass Trial, 88 of the 134 cultivars and experimental lines entered had endophyte levels above 70% (Mohr et al., 2002). Many additional entries had levels above 60%, which may be higher in different seedlots than that submitted to the test, and only one had levels below 5%. Maintaining high levels in this species is not as hard due to greater longevity in the seed than the fine fescues. However, a high level of endophyte in one test does not ensure the cultivar will always have high levels in the marketplace. A company must still ensure that its seedstock is stored under cool, dry conditions or is regularly replaced with fresh seed. At the time of these surveys, the market had become saturated with perennial ryegrass seed due to abnormal market conditions, and many seedlots were stored for significant time periods with no monitoring of endophyte viability. Growth conditions of the seed crop and harvest conditions can also influence endophyte seed levels. High stress during seed filling and maturation can reduce the level of viable endophyte. Recently, it has been noted that new harvesting equipment that allows the seed to be combined sooner may be shortening the period for movement of the endophyte into the maturing seed and reducing seed percentage levels. Monitoring individual lines for endophyte levels during breeding has shown us much lower levels in seed examined shortly after cutting than after full seed filling and dry down. An additional problem is that varieties which are grown for seed in parts of the European Union, even if the seed is for turf usage, are not allowed to contain endophytes due to animal problems. Seed growers in the Williamette Valley of Oregon have also learned they need to analyze the seed straw for alkaloids before marketing the straw for fiber use in certain areas.

A significant portion of the perennial ryegrass crop is used by professional turf managers, either at golf courses, sports fields, or by landscape management companies, who know the benefits of endophytes. These customers often request endophyte-enhanced seed, which would explain why a large percentage of varieties contain high levels of endophytes. Unless an endophyte test is requested, and usually this is a seed test that does not test for viability, end users rely on the tests of the seedlot supplied to the National Test performed by Rutgers University, or a publication by Oregon State of endophyte levels, which is self reporting by companies to determine if a variety has endophytes.

Approximately 60 to 70% of the perennial ryegrass crop is sold into the temporary overseeding market in the southern USA. The endophyte has not been documented to provide advantages to this market, and some individuals have extrapolated from greater heat and drought tolerance data in permanent turf that it may slow the transition back to bermudagrass from the overseeded turf. However, the only trial to examine this was when a variety with and without the endophyte was entered in an overseeding trial for one year at the University of Arizona. The seedlot with the endophyte transitioned quicker than the one without, but the difference was not statistically significant (Brilman, 1995, unpublished data). This type of trial needs to be repeated for multiple locations and for different varieties to document benefits or problems with endophytes in perennial ryegrasses used for overseeding.

Tall fescue varieties for turf with endophytes present some additional problems. In the 2001 NTEP Tall Fescue Trial, only 50 out of 160 entries had endophyte levels above 70%, and 25 had levels below 6%, with 14 of these listed as 0% (Mohr et al., 2003). Certain seed growers prefer tall fescue cultivars with no endophyte because they want to sell or use the straw without having to test for alkaloid content. In certain markets, primarily where irrigation is utilized and insect pressure is low, no benefits from endophytes in tall fescue turf have been documented. Additional markets want at least some cultivars available with no endophyte for the homeowner who may also graze animals such as a horse. Selection and evaluation of tall fescue cultivars during high-stress periods show the benefits of endophytes, and greater persistence has been documented. However, additional breeding goals must also be integrated to realize these benefits in turf, with brown patch resistance being the primary concern in many markets.

The species of fine fescues present many unique problems in consistently selling seed with high endophyte levels or even maintaining it in varieties. The *Epichloë* endophytes can have varying levels of stromata choke produced in different years and under different management regimes, which can influence seed yields. The endophytes do not persist in fine fescue seed as long as in perennial ryegrass or tall fescue seed, so long-term storage becomes more difficult and maintenance of high endophyte levels can be a problem in some locations. The 1998 NTEP Fine Fescue Trial had 16 of 79 entries with levels above 70%, 17 entries with levels < 25%, and 18 entries with no infected seed (DaCosta et al., 1999). The hard fescues exhibited the highest levels, with 22 of 24 cultivars showing some level, and are also the species with less choke problems. In the Chewings fescues, 20 of 24 were infected, and in the strong creeping red fescues, 15 of 22 were infected. One of four slender creeping red fescues was infected by an endophyte. Comparisons of this trial with the 1993 National Trial showed a greater percentage of infected cultivars, but many with only moderate levels of infection. In this trial, the authors stated at least 25% infection is necessary to obtain benefits, but no data was supplied.

Fine fescues are usually combined with other grass species in turf mixtures including Kentucky bluegrass, perennial ryegrass, and multiple species of fine fescues. Much of this seed is sold into the homeowner market, where the benefits would be great but often the price becomes the primary criteria. The level of endophyte in different components of a mixture necessary to provide benefits has not been adequately studied. The extra expense associated with monitoring and maintaining high levels of endophyte in fine fescues when faced with market pressures is difficult to justify unless the customer sees true advantages and requests this seed. Even when trying to maintain high levels, conditions may occur that reduce the endophyte level that reaches the customer.

FUTURE RESEARCH AND DIRECTION

A rapid test for viable endophyte that would not delay marketing would be one of the most important steps in moving the market towards the ability to again obtain a premium for endophyte-enhanced seed. Dr. Ron Welty, formerly of the USDA working with Seed Research of Oregon (1997, unpublished data), developed a technique based on examining perennial ryegrass and tall fescue seed just as it germinated for endophyte growth that we have utilized to monitor viable levels in all species. It was generally found that in sprouted hypocotyls, the endophyte, if viable, had started to uncoil, and the diameters of the hyphae were small. Endophytes were also observed to not be exclusively associated with the aleurone. This viability test could be further refined and submitted for trial with seed analysts.

Further documentation of the percentage infection is needed in different turfgrass species, and with different endophyte genotypes in each species, to obtain benefits. This may enable the turf seed industry to establish a minimal endophyte level necessary to claim benefits in each species or in turfgrass mixtures. As the turf industry moves in the direction of reduced inputs, including chemicals and water, this will become increasingly important. National standards would be the most effective, as opposed to standards developed at the state level. Alternatively, a new type of voluntary certification, similar to a sod test, may also be effective. As in any program of this type, a statistical margin for the error in different samples must be taken into account.

Research on the conditions necessary to maintain endophytes in the seed needs to be further refined to see if improved packaging could increase the likelihood of the benefits reaching consumers. Mature, well-dried seed kept in moistureproof containers may maintain levels longer, even in hotter areas of the world. In Oregon, the lowest moisture levels are achieved during the late summer and early fall, with the seed gaining water during the winter months in woven poly bags or open containers. Further work is also justified in obtaining endophytes, or endophyte-plant combinations, that persist longer in the seed under less-than-ideal conditions.

The benefits of endophytes in turfgrass species are well documented, and many customers request them. More research needs to be done to more reliably deliver these benefits to the customers.

REFERENCES

Ahmed, S., S. Govindaraian, J.M. Johnson-Cicalese, and C.R. Funk. 1987. Association of a fungal endophyte in perennial ryegrass with antibiosis to larvae of the southern armyworm, *Spodoptera eridania*. Entomol. Exp. Appl. 43:287–294.

Ahmed, S., J.M. Johnson-Cicalese, W.K. Dickson, and C.R. Funk. 1986.Endophyte-enhanced resistance in perennial ryegrass to the bluegrass billbug, *Sphenophorus parvulus*. Entomol. Exp. Appl. 41:3–10.

Breen, J.P. 1993a. Enhanced resistance to fall armyworm (Lepidoptera: Noctudiae) in *Acremonium* endophyte-infected turfgrass. J. Econ. Entomol. 86:621–629.

Breen, J.P. 1993b. Enhanced resistance to three species of aphids (Homoptera: Aphididae) in *Acremonium* endophyte-infected turfgrass. J. Econ. Entomol. 86:1279–1286.

Carriere, Y., A. Brouchard, S. Bourassa, and J. Broudeur. 1998. Effect of endophyte incidence in perennial ryegrass on distribution, host-choice, and performance of the hairy chinch bug (Hemiptera: Lygaeidae). J. Econ. Entomol. 91:324–328.

Clark, B.B., D.R. Huff, D.A. Smith, and C.R. Funk. 1994. Enhanced resistance to dollar spot in endophyte-enhanced fine fescues. p. 187. *In* 1994 Agronomy abstracts. ASA, Madison, WI.

DaCosta, M., B. Bhandari, J. Carson, J. Johnson-Cicalese, and W.A. Meyer. 1999. Incidence of endophytic fungi in seed of cultivars and selections in the 1998 National Fine Fescue Test. Rutgers Turfgrass Proc. 30:185–190.

Funk, C.R., D.C. Saha, and J.M. Johnson-Cicalese. 1984. Association of endophytic fungi with increased persistence and improved performance of perennial ryegrass and tall fescue in low maintenance turfs seeded in 1976. Rutgers Turfgrass Proc. 15:134–146.

Funk, C.R., and J.F. White, Jr. 1997. Use of natural and transformed endophytes for turf improvement. p. 229–239. *In* C.W. Bacon and N.S. Hill (ed.) *Neotyphodium*/Grass Interactions. Proc. Int. Symp. *Acremonium*/Grass Interactions, 3rd, Athens, GA, USA. 28–31 May 1997. Plenum Press, New York.

Grewal, P.S., and D. Richmond. 2004. New benefits of endophyte-infected grasses emerge [Online]. Available at www.turfgrasstrends.com. Turfgrass Trends 1 Mar. 2004.

Mohr, M., W.A. Meyer, and C. Mansue. 2002. Incidence of *Neotyphodium* endophyte in seed lots of cultivars and selections of the 1999 National Perennial Ryegrass Test. Rutgers Turfgrass Proc. 33:185–188.

Mohr, M., W.A. Meyer, and C. Mansue. 2003. Incidence of *Neotyphodium* endophyte in seed lots of cultivars and selections of the 2001 National Tall Fescue Test. Rutgers Turfgrass Proc. 34:199–204.

Prestidge, R.J., R.P. Pottinger, and G.M. Barker. 1982. An association of *Lolium* endophyte with ryegrass resistance to Argentine stem weevil. Proc. N. Z. Weed Pest Control Conf. 35:119–122.

Richmond, D.S., P.S. Grewal, and J. Cardina. 2004. Influence of Japanese beetle *Popillia japonica* larvae and fungal endophytes on competition between turfgrasses and dandelion. Crop Sci. 44:600–606.

Richmond, D.S., and D.J. Shetlar. 1999. Larval survival and movement of bluegrass webworm in mixed stands of endophytic perennial ryegrass and Kentucky bluegrass. J. Econ. Entomol. 92:1329–1334.

Richmond, D.S., and D.J. Shetlar. 2000. Hairy chinch bug (Hemiptera: Lygaeidae) damage, population density, and movement in relation to the incidence of perennial ryegrass infected by *Neotyphodium* endophytes. J. Econ. Entomol. 93:1167–1172.

Saha, D.C., J.M. Johnson-Cicalese, and C.R. Funk. 1988. Endophyte content of perennial ryegrass cultivars and selections in the National Ryegrass Evaluation Program. Rutgers Turfgrass Proc. 18:73–78.

Saha, D.C., J.M. Johnson-Cicalese, P.M. Haliskey, M.I. Van Heemstra, and C.R. Funk. 1987. Occurrence and significance of endophytic fungi in the fine fescues. Plant Dis. 71:1021–1024.

Salminen, S.O., and P.S. Grewal. 2002. Does decreased mowing frequency enhance alkaloid production in endophytic tall fescue and perennial ryegrass? J. Chem. Ecol. 28:939–950.

Salminen, S.O., P.S. Grewal, and M.S. Quigley. 2003. Does mowing height influence alkaloid production in endophytic tall fescue and perennial ryegrass? J. Chem. Ecol. 29:1319–1328.

Yue, O., J. Johnson-Cicalese, T.J. Gianfagna, and W.A. Meyer. 2000. Alkaloid production and chinch bug resistance in endophyte-inoculated Chewings and strong creeping red fescue. J. Chem. Ecol. 26:279–292.

Zaurov, D.E., S. Bonos, J.A. Murphy, M. Richardson, and F.C. Balanger. 2001. Endophyte infection can contribute to aluminum tolerance in fine fescues. Crop Sci. 41:1981–1984.

ENDOPHYTES, QUALITY ASSURANCE, AND THE SEED TRADE IN EASTERN AUSTRALIA

W. M. Wheatley[1]

Climatic conditions vary widely across eastern Australia. Coastal areas are characterized by warm temperatures with high humidity for much of the year, particularly from mid-spring to mid-autumn. Tableland areas experience cold winters, warm to hot summers, with humidity often high in northern New South Wales (NSW) and Queensland (QLD) throughout summer. The slopes and plains experience cool to cold winters and hot summers, with a prolonged dry summer in the south and dry winter in the north. Rainfall is summer dominant north of an east-to-west line across central NSW, winter dominant to the south, with a narrow nonseasonal belt on either side. Temperate pasture species are mainly confined to the coast and tablelands south of Brisbane, QLD, but restricted to the higher rainfall zones and irrigation areas on the slopes and plains in southern states. Annual rainfall of the temperate zone is >550 mm, with a minimum growing season of 7 mo (Cunningham et al., 1994).

Fig. 17.1. *The higher rainfall (>550 mm per annum) temperate zone of eastern Australia. Reprinted with permission from Kemp and Michalk (2004).*

---

[1] The University of Sydney, Orange, NSW, 2800, Australia

Perennial Ryegrass

Perennial ryegrass (*Lolium perenne*) is a temperate, introduced grass in Australia and even though it is widely sown for pasture, it has also become naturalized throughout this area. It is recommended for pastures because of its ease of establishment and contribution to higher levels of animal production, but its persistence is less than alternative grasses such as phalaris (*Phalaris aquatica*) or Kikuyu (*Pennisetum clandestinum*). Estimates on the area of established perennial ryegrass vary from more than 6 million ha across Australia (Cunningham et al., 1994) to 4.9 million ha in southeastern Australia, with a modeled potential area of adaption of 14.4 million ha in this latter zone (Hill and Donald, 1998). The latter estimate represents 34% of the modeled potential area.

Fig. 17.2. *Area established to perennial ryegrass in eastern Australia. Reprinted with permission from Hill and Donald (1998).*

Fig. 17.3. *Area established to tall fescue in eastern Australia. Reprinted with permission from Hill and Donald (1998).*

Perennial ryegrass is usually infected with the endemic fungal endophyte *Neotyphodium lolii*. A number of secondary metabolites are produced within the plant as a result of this grass host and endophyte association (Lane, 1999). These result in beneficial effects to the grass, such as enhanced ability to resist insect damage (H.W. Kemp, unpublished data; Prestige et al., 1982; Quigley and Reed, 1999) and improved plant establishment and persistence (Cunningham et al., 1993; H.W. Kemp, unpublished data) or a range of animal disorders (Fletcher and Easton, 2000), of which ryegrass staggers is the most commonly identified in Australia.

TALL FESCUE

Tall fescue (*Festuca arundinacea*) is a temperate, introduced perennial grass that is usually included in a pasture mixture. While it is productive and persistent in summer rainfall areas and contributes to higher levels of animal production, its use has often been limited by weak seedling vigor and insect damage, particularly from African black beetle (*Heteronychus arator*) (Wheatley et al., 2003). Hill and Donald (1998) estimate the area of tall fescue in southeastern Australia at 1.1 million ha, only 6.5% of its modeled potential area of 16.8 million ha.

In Australia, tall fescue cultivars sold for grazing systems do not contain the endemic fungal endophyte *Neotyphodium coenophialum* (Easton et al., 1994; Valentine et al., 1993) found in turf cultivars and present in grazing systems in other parts of the world. However, there is at least one small area of tall fescue on the south coast of NSW which has been established for a long period. This stand does not appear to be consumed by livestock if alternative feed is available and contains an endemic endophyte (D.E. Hume, H.W. Kemp, and W.M. Wheatley, unpublished data). Anecdotal evidence suggests that similar situations exist in other areas.

SPREAD OF ENDOPHYTE

Neotyphodium endophyte is only spread via seed (Philipson and Christey, 1986; Rolston et al., 1994; White et al., 1993). In perennial ryegrass, nearly 100% of tillers are naturally infected and a high percentage of fresh perennial ryegrass seed sold in Australia is infected with endemic endophyte (Quigley and Reed, 1999; Valentine et al., 1993).

ENDOPHYTE VIABILITY IN SEED

A number of factors are involved in maintaining endophyte viability in seed:

- Seed production practices (Rolston et al., 1994)
- Age of seed (Neill, 1940; Reed et al., 1987; Siegel et al., 1985)

- Temperature and humidity during seed storage (Latch and Christensen, 1982; Mika and Bumerl, 1995; Rolston et al., 1986; Siegel et al., 1985; Welty et al., 1987)
- Seed moisture content during storage/packaging materials (Mika and Bumerl, 1995; Rolston et al., 1986, 1994; Welty et al., 1987)

In summary, if seed was stored under ambient conditions in polyweave bags, virtually no viable endophyte was found after 18 to 24 months from harvest. However, if different packaging materials and storage conditions were used, endophyte viability was considerably enhanced.

GOVERNMENT POLICY AND ENDOPHYTES

To ensure the quality of seed offered for sale in Australia, state governments have legislated a Seeds Act to ensure genetic purity, germination percentage, and impurities. However, there is no requirement to provide details on the endophyte status of the seed. In the past, state government departments enforced their Seeds Act, but in recent times, there has been a shift towards industry deregulation, in so much as the Act has remained, but compliance has become the responsibility of the seed industry. Victoria (VIC) and Tasmania (TAS) have already deregulated (K. Reed and S. Smith, personal communication) and NSW is moving in that direction (R. Walker, personal communication). There also appears to be reluctance on the part of government officials to develop a policy on endophytes, preferring to leave it to the seed industry.

ENDEMIC ENDOPHYTE IN SEED LINES

In an early study (M. Mebalds, personal communication cited by van Heeswijck and McDonald, 1992), endemic endophyte was detected in 63 to 90% of stored seed lines of the perennial ryegrass ecotype 'Victorian'.

Valentine et al. (1993) collected 154 samples of perennial ryegrass seed lines from seed certification laboratories and seed companies in VIC, NSW, and TAS in autumn 1991 and examined each for endophyte infection. One-hundred seeds from each sample were germinated and assessed after 7 d by microscopy of leaf sheaths stained in accordance with Di Menna and Waller (1986). Mean infection levels ranged from 16% in 'Nui' to 71% in 'Ellett'. Only eight perennial ryegrass seed lines were endophyte free. Of the eight cultivars and ecotypes included, only 'Brumby' (seven samples) had a narrow infection range (31 to 46%), with the ecotypes Victorian (58 samples) and 'Kangaroo Valley' (7 samples) having ranges of 0 to 85% and 0 to 86%, respectively. Endophyte was detected in 56 of the 58 seed lines of Victorian. The results from this study indicate variable levels of viable endemic endophyte in fresh perennial ryegrass seed.

Wheatley et al. (unpublished data) sought information on the status of endemic endophyte in perennial ryegrass seed in retail/wholesale outlets in 15 different locations across the coast and tablelands of NSW and the south coast of VIC in spring 2002. A total of 53 samples from 13 diploid perennial ryegrass cultivars and ecotypes was collected. Seed-borne endophyte levels were determined by the seed squash method described by Latch et al. (1987) using 50 seeds per sample. An endophyte grow-out test, to determine viable endophyte levels at the time of sampling, was immediately commenced by planting 100 seeds from each sample. One tiller from each seedling was blotted onto nitrocellulose membrane and then assayed using the tissue-print immunoblot technique described by Gwinn et al. (1991) with modifications by Wheatley and Simpson (2001). Age of seed was available from respective seed companies for 46 samples. Two-thirds of the retail facilities in this study were primarily constructed of metal, facilitating a high heat transfer, and one quarter were constructed of concrete walls with a metal roof, facilitating high heat transfer and retention—an important factor in endophyte viability under ambient conditions, given the extremely high temperatures and humidity that can occur in Australia. The impact of this on endophyte viability requires further investigation. Two-thirds of the seed samples were 2 yr or less since harvest, but seed up to 7 yr old was being offered for sale. In general, seed viability remained high, although viability dropped below 50% when samples were 3 yr and older. Seed-borne endophyte levels were similar to the way in which each cultivar was promoted for sale, although one 4-yr-old seed sample promoted as low endophyte had a seed-borne level of 73%, but had 0% viable endophyte. The ecotypes Victorian and Kangaroo Valley had seed-borne infections ranging from 10 to 68%, similar to the findings of Valentine et al. (1993). Viable endophyte remained at acceptable levels for 2 yr from harvest, although samples were present that had viability levels of 2, 10, and 30% at this stage. The loss in endophyte viability increased rapidly after 2 yr, with 75% of the known samples having viable levels near 0%.

Table 17.1. Relationship between seedling establishment (seed viability) and endophyte viability.†‡

| Age of seed | Seedling establishment | | Viable endophyte | |
	Average	Range	Average	Range
yr		%———		
1	84	73–95	74	32–99
2	85	59–98	61	2–98
3	73	50–90	18	0–58
4	71	30–92	2	0–8
5	41	0–81	0.5	0–1
6	ns§		ns	
7	50	50	0	0

† Four samples whose seed-borne endophyte < 50% were not included. Seven samples whose age was not known were not included.

‡ Wheatley et al., unpublished data.

§ ns, no samples collected.

The decline in endophyte viability at a much faster rate than seed viability supports earlier studies (Rolston et al., 1986, 1994; and Welty et al., 1987). The loss in endophyte viability in this study was significantly influenced by cultivar, age of seed, and the interaction between cultivar and age of seed. The study has established that viable levels of endemic endophyte currently in the marketplace are a major cause for concern and has important implications for the sale of selected endophytes in perennial ryegrass and tall fescue.

SELECTED ENDOPHYTES

Selected, novel, or nontoxic endophytes are those which, in association with their grass host, produce metabolites that result in beneficial effects to the grass, such as insect resistance, but eliminate or significantly reduce the range of animal disorders described by Fletcher and Easton (2000). In Australia, there are numerous perennial ryegrass cultivars commercially available with endophyte AR1 (marketed as AR1) and five cultivars of tall fescue with endophyte AR542 (marketed as MaxP). To be marketed as such, there must be a minimum viable endophyte level of 70% for cultivars with AR1 at the time of sale and 65% for MaxP cultivars.

Selected ryegrass endophytes that are free of lolitrem B, but produce ergovaline and peramine, are also available in the cultivars Tolosa (endophyte NEA2) and Extreme (endophyte AR6).

PACKAGING AND STORAGE OF SEED WITH ENDOPHYTES

Perennial ryegrass seed with either endemic or AR1 endophyte is packaged in a standard polyweave bag and stored under ambient conditions until sold to the producer. AR1 is clearly marked on each seed bag. Some cultivars with AR1 are only being delivered to the retailer on an exact-quantity-ordered basis to reduce seed movement and storage around the country. The level of viable endophyte in fresh seed in storage is monitored on a regular basis (seed company representative, personal communication).

Tall fescue seed with MaxP endophyte is packaged in either a 70-μ plastic bag inside a standard polyweave bag or in an Al-foil bag, depending upon cultivar. MaxP is clearly marked on each bag. As soon as the seed has been packaged, it remains in cool storage (5°C) until an order is received and a preferred delivery date is determined. The exact quantity of seed is delivered to the retailer as close to this date as possible and no later than 14 d beforehand (seed company representatives, personal communication). This minimizes the time the seed is stored under less-than-ideal conditions.

Selected endophytes attract a premium of $1.50AU to $2.10 per kg seed, which carries the obligation and expectation that those fungi are viable in the seed purchased. These endophytes are currently being marketed in fresh seed, so loss of viable endophyte should be minimal as long as demand is high. As the demand for a cultivar or endophyte declines, the potential for older seed remaining in the marketplace will become a greater problem, and packaging and storage of AR1 (or any selected endophyte) seed may have to be reviewed. The viability of endemic endophyte in the Australian marketplace has been highlighted by the introduction of selected endophytes. Continuing with the existing procedure for endemic endophyte may invite potential litigation.

ENDOPHYTE-FREE SEED

Where the opportunity exists to purchase a perennial ryegrass cultivar either infected with endophyte or endophyte-free, demand for endophyte-free seed is extremely low (seed company representatives, personal communication). One seed company has developed and retails a number of perennial ryegrass cultivars as low or nil endemic endophyte.

PRODUCER EXPECTATIONS

Producers who require high endemic endophyte infection levels in perennial ryegrass to ensure plant persistence, particularly against African black beetle, rely on seed labeling that indicates high endophyte (HE) and/or the verbal advice of the seed retailer. It is extremely rare for an infection level to be known or provided. In addition, there have been many examples where the endemic endophyte infection level of seed at retail outlets has not been tested. Variable results in the field have led to questions regarding the levels of viable endemic endophyte in perennial ryegrass cultivars at the time of sale.

Quality assurance in the marketplace remains the responsibility of seed companies, wholesalers, and retailers. While state government Seeds Acts do not protect the consumers of endophyte-infected seed, other existing consumer legislation does. Producers need to be made aware of the many differences in packaging, storage, and sale of different endophytes, based on research information that is available.

SEED INDUSTRY

In response to the information presented on the quality of seed for sale at wholesale and retail outlets, the Plant Breeders and Proprietory Marketers Group (PB&PMG) within the Australian Seed Federation (ASF) initiated a meeting in early February 2004. As a result, a working group was formed to develop an education/information package for the industry. As well as ASF, the group, coordinated initially by W. Wheatley, includes a representative from member seed companies, AgResearch Grasslands New Zealand, and the state agricultural departments of NSW, VIC, and TAS, ensuring that a unified message is compiled. Key areas being addressed by the group include:

- Awareness of endophyte throughout the pastoral industry
- A common language on endophyte related matters
- Role of endophytes in perennial ryegrass and tall fescue pastures
- Maintaining endophyte viability through packaging, storage and handling of seed
- Standardization and clarity of endophyte testing methods
- Feasibility of seed being regularly tested to ensure the endophyte status is known at or about the time of sale, with the level being indicated on a tag on the bag

PB&PMG has stated that endophyte levels of high, medium, or low will not be used in relation to endophyte. The viable endophyte percentage will be stated and the information package will develop an understanding of efficacy of endophytes at certain percentage levels in certain environments. The distribution of the package will be the responsibility of the ASF Retail Group and will target all stakeholders, including retailers, producers, state government agricultural departments, and the major research funding bodies. The ASF is to be commended for their initiative, and illustrates how industry responsibility can be an alternative to government legislation. In the long term, industry self-regulation may be more effective in enhancing the seeds sector and protecting consumers than acts of state or federal parliaments. However, compliance will be the responsibility of members and nonmembers of ASF.

ACKNOWLEDGMENTS

The author thanks representatives of a number of seed companies for supplying information on the packaging and marketing of their endophytes and cultivars. Due to the sensitivity of information provided, the contributors have been kept anonymous.

REFERENCES

Cunningham, P.J., M.J. Blumenthal, M.W. Anderson, K.S. Prakash, and A. Leonforte. 1994. Perennial ryegrass improvement in Australia. N. Z. J. Agric. Res. 37:295–310.

Cunningham, P.J., J.Z. Foot, and K.F.M. Reed. 1993. Perennial ryegrass (*Lolium perenne*) endophyte (*Acremonium lolii*) relationships: the Australian experience. Agric. Ecosyst. Environ. 44:157–168.

Di Menna, M.E., and J.E. Waller. 1986. Visual assessment of seasonal changes in the amount of mycelium of *A. lolii* in leaf sheaths of perennial ryegrass. N. Z. J. Agric. Res. 29:111–116.

Easton, H.S., C.K. Lee, and R.D. Fitzgerald. 1994. Tall fescue in Australia and New Zealand. N. Z. J. Agric. Res. 37:405–417.

Fletcher, L.R., and H.S. Easton. 2000. Using endophytes for pasture improvement in New Zealand. p. 149–162. *In* V. Paul and P. Dapprich (ed.) Proc. 4th Int. *Neotyphodium*/Grass Interactions Symp., Soest, Germany. 27–29 Sept. 2000.Univ. of Paderborn, Soest.

Gwinn, K.D., M.H. Collins-Shepard, and B.B. Reddick. 1991. Tissue print-immunoblot, an accurate method for the detection of *Acremonium coenophialum* in tall fescue. Phytopathology 81:747–748.

Hill, M.J., and G.E. Donald. 1998. Determination of benefits from pasture improvement. Final report for Australian Meat Research Corporation. Revised ed. CSIRO, Australia.

Kemp, D.R., and D.L. Michalk. 2004. Australian temperate grasslands: Changing philosophies and future prospects. Chapter 20 (in press). *In* S. Reynolds (ed). Grasslands—Future perspectives. Food Agric. Org., Rome.

Lane, G.A. 1999. Chemistry of endophytes: Patterns and diversity. Ryegrass endophyte: An essential New Zealand symbiosis. Grassland Res. Practice Ser. 7:85–94.

Latch, G.C.M., and M.J. Christensen. 1982. Ryegrass endophyte, incidence and control. N. Z. J. Agric. Res. 25:443–448.

Latch, G.C.M., L.R. Potter, and B.F. Tyler. 1987. Incidence of endophytes in seeds from collections of *Lolium* and *Festuca* species. Ann. Appl. Biol. 111:59–64.

Mika, V., and J. Bumerl. 1995. Decline of endophyte viability during seed storage. p. 453–455. *In* Proc. 3rd Int. Herbage Seeds Conf., Halle, Germany. 18–23 June 1995.

Neill, J.C. 1940. The endophyte of rye-grass (*Lolium perenne*). N. Z. J. Sci. Technol. 21A:280–291.

Philipson, M.N., and M.C. Christey. 1986. The relationship of host and endophyte during flowering, seed formation and germination of *Lolium perenne*. N. Z. J. Bot. 24:125–134.

Prestige, R.A., R.P. Pottinger, and G.M. Barker. 1982. An association of *Lolium* endophyte with ryegrass resistance to Argentine stem weevil. Proc. N. Z. Weed Pest Control Conf. 35:119–122.

Quigley, P.E., and K.F.M. Reed. 1999. Endophyte in perennial ryegrass: Effect on host plants and livestock. *In* Agriculture Notes Series. No. AG0202. Dep. Natural Resources and Environment, VIC, Australia.

Reed, K.F.M., P.J. Cunningham, J.T. Barrie, and J.F. Chin. 1987. Productivity and persistence of cultivars and introductions of perennial ryegrass (*Lolium perenne*) in Victoria. Aust. J. Exp. Agric. 27:267–274.

Rolston, M.P., M.D. Hare, G.C.M. Latch, and M.J. Christensen. 1994. *Acremonium* endophyte viability in seeds and effects of storage. Seeds symposium: Seed development and germination. Spec. Publ. 9:77–82. Seed Symp. Aug. 1991. Agron. Soc. of New Zealand, Christchurch.

Rolston, M.P., M.D. Hare, K.K. Moore, and M.J. Christensen. 1986. Viability of *Lolium* endophyte fungus in seed stored at different moisture contents and temperatures. N. Z. J. Exp. Agric. 14:297–300.

Siegel, M.R., G.C.M. Latch, and M.C. Johnson. 1985. *Acremonium* fungal endophytes of tall fescue and perennial ryegrass: Significance and control. Plant Dis. 69:179–183.

Valentine, S.C., B.D. Bartsch, K.G. Boyce, M.J. Mathison, and T.R. Newbery. 1993. Incidence and significance of endophyte in perennial ryegrass seed lines. Final report on project DAS33 prepared for the Dairy Research and Development Corporation. South Australian Research and Development Institute.

van Heeswijck, R., and G. McDonald. 1992. *Acremonium* endophytes in perennial ryegrass and other pasture grasses in Australia and New Zealand. Aust. J. Agric. Res. 43:1683–1709.

Welty, R.E., M.D. Azevedo, and T.M. Cooper. 1987. Influence of moisture content, temperature and length of storage on seed germination and survival of endophytic fungi in seeds of tall fescue and perennial ryegrass. Phytopathology 77:893–900.

Wheatley, W.M., D.E. Hume, H.W. Kemp, M.S. Monk, K.F. Lowe, A.J. Popay, D.B. Baird, and B.A. Tapper. 2003. Effects of fungal endophyte on the persistence and productivity of tall fescue at 3 sites in eastern Australia. In Proc. 11th Australian Agron. Conf. [CD-ROM—Pastures]. Geelong, Australia.

Wheatley, W.M., and W.R. Simpson. 2001. Tissue immunoblot procedure—Stability of *Neotyphodium* endophyte antigen on nitrocellulose membrane. p. 145–147. *In* V. Paul and P. Dapprich (ed.) Proc. 4th Int. *Neotyphodium*/Grass Interactions Symp., Soest, Germany. 27–29 Sept. 2000. Univ. of Paderborn, Soest.

White, J.F., G. Morgan-Jones, and A.C. Morrow. 1993. Taxonomy, life cycle, reproduction and detection of *Acremonium* endophytes. Agric. Ecosyst. Environ. 44:13–37.

PUBLIC EDUCATION ON TALL FESCUE TOXICOSIS

Craig Roberts[1] and John Andrae[2]

During the past 25 to 30 yr, agricultural researchers have discovered the general cause of fescue toxicosis (associated with tall fescue, *Festuca arundinacea*) (Bacon et al., 1977), and have shown that it can be managed. Researchers demonstrated that fescue toxicosis can be eliminated by replacing toxic cultivars, such as endophyte-infected 'Kentucky 31', with nontoxic forage species (Bouton et al., 2002; Hoveland et al., 1983; Nihsen et al., 2004; Parish et al., 2003). They have also reported that toxicosis can be alleviated with careful management of infected Kentucky 31, should replacement not be an option (Chestnut et al, 1991; Roberts and Andrae, 2004).

Such management requires some level of scientific knowledge. This knowledge must first include an understanding of mechanisms that cause fescue toxicosis—the endophyte and its alkaloids. Knowledge must also include the ability to separate effective, science-based management practices from unsound, anecdotal remedies. The gathering, packaging, and transferring of this knowledge to the public is the duty of extension specialists, crop consultants, and other educators.

Management practices for tall fescue toxicosis have been presented extensively in international symposia (Ball, 1997; Cross, 1997; De Battista et al, 1997; Foot, 1997; Lewis, 1997). These practices, along with recently developed recommendations, have been published in an electronic guide (Roberts and Andrae, 2004). The reader is referred to these articles for detailed reviews of practices used to alleviate fescue toxicosis in various areas of the world.

This chapter will address public education in fescue toxicosis. It will discuss general management recommendations as offered by university extension specialists and the adoption of these practices by producers. It will also list some questionable practices. Finally, the chapter will offer suggestions for educational efforts on tall fescue toxicosis, including a focused, multiday workshop with presentations, demonstrations, and the layout of a grazing system. While the

[1] University of Missouri, Columbia, MO USA
[2] University of Georgia, Athens, GA USA

chapter focuses on tall fescue toxicosis in the USA, we hope the principles discussed apply to approaches in dealing with ryegrass staggers in Australia and New Zealand.

CO-WORKERS AND COW-WORKERS: WHAT DO SPECIALISTS RECOMMEND, AND WHAT WILL PRODUCERS ADOPT?

What are extension specialists saying? How consistent is their message throughout the Fescue Belt? What is the most effective delivery method for these messages? Are the producers adopting these recommendations?

In the spring of 2004, we conducted a survey among state extension specialists in the tall fescue growing region of the USA. This survey asked questions regarding current university extension recommendations and producer practices. It was completed by 15 state-level forage and livestock extension specialists. Results are summarized as a compilation of total points assigned to a response.

What Do the Specialists Say?

The first question was simply a classifier to determine whether the specialist was trained in agronomy, animal sciences, or another discipline. The next question was the first real question of substance. It asked, "If a producer continues to manage E+ KY31 (toxic endophyte-infected tall fescue), which practices would YOU (the specialist) recommend most? (Rank 1 = least important practice, 10 = most important.)" The categories were

__Dilution of pasture with other forages
__Feeding of supplements
__Rotation to other nontoxic pastures (either tall fescue or another species)
__Reduction in N fertilizer or poultry litter application rates
__Controlling seedheads (mainly referring to clipping)
__Feeding mineral mixes
__Feeding seaweed product
__Administering ivermectin
__Breeding for animal resistance
__Other (_____)

The results indicate that extension specialists are keeping abreast of current scientific data, interpreting and applying the literature well, and are recommending proven practices from controlled experiments (Fig. 18.1). The extension specialists' top two recommendations are dilution of the pasture (mainly with legumes) and rotating livestock to nontoxic pastures.

Several specialists did not rank all practices listed in the question. Some of these unranked practices included feeding mineral mixes, feeding seaweed product, administering ivermectin, and breeding for animal resistance. Specialists who refused to rank practices commented that ranking would give the impression that all practices were recommended, albeit less than other practices. Several other specialists who did assign a number to all practices commented that they did so only to fulfill survey requirements.

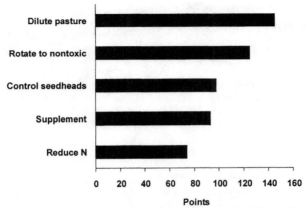

Fig. 18.1. *Current management recommendations for alleviation of tall fescue toxicosis. Recommendations are from state extension specialists in the USA.*

What Will the Producers Implement?

The third question in the survey was, "If a producer continues to manage E+ KY31, which practices are PRODUCERS most likely to adopt (regardless of your perceived effectiveness)? Please rank 1 = least likely to adopt, 10 = most likely to adopt." The categories were the same as those for specialists' recommendations.

According to the specialists' responses, producers are diluting pastures with legumes, as has been a proven recommendation for many years (Fig. 18.2). But the responses also indicate that producers may be attempting to feed their way out of this problem rather than managing their pastures. Supplying minerals and mineral mixes, a nutritional practice, was ranked above livestock rotation to nontoxic pastures or seedhead control, two practices that can drastically reduce the amount of ergot alkaloids ingested by the grazing animal.

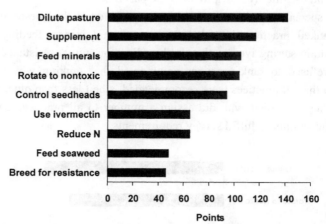

Fig. 18.2. *Practices of producers attempting to alleviate tall fescue toxicosis as reported by state extension specialists in the USA.*

Fire at Will!

A fourth question invited the specialists to comment on any of the practices listed in the previous question. As noted earlier, several specialists commented that they do not recommend some of the feeding practices and proposed cures. Others commented that many of the practices are recommended, but not as a way to alleviate toxicosis; administration of ivermectin for parasite control and feeding of minerals were cited as examples of sound management that are recommended with or without a toxicosis problem.

Several specialists responded that the most successful producers in their regions dilute pastures with legumes and move livestock to summer pastures. One specialist commented that rotational grazing was the most effective method for alleviating fescue toxicosis. Another respondent reminded us that reducing N fertilization is not always feasible in areas with heavily concentrated poultry production. Two specialists ranked seedhead control high for reasons related both to toxin concentration and forage quality. One specialist commented that too many of the recommendations are based on testimonials rather than scientifically proven practices. Although specialists did not unanimously agree on a single, most effective practice to alleviate fescue toxicosis, it is clear that a combination of practices is likely the most effective overall approach to capitalize on potential additive effects.

How Can Information Be Effectively Delivered to Producers?

The survey also asked about program delivery. The question was, "When you deliver information related to tall fescue toxicosis and management, which delivery method is most effective in causing a change in producer's management? Please rank 1 = least effective, 9 = most effective." The categories were

__Multiday school dedicated solely to tall fescue toxicosis and management
__Portion of a multiday meeting (e.g., a breakout session)
__Extension talks at routine county meetings (e.g., winter soil and crop meeting)
__Extension guide sheets and bulletins
__Newsletters from producer groups (e.g., Forage and Grassland Council)
__Popular press articles where you are quoted (i.e., trade magazines, hard copy and electronic)
__Forage and livestock web pages
__Demonstration plots/field days
__Other (_____)

The specialists stated that demonstrations and field days are the most effective ways of getting the information to the producers. They also stated that breakout sessions at multiday events are effective. Some specialists commented that tall fescue schools would be employed, but there were currently no such schools in existence; this is why that mode of delivery ranks low in Fig. 18.3.

It is worth noting that all direct, person-to-person methods of delivery were considered more effective at changing producer practices than all indirect, written methods of delivery. The least effective of the indirect methods of delivery was considered to be newsletters from producer groups.

Are Producers Aware of the Endophyte Problem?

Another question asked the specialists to offer their best estimate on the proportion of producers who realized they had an endophyte problem. The question

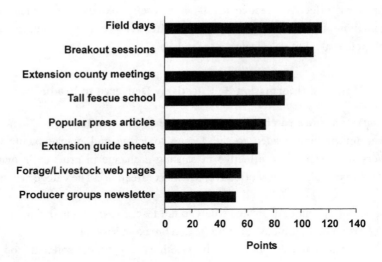

Fig. 18.3. *Most effective mode of delivery in tall fescue toxicosis extension as ranked by U.S. extension specialists.*

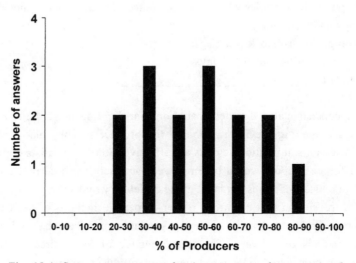

Fig. 18.4. *State extension specialists' estimations of proportion of producers who realize they have an endophyte problem.*

read, "An educated guess is required here. Of the producers in your state who grow infected tall fescue (such as KY31), what percentage of them understand they have an endophyte problem?" The ranges were

__0 to 10%   __50 to 60%
__10 to 20%   __60 to 70%
__20 to 30%   __70 to 80%
__30 to 40%   __80 to 90%
__40 to 50%   __90 to 100%

The results varied greatly, indicating either that specialists had no idea, or that producer knowledge varies widely from state to state (Fig. 18.4). Averaged across states, the specialists estimated that half of producers realize an endophyte problem exists on their farm.

What Are the Concerns about Replacing Toxic Endophytes with Novel Endophytes?

Two questions dealt with replacement of toxic KY31 with a tall fescue containing a novel endophyte. The first question addressed the specialists' opinions on replacement methods. This question was worded, "In YOUR opinion, what are the main drawbacks in replacing toxic tall fescue with a cultivar that contains a novel endophyte? Rank as 1 = least important drawback, 7 = most important drawback." The categories were

__Seed cost
__Process of killing old and planting new (spray, smother crop, etc.)
__Inability to harvest seed for own use
__Not good investment on rented land
__Less persistent than E+ KY31
__Pasture reverting back to E+ KY31
__Other (_____)

The second question was the same, except that it asked about producers' main concerns. According to the results (Fig. 18.5 and 18.6), the main drawbacks stated by the specialists were similar to those expected from producers—the replacement process itself and the seed cost. The specialists also assumed that producers are concerned about newly established pastures (with novel endophytes) becoming reinfected with the old, toxic endophyte. It is important to note that concerns with persistence when tall fescue contains a novel endophyte ranked relatively low in both specialist and perceived producer responses.

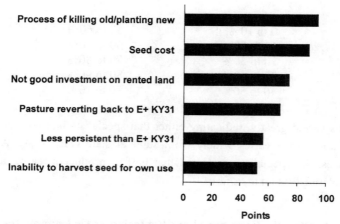

Fig. 18.5. *State extension specialists' view: Main drawbacks to replacing toxic tall fescue with a cultivar containing a novel endophyte.*

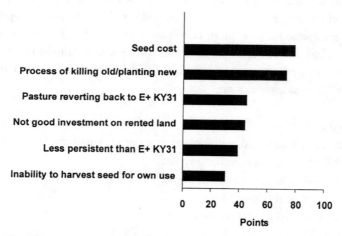

Fig. 18.6. *Predicted producers' view: Main drawbacks to replacing toxic tall fescue with a cultivar containing a novel endophyte.*

Will Producers Test Pastures?

Perhaps the most predictable response came from the last question. It read, "An educated guess is required here. Of the producers in your state who grow infected tall fescue (such as KY31), what percentage of them will test for the endophyte in the next 5 years?" The ranges were

___ 0 to 25%	___ 50 to 75%
___ 25 to 50%	___ >75%

The state specialists unanimously predicted that only 0 to 25% of the tall fescue growers will test for the endophyte. This is not surprising, as many producers do not believe the test is necessary, and others are still unaware of an endophyte problem. Although this proportion is low, testing tall fescue fields is a useful practice. Fields with modest endophyte infection rates are not uncommon, particularly in more northern areas of the U.S. tall fescue region. If accurate infection rates of all fields on a farm were known, producers could prioritize fields for renovation and identify pastures that would provide adequate animal performance with addition of legumes.

THE WITCH'S BREW

In general, the survey results indicate some agreement between extension specialist recommendations and production practices. According to the responses, more than one of the primary management practices is being adopted.

However, it should be noted that the above results are not from a survey of random producers. They are responses from a questionnaire, albeit a representative questionnaire, as it sampled more than half of the state extension specialists who are involved in tall fescue toxicosis work. It should also be noted that the questionnaire only addresses mainstream recommendations and practices. It does not address the wide range of practices, many of which include the bizarre. This next section will do just that—address, even document, both the normal and the unusual.

The following is a curious list compiled by Eldon Cole, Extension Livestock Specialist in Mt. Vernon, Missouri. It is a collection of suggestions for alleviating tall fescue toxicosis. It is not a list of practices that Mr. Cole endorses, but one that he maintains as a record. Items on this list were offered by farmers, feed dealers, researchers, extension staff, and others as practices they had seen or heard about that helped alleviate the severity of fescue toxicosis in beef cattle. Some suggestions are scientific practices while others are home remedies. Many of these practices are not legally approved for use. A few are rather macabre.[1]

---

[1] This list contains trade names of several products. These names are part of the common vernacular of forage–livestock producers in the USA. Mention of these trade names and products does not constitute endorsement on the part of the authors, editors, or publishers.

1. Dilute by adding legumes to the toxic pastures.
2. Graze fungus-free or novel endophyte pastures, especially in the summer.
3. Rotate among pastures.
4. Use warm-season grass pastures in June, July, and August (johnsongrass, big bluestem, bermudagrass, etc.).
5. Graze wheat pasture in the spring.
6. Prevent cattle from grazing the tall fescue stem and seedhead (possibly ergot) by early clipping or by using a plant growth regulator, such as Embark.
7. Graze pastures short.
8. Stock pasture lightly.
9. Sometime during the year, mow the fescue as close to the ground as possible.
10. Burn the fescue pasture.
11. Be sure there is adequate shade in the pasture.
12. Fill in or fence ponds, let cattle drink fresh, clean, cool water.
13. Feed some alfalfa hay (3–7 lbs.) during the winter and midsummer.
14. Do not allow cattle to eat big round bales that have been stored outside.
15. Ammoniate mature fescue hay or fescue stubble hay.
16. Topdress pasture with sulfur.
17. Lime pasture more often.
18. Use more magnesium in lime and fertilizer.
19. Cut back on nitrogen fertilizer or poultry litter.
20. Apply seaweed extract to growing tall fescue.
21. Feed 2 to 4 pounds of grain per day in the summer.
22. Feed a grain sorghum that contains a relatively high level of antifungal activity.
23. Feed a protein supplement.
24. Feed two pounds a day of 20% liquid supplement.
25. Use ionophores (Rumensin or Bovatec).
26. Supplement with a high oil/fat feed like whole soybean or cottonseed.
27. Include walnut by-product supplement.
28. Feed MFA no. 1 salt mix year-round.
29. Do not use salt to limit grain or protein intake.
30. Feed cattle some kind of alcohol-containing product, such as an alcohol–molasses mix.

31. Feed "special" enzyme compound.
32. Feed fermentation extracts.
33. Feed an antioxidant.
34. Feed extra selenium.
35. Feed extra sulfur.
36. Feed extra potassium.
37. Feed extra copper.
38. Feed extra zinc.
39. Feed extra calcium.
40. Feed extra phosphorus.
41. Feed extra manganese.
42. Feed extra cobalt.
43. Give vitamin B12.
44. Give extra vitamins A and E.
45. Administer thiamine.
46. Feed niacin.
47. Administer nitric oxide.
48. Vaccinate for BVD and IBR.
49. Vaccinate for chlamydial abortion.
50. Use Ivomec (injectable or bolus).
51. Use thiabendazole.
52. Feed phenothiazine.
53. Administer growth-promoting implants to stocker cattle.
54. Treat for coccidiosis.
55. Give aspirin to reduce temperature.
56. Give dipyrone to reduce temperature.
57. Administer cimetidine.
58. Administer metoclopramide.
59. Administer domperidone.
60. Vaccinate against ergot alkaloids.
61. Feed tea leaves.
62. Feed caffeine.
63. Feed a brown seaweed product (Tasco).
64. Give a shot of Pet milk under the skin to make cattle shed their hair.
65. Calve in the fall.
66. Calve in January and February.
67. Use bulls that sire shorter-haired, easy shedding calves.
68. Select cattle that show heat tolerance.

69. Clip the hair off the affected cattle in hot weather.
70. Sell long, rough-haired cows.
71. Sell cows that pond-stand.
72. Sell bob-tailed cows.
73. Crossbreed.
74. Add Brahman breeding.
75. Add Longhorn breeding.
76. Add Limousin breeding.
77. Add Jersey breeding.
78. Add Gelbvieh breeding.
79. Add Senepol breeding.
80. Reduce percentage of Hereford.
81. Reduce percentage of Simmental.
82. Reduce percentage of Angus.
83. Reduce percentage of Shorthorn.
84. Reduce percentage of Charolais.
85. Reduce percentage of Salers.
86. Select for early shedding ability in the above breeds.
87. Use bulls that have EPDs that resulted from progeny raised on fescue.
88. Keep cattle in fleshy condition.
89. Keep first-calf heifers off toxic pastures in spring and summer.
90. Let cattle have access to brushy pastures where there are sassafras, persimmon, and sumac sprouts to browse.
91. Have curly dock in the pasture for cattle to graze.
92. Drench affected cattle with a chelated copper solution.
93. Inject copper solution into their briskets.
94. Feed zeolite, a clay-like compound.
95. Feed hydrated sodium calcium aluminosilicate.
96. Pour diesel fuel on the lower leg and hoof.
97. Chase sorefooted animals with pickup, horse, dog, or four-wheeler to stimulate circulation.

Even after 25 yr of scientific research and extension efforts, tall fescue toxicosis is still regarded, on occasion, as a mysterious plague from the dark abyss. It is only fitting, therefore, that some management practices resemble ghoulish potions more than good farming.

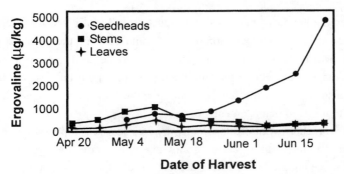

Fig. 18.7. *Ergovaline concentration in tall fescue pastures during the spring of 1987. Ergovaline is one of several ergot alkaloids. From Rottinghaus et al., 1991. Reprinted with permission from Journal of Agricultural and Food Chemistry. Copyright 1991 American Chemical Society.*

PROPOSED EDUCATIONAL APPROACHES

Education about tall fescue toxicosis should include information about ergot alkaloids, since these alkaloids cause the problem. This information should describe the alkaloids and the animal symptoms caused by high alkaloid concentrations. In addition, information should explain how concentrations are affected by environment, management, plant physiology, and endophyte strain. An example of data that could be presented is seen in Fig. 18.7, which shows that ergovaline is most highly concentrated in the late spring, when seedheads are mature.

Knowledge about ergot alkaloids will help the producer make day-to-day management decisions to limit toxin consumption. Such a management approach, a sort of *alkaloid management*, would employ a series of practices known to reduce consumption; some of these practices were mentioned in the survey above and were described in the electronic guide cited earlier (Roberts and Andrae, 2004).

Tall Fescue Toxicosis and Management Workshop

A multiday workshop, entitled "Tall Fescue Toxicosis and Management," might be the most effective way to present an integrated, yet focused, message about this syndrome and how to deal with it. In many regions, such a workshop would be easily justified because tall fescue toxicosis represents the main livestock disorder, the impact of toxicosis on production and profit is severe, and there are many effective management options now available to aid the livestock producer.

Such a workshop would have a goal—to provide producers with sound, sensible management options for eliminating or alleviating tall fescue toxicosis. It would also have an expected outcome—producers would increase their profits through improved management.

This workshop would follow a step-by-step learning process, beginning with an understanding of toxicosis and following with a series of suggested practices, each practice coupled with a field demonstration. The workshop might conclude with the presentation of a grazing system that incorporated the suggested management practices. Because these practices often involve commercial products, the workshop should receive broad support from industry. A sample agenda of a workshop utilizing this step-by-step learning process is proposed below.

Example Agenda: Tall Fescue Toxicosis and Management Workshop

I. Introduction: Why is tall fescue such a popular forage species?
Benefits of Tall Fescue: ease of establishment, persistence and adaptation, yield distribution, forage quality, and lack of better alternative pasture forage species

II. Identification of the Problem: Tall fescue toxicosis
An Animal Science or Veterinary Perspective.
Describe toxicosis symptoms and resulting performance of various livestock classes

A Plant Science or Mycology Perspective.
Describe endophyte, ergot alkaloids, seasonal fluctuation in alkaloid concentration, endophyte reproduction, endophyte/plant mutualism, and plant persistence
Field Demo: Steers grazing tall fescue infected with toxic endophyte, novel endophyte, and no endophyte

III. Prerequirement to Management: Endophyte testing
Discuss importance of testing for endophyte vs. ergot alkaloids, testing tillers and seed, sampling procedures, immunochemical procedures vs. microscopic procedures
Field Demo: When and how to sample
Lab Demo: How test is performed

IV. Management Option 1: Eradicating toxic tall fescue

Discuss replacing toxic tall fescue with other cool-season grasses, including tall fescue with novel endophytes; establishment process of "spray-smother-spray"; pros and cons of replacing existing stand with new stand

Field Demo: Spray-smother-spray with various available herbicides

V. Management Option 2: Managing existing pastures of toxic tall fescue

Introduce two concepts (see Roberts and Andrae, 2004):

- Alkaloid Management (techniques to minimize alkaloid intake)
- Incremental Alleviation (Fig. 18.8)

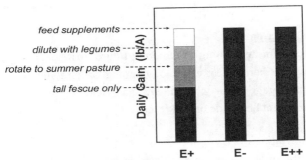

Fig. 18.8. *The principle of incremental improvement as popularized by George Garner at the University of Missouri. Producers who manage toxic tall fescue should employ a series of practices that produce incremental alleviation in livestock performance. These practices represent annual inputs. When employed simultaneously, these practices can have an additive effect on livestock, with performance that approaches that from consuming nontoxic tall fescue. E+ = toxic tall fescue, E- = endophyte-free tall fescue and E++ = nontoxic endophyte-infected tall fescue. (From Roberts and Andrae, 2004.)*

Step 1: Livestock Rotation to Summer Pastures

Discuss importance of rotating livestock in late spring; the use of summer pastures; interseeding wheat (*Triticum aestivum*) into bermudagrass (*Cynodon dactylon*) for alternative winter grazing; management of overly mature tall fescue. Also discuss "double-crop" result of a pasture that contains tall fescue mixed with a warm-season forage, such as annual lespedeza (*Kummerowia* spp.), crabgrass (*Digitaria* spp.), or bermudagrass (depending upon location)

Field Demo: Summer perennial pastures overseeded with wheat

Step 2: Dilution of Pasture

Discuss legumes and blends of cool-season grasses; nutritional effect of legumes, as well as their effect on reducing ingestion of alkaloids; benefits of legumes in pasture yield distribution and N fixation; establishment of legumes in existing sod

Field Demo: Tall fescue pasture mixed with cool-season legumes and grasses

Step 3: Diet Supplementation

Discuss impact of various supplements on animal performance, including standard supplements and by-product feeds; pros and cons of various feedstuffs including substitution, handling, costs, etc.

Field Demo: Supplementation demo

Step 4: Seedhead Control

Discuss reasons for and timing of clipping; future treatments of growth regulators and low rate herbicides; effect of clipping or spraying on ergot alkaloid concentrations, forage quality components, and yield

Field Demo: Select herbicide on tall fescue

Step 5: Ammoniation of Hay

Discuss process of ammoniation as related to toxicology vs. nutrition; effect of ammoniation on ergot alkaloid concentrations in plant and animal performance; effects of ammoniation vs. ensiling or greenchopping; cost of ammoniation

Field Demo: Ammoniation process

Step 6: Winter Grazing Management

Discuss rationale of deferred grazing; practice of stockpiling tall fescue; high forage quality; seasonal fluctuation of ergovaline; strip-grazing issues; cost of deferred grazing vs. hay

Field Demo: Strip-grazing stockpiled tall fescue

Step 7: Other Practices

An overview of other management practices proposed to alleviate fescue toxicosis. Overview would include both proven and speculative methods

Step 8: Designing a System (see example below)

Base Pasture

Spring/Fall (70% acres)	Summer/Winter (30% acres)
Perennials: E+ tall fescue Red clover White clover **Reseeding Annual:** Annual lespedeza	**Perennial:** Bermudagrass **Summer Annual:** Crabgrass **Winter Annual:** Wheat or Rye

Management Calendar (for the Ozarks)

Spring (March - May):
- Graze fescue/clover pasture at green up
- Rotate to bermudagrass pasture mid-May or as soon as possible
- Manage ungrazed pasture for growth of summer species

Summer (June - August):
- Strip graze bermudagrass
- Cut tall fescue in early head stage if weather permits
- If hay is cut late, cover and ammoniate it
- Graze annual lespedeza in July, as it appears in fescue/clover field
- Fertilize some portion of fescue acreage for late winter stockpile
    (This can be done until mid-September in Arkansas)

Fall (September - November)
- Rotate back to fescue/clover pastures
- Drill wheat or rye into bermudagrass pastures, fertilize with N
- Graze fescue/clover until wheat can be grazed
- Defer grazing tall fescue until later winter

Winter (January - February)
- Feed ammoniated hay
- Graze tall fescue pasture and stockpile
- Graze wheat as soon as it is ready

Fig. 18.9. *A low-cost, productive grazing system for Ozark beef producers who have toxic Kentucky 31 tall fescue. One important clarification is in order. For nutritional reasons, recommendations usually advise to cut hay before the plant is mature. The cutting management suggested in the above system—cutting when hay is mature and as weather permits—is an exception and only possible because (i) cattle are grazing a warm-season pasture, and (ii) the hay will be ammoniated.*

Example: A Simple, Low-Cost Grazing System. Presenting a layout of a good grazing system would be an excellent way to conclude a focused, multiday workshop on tall fescue toxicosis. Such a system is offered below (Fig. 18.9). It was designed for the Ozark (northern Arkansas or southern Missouri, USA) beef producer. It is a simple, low-cost grazing plan that uses two base pastures—tall fescue that contains the toxic endophyte and is mixed with legumes, and bermudagrass that can be overseeded with a winter annual. The system is based on data, not superstition. It depends on management, particularly the timing of livestock rotation. It calls for ordinary forage cultivars rather than rare plants from mystic, far-off realms. Though it was designed for Ozark producers, it would likely apply to other regions with similar temperature and rainfall patterns.

SUMMARY

Survey results reveal both positive and negative aspects of fescue toxicosis education. On a positive note, extension specialists are delivering relevant, current information concerning fescue toxicosis treatments. On the basis of this survey, both livestock and forage extension specialists tend to recommend pasture management strategies; however, these same specialists perceive that farmers are more likely to favor direct supplementation of feedstuffs to animals. Information is most effectively delivered via oral presentations and field demonstrations. On a more negative note, specialists report that the spray-smother-spray process to replace toxic tall fescue as well as relatively high seed costs of novel endophyte-infected cultivars are the main impediments to replacing toxic tall fescue with nontoxic cultivars. Specialists also estimate that only 50% of producers in the tall fescue region are aware that an endophyte problem exists, and 25% or less are likely to test their pastures for endophyte. There is clearly a need for continued public education. This need can be efficiently addressed in a multiday tall fescue toxicosis and management workshop. This workshop capitalizes on classroom and in-field demonstrations, two effective delivery techniques, to educate producers on tall fescue toxicosis causes and a multitude of remedies.

REFERENCES

Bacon, C.W., J.K. Porter, J.D. Robbins, and E.S. Luttrell. 1977. *Epichloe typhina* from tall fescue grasses. Appl. Environ. Microbiol. 35:576–581.
Ball, D.M. 1997. Significance of endophyte toxicosis and current practices in dealing with the problem in the United States. p. 395–410. *In* C.W. Bacon and N.S. Hill (ed.) *Neotyphodium*/Grass Interactions. Plenum Press, New York.
Bouton, J.H., G.C.M. Latch, N.S. Hill, C.S. Hoveland, M.A. McCann, R.H. Watson, J.A. Parish, L.L. Hawkins, and F.N. Thompson. 2002. Reinfection of tall fescue cultivars with non-ergot alkaloid-producing endophytes. Agron. J. 94:567–574.

Chestnut, A.B., H.A. Fribourg, J.B. McClaren, D.G. Keltner, B.B. Reddick, R.J. Carlisle, and M.C. Smith. 1991. Effects of *Acremonium coenophialum* infestation, bermudagrass and nitrogen or clover on steers grazing tall fescue pastures. J. Prod. Agric. 4:208–213.

Cross, D.L. 1997. Fescue toxicosis in horses. p. 289–310. *In* C.W. Bacon and N.S. Hill (ed.) *Neotyphodium*/Grass Interactions. Plenum Press, New York.

De Battista, J., N. Altier, D.R. Galdames, and M. Dall'Agnol. 1997. Significance of endophyte toxicosis and current practices in dealing with the problem in South America. p. 383–388. *In* C.W. Bacon and N.S. Hill (ed.) *Neotyphodium*/Grass Interactions. Plenum Press, New York.

Foot, J.Z. 1997. Significance of endophyte toxicosis and current practices in dealing with the problem in Australia and New Zealand. p. 389–394. *In* C.W. Bacon and N.S. Hill (ed.) *Neotyphodium*/Grass Interactions. Plenum Press, New York.

Hoveland, C.S., S.P. Schmidt, C.C. King, Jr., J.W. Odom, E.M. Clark, J.A. McGuire, L.A. Smith, H.W. Grimes, and J.L. Holliman. 1983. Steer performance and association of *Acremonium coenophialum* fungal endophyte on tall fescue pasture. Agron. J. 75:821–824.

Lewis, G.C. 1997. Significance of endophyte toxicosis and current practices in dealing with the problem in Europe. p. 377–382. *In* C.W. Bacon and N.S. Hill (ed.) *Neotyphodium*/Grass Interactions. Plenum Press, New York.

Nihsen, M.E., E.L. Piper, C.P. West, R.J. Crawford, T.M. Denard, Z.B. Johnson, C.A. Roberts, D.A. Spiers, and C.F. Rosenkrans, Jr. 2004. Growth rate and physiology of steers grazing tall fescue inoculated with novel endophytes. J. Anim. Sci. 82:878–883.

Parish, J.A., M.A. McCann, R.H. Watson, N.N. Paiva, C.S. Hoveland, A.H. Parks, B.L. Upchurch, N.S. Hill, and J.H. Bouton. 2003. Use of non-ergot alkaloid-producing endophytes for alleviating tall fescue toxicosis in stocker cattle. J. Anim. Sci. 81:2856–2868.

Roberts, C., and J. Andrae. 2004. Tall fescue toxicosis and management [Online]. Available at www.plantmanagementnetwork.org/cm/. Crop Manage. DOI 10.1094/CM-2004-0427-01-MG.

Rottinghaus, G.E., G.B. Garner, C.N. Cornell, and J.L. Ellis. 1991. HPLC method for quantitating ergovaline in endophyte-infested tall fescue: Seasonal variation of ergovaline levels in stems with leaf sheaths, leaf blades, and seed heads. J. Agric. Food Chem. 39:112–115.